# RELATIVE INVARIANTS OF RINGS

**PURE AND APPLIED MATHEMATICS**

*A Program of Monographs, Textbooks, and Lecture Notes*

EXECUTIVE EDITORS

Earl J. Taft
*Rutgers University*
*New Brunswick, New Jersey*

Zuhair Nashed
*University of Delaware*
*Newark, Delaware*

CHAIRMEN OF THE EDITORIAL BOARD

S. Kobayashi
*University of California, Berkeley*
*Berkeley, California*

Edwin Hewitt
*University of Washington*
*Seattle, Washington*

EDITORIAL BOARD

*M. S. Baouendi*
*Purdue University*

*Jack K. Hale*
*Brown University*

*Marvin Marcus*
*University of California, Santa Barbara*

*W. S. Massey*
*Yale University*

*Leopoldo Nachbin*
*Centro Brasileiro de Pesquisas Físicas*
*and University of Rochester*

*Anil Nerode*
*Cornell University*

*Donald Passman*
*University of Wisconsin*

*Fred S. Roberts*
*Rutgers University*

*Gian-Carlo Rota*
*Massachusetts Institute of*
*Technology*

*David Russell*
*University of Wisconsin-Madison*

*Jane Cronin Scanlon*
*Rutgers University*

*Walter Schempp*
*Universität Siegen*

*Mark Teply*
*University of Florida*

## MONOGRAPHS AND TEXTBOOKS IN PURE AND APPLIED MATHEMATICS

1. *K. Yano,* Integral Formulas in Riemannian Geometry (1970)*(out of print)*
2. *S. Kobayashi,* Hyperbolic Manifolds and·Holomorphic Mappings (1970) *(out of print)*
3. *V. S. Vladimiorv,* Equations of Mathematical Physics ( A. Jeffrey, editor; A. Littlewood, translator) (1970) *(out of print)*
4. *B. N. Pshenichnyi,* Necessary Conditions for an Extremum (L. Neustadt, translation editor; K. Makowski, translator) (1971)
5. *L. Narici, E. Beckenstein, and G. Bachman,* Functional Analysis and Valuation Theory (1971)
6. *D. S. Passman,* Infinite Group Rings (1971)
7. *L. Dornhoff,* Group Representation Theory (in two parts). Part A: Ordinary Representation Theory. Part B: Modular Representation Theory
8. (1971, 1972)
8. *W. Boothby and G. L. Weiss (eds.),* Symmetric Spaces: Short Courses Presented and Washington University (1972)
9. *Y. Matsushima,* Differentiable Manifolds (E. T. Kobayashi, translator) (1972)
10. *L. E. Ward, Jr.* Topology: An Outline for a First Course (1972) *(out of print)*
11. *A. Babakhanian,* Cohomological Methods in Group Theory (1972)
12. *R. Gilmer,* Multiplicative Ideal Theory (1972)
13. *J. Yeh,* Stochastic Processes and the Wiener Integral (1973) *(out of print)*
14. *J. Barros-Neto,* Introduction to the Theory of Distributions (1973)
15. *(out of print)*
15. *R. Larsen,* Functional Analysis: An Introduction (1973) *(out of print)*
16. *K. Yano and S. Ishihara,* Tangent and Cotangent Bundles: Differential Geometry (1973) *(out of print)*
17. *C. Procesi,* Rings with Polynomial Identities (1973)
18. *R. Hermann,* Geometry, Physics, and Systems (1973)
19. *N. R. Wallach,* Harmonic Analysis on Homogeneous Spaces (1973) *(out of print)*
20. *J. Dieudonné,* Introduction to the Theory of Formal Groups (1973)
21. *I. Vaisman,* Cohomology and Differential Forms (1973)
22. *B. -Y. Chen,* Geometry of Submainfolds (1973)
23. *M. Marcus,* Finite Dimensional Multilinear Algebra ( in two parts) (1973, 1975)
24. *R. Larsen,* Banach Algebras: An Introduction (1973)
25. *R. O. Kujala and A. L. Vitter (eds.),* Value Distribution Theory: Part A; Part B. Deficit and Bezout Estimates by Wilhelm Stoll (1973)
26. *K. B. Stolarsky,* Algebraic Numbers and Diophantine Approximation (1974)
27. *A. R. Magid,* The Separable Galois Theory of Commutative Rings (1974)
28. *B. R. McDonald,* Finite Rings with Identity (1974)
29. *J. Satake,* Linear Algebra (S. Koh, T. A. Akiba, and S. Ihara, translators) (1975)

30. *J. S. Golan,* Localization of Noncommutative Rings (1975)
31. *G. Klambauer,* Mathematical Analysis (1975)
32. *M. K. Agoston,* Algebraic Topology: A First Course (1976)
33. *K. R. Goodearl,* Ring Theory: Nonsingular Rings and Modules (1976)
34. *L. E. Mansfield,* Linear Algebra with Geometric Applications: Selected Topics Topics (1976)
35. *N. J. Pullman,* Matrix Theory and Its Applications (1976)
36. *B. R. McDonald,* Geometric Algebra Over Local Rings (1976)
37. *C. W. Groetsch,* Generalized Inverses of Linear Operators: Representation and Approximation (1977)
38. *J. E. Kuczkowski and J. L. Gersting,* Abstract Algebra: A First Look (1977)
39. *C. O. Christenson and W. L. Voxman,* Aspects of Topology (1977)
40. *M. Nagata,* Field Theory, (1977)
41. *R. L. Long,* Algebraic Number Theory (1977)
42. *W. F. Pfeffer,* Integrals and Measures (1977)
43. *R. L. Wheeden and A. Zygmund,* Measure and Integral: An Introduction to Real Analysis (1977)
44. *J. H. Curtiss,* Introduction to Functions of a Complex Variable (1978)
45. *K. Hrbacek and T. Jech,* Introduction to Set Theory (1978)
46. *W. S. Massey,* Homology and Cohomology Theory (1978)
47. *M. Marcus,* Introduction to Modern Algebra (1978)
48. *E. C. Young,* Vector and Tensor Analysis (1978)
49. *S. B. Nadler, Jr.* Hyperspaces of Sets (1978)
50. *S. K. Segal,* Topics in Group Rings (1978)
51. *A. C. M. van Rooij,* Non-Archimedean Functional Analysis (1978)
54. *L. Corwin and R. Szczarba,* Calculus in Vector Spaces (1979)
53. *C. Sadosky,* Interpolation of Operators and Singular Integrals: An Introduction to Harmonic Analysis (1979)
54. *J. Cronin,* Differential Equations: Introduction and Quantitative Theory (1980)
55. *C. W. Groetsch,* Elements of Applicable Functional Analysis (1980)
56. *I. Vaisman,* Foundations of Three-Dimensional Euclidean Geometry, (1980)
57. *H. I. Freedman,* Deterministic Mathematical Models in Population Ecology (1980)
58. *S. B. Chae,* Lebesgue Integration (1980)
59. *C. S. Rees, S. M. Shah, and C. V. Stanojević,* Theory and Applications of Fourier Analysis (1981)
60. *L. Nachbin,* Introduction to Functional Analysis: Banach Spaces and Differential Calculus (R. M. Aron, translator) (1981)
61. *G. Orzech and M. Orzech,* Plane Algebraic Curves: An Introduction Via Valuations (1981)
62. *R. Johnsonbaugh and W. E. Pfaffenberger,* Foundations of Mathematical Analysis (1981)
63. *W. L. Voxman and R. H. Goetschel,* Advanced Calculus: An Introduction to Modern Analysis (1981)
64. *L. J. Corwin and R. H. Szcarba,* Multivariable Calculus (1982)
65. *V. I. Istrătescu,* Introduction to Linear Operator Theory (1981)
66. *R. D. Järvinen,* Finite and Infinite Dimensional Linear Spaces: A Comparative Study in Algebraic and Analytic Settings (1981)

67. *J. K. Beem and P. E. Ehrlich,* Global Lorentzian Geometry (1981)
68. *D. L. Armacost,* The Structure of Locally Compact Abelian Groups (1981)
69. *J. W. Brewer and M. K. Smith, eds.,* Emmy Noether: A Tribute to Her Life and Work
70. *K. H. Kim,* Boolean Matrix Theory and Applications (1982)
71. *T. W. Wieting,* The Mathematical Theory of Chromatic Plane Ornaments (1982)
72. *D. B. Gauld,* Differential Topology: An Introduction (1982)
73. *R. L. Faber,* Foundations of Euclidean and Non-Euclidean Geometry (1983)
74. *M. Carmeli,* Statistical Theory and Rnadom Matrices (1983)
75. *J. H. Carruth, J. A. Hildebrant, and R. J. Koch,* The Theory of Topological Semigroups (1983)
76. *R. L. Faber,* Differential Geometry and Relativity Theory: An Introduction (1983)
77. *S. Barnett,* Polymonials and Linear Control Systems (1983)
78. *G. Karpilovsky,* Commutative Group Algebras (1983)
79. *F. Van Oystaeyen and A. Verschoren,* Relative Invariants of Rings: The Commutative Theory (1983)
80. *I. Vaisman,* A First Course in Differential Geometry (1984)
81. *G. W. Swan,* Applications of Optimal Control Theory in Biomedicine (1984)
82. *T. Petrie and J. D. Randall,* Transformation Groups on Manifolds (1984)
83. *K. Goebel and S. Reich,* Uniform Convexity, Hyperbolic Geometry, and Nonexpansive Mappings (1984)
84. *T. Albu and C. Năstăsescu,* Relative Finiteness in Module Theory (1984)
85. *K. Hrbacek and T. Jech,* Introduction to Set Theory, Second Edition, Revised and Expanded (1984)
86. *F. Van Oystaeyen and A. Verschoren,* Relative Invariants of Rings: The Noncommutative Theory (1984)

# RELATIVE INVARIANTS OF RINGS

## The Noncommutative Theory

F. Van Oystaeyen • A. Verschoren
Department of Mathematics
University of Antwerp, UIA
Antwerp-Wilrijk, Belgium

MARCEL DEKKER, INC. New York and Basel

Library of Congress Cataloging in Publication Data

Oystaeyen, F. van, [date]
Relative invariants of rings.

(Monographs and textbooks in pure and applied mathematics ; 86)
Bibliography: p.
Includes index.
1. Noncommutative rings. 2. Invariants.
I. Verschoren, A., [date]. II. Title. III. Series.
QA251.4.098 1984 512'.4 84-14983
ISBN 0-8247-7281-4



MARCEL DEKKER, INC.
270 Madison Avenue, New York, New York 10016

Current printing (last digit):
10 9 8 7 6 5 4 3 2 1

PRINTED IN THE UNITED STATES OF AMERICA

Since I did most of the work in writing this book, it may be justified to dedicate it to my co-author for being able to cope with my cooperation in his usual modest and friendly way.

# PREFACE

The theory of relative invariants of commutative rings has been treated from a ring theoretical point of view in the first part of this work [162]. One of the fundamental facts on which that theory has been built is that one can obtain ring theoretical interpretations for invariants otherwise defined in terms of geometric objects. Sheaf theoretical objects like the Brauer group (Picard group) of a scheme or of a locally ringed space, in particular their restriction to not-necessarily affine subsets, may be related to relative Brauer groups (relative Picard groups). The subsets considered need not be open subsets and it turned out to be very rewarding to allow the consideration of geometrically stable subsets. These geometrically stable subsets of the prime spectrum correspond exactly to localizations of the ring, determined by Gabriel filters which are defined in terms of certain sets of prime ideals; this accounts for the utility of localization functors in the commutative theory of relative invariants. An important example is obtained by considering the geometrically stable set $X^{(1)}$ of prime ideals of height one, which leads to the study of Krull domains and orders over these in a very natural way. If R is a Krull domain, then the relative Picard group corresponding to the set $X^{(1)}(R)$ is nothing but the divisorial class group of the Krull domain.

Although in the non-commutative situation we cannot expect a geometrical context for the study of relative invariants, it is still plausible to expect that relative Picard groups or similar objects may provide a tool for deriving structural properties of the ring. This conviction is based upon the knowledge that a useful theory does exist in the absolute case. Indeed, A. Fröhlich developed the general theory of Picard groups of non-commutative rings and in particular of orders in central (semi)simple algebras with an eye to application in the particular case of integral group rings. The method provided a useful tool in the study of Galois modules in Number Theory and it became an ingredient in the proof of Fröhlich's conjecture concerning the existence of an integral basis (with further consequences in the theory of L-series). Because of the rather specific aims, A. Fröhlich restricts attention to orders over Dedekind rings in the main part of his basic paper [52]. Nevertheless, it is pointed out in this paper that many results remain valid over more general rings. Guided by an interest in maximal orders, one is lead to focus on orders over completely integrally closed rings or over Krull domains. But even in the commutative case, it is clear that the Picard group is not the right invariant for studying Krull domains. Indeed the philosophy behind the introduction of a central invariant associated to a ring is that the vanishing of the invariants should be equivalent to specific properties of the ring. From this point of view, the class group is more intrinsically related to the notion of a Krull domain than the Picard group and so we are motivated to study certain class groups of orders or more general non-commutative rings. Needless to say, there are many possibilities to extend the notion of "class group" to the non-commutative situation and indeed we will consider several useful "non-commutative class groups" in this book. In view of the commutative theory, however, it is the relative Picard group which presents itself as a first possible candidate for extension to non-commutative rings.

Chapter B is devoted to the theory of relative Picard groups. Exact sequences, much in the vein of A. Fröhlich's results for the absolute Picard group, as well as Mayer-Vietoris sequences, will be derived. In this chapter the particular case of maximal orders over Krull domains receives special consideration only occasionally.

Chapter C is partly a return to the commutative situation because it is concerned with the study of relative Azumaya algebras, i.e., the algebras representing elements of relative Brauer groups. However, the first section deals with the structure of the algebras themselves and so we thought it appropriate to include it in the non-commutative part. In this section graded methods reappear because of the generalized crossed product structure of certain relative Azumaya algebras. The second section, on relative cohomology, has been included as an application and it may be viewed as an addendum to the theory of relative Brauer groups.

Graded ring theory and so-called graded invariants entered the picture when projective varieties and schemes were considered in the commutative theory. In Chapter C the non-vanishing of certain central Picard groups forces us to use generalized crossed products (strongly graded rings in the sense of [107]) instead of common crossed products and this accounts for the reappearance of graded techniques in this book. But there is a significant additional benefit to this because, apart from their own ring theoretical interests, constructions of graded nature like the generalized Rees rings over arithmetical orders may be utilized to calculate some of the invariants we have taken an interest in. To make this application possible, it is necessary to relate the graded invariants of graded rings of some suitable type to the ungraded equivalents; the second step is then to develop methods to calculate the graded invariants.

In Chapter D the main topic is the construction of new graded orders starting from a given one in such a way that the calculation of a certain invariant or its graded equivalent becomes easier. Links between the new graded orders' invariants and the corresponding invariants of the original order are then obtained by standard techniques in graded ring theory. Actually, the construction of well-chosen generalized Rees rings over a given order allows to kill-off a chosen subgroup of the "class group" of the given order and on many occasions one is able to do this in such a way that the "class group" of the new order vanishes. This brings us back to the point that structural properties of the ring resulting from the vanishing of a certain class group should be traceable. Another delicate point which presses forward here is that the structure of the subgroup of the class group which one is going to kill-off in the construction has vital importance because it will be (linked to) the grading group used in the construction. Separate sections in chapter D deal with different possibilities, e.g., $\mathbb{Z}$, $\mathbb{Z}^n$ or free abelian groups, torsion free abelian groups satisfying the ascending chain condition on cyclic subgroups, and also arbitrary finite groups (although we did not include the description of the behavior of the central class group in the finite case).

Before we can be more precise, we must point out an obvious weakness in the theory of the relative Picard groups. If $\Lambda$ is an R-order in a central simple algebra $\Sigma$ and if $\kappa$ is a kernel functor on R-modules corresponding to some geometrically stable subset Y in Spec(R), then a first impulse would be to try to relate Picent$(\Lambda,\kappa)$ and Pic$(R,\kappa)$. By just looking up the definitions it will be clear that the existence of nontrivial R-automorphisms of $\Lambda$ may obstruct the existence of a descent link between Picent$(\Lambda,\kappa)$ and Pic$(R,\kappa)$; actually Picent$(\Lambda,\kappa)$ need not be abelian in general. To remedy this we introduce the central class group of a maximal order over a Krull domain in Chapter E.

If the central class group of $\Lambda$ equals the class group of R, then $\Lambda$ is a reflexive Azumaya algebra (excluding some exceptional cases arising from undesired inseparability phenomena). In combination with the constructive tecnniques of Chapter D, the forementioned property allows to reduce problems about the central class group to problems about the class group of the center of a suitably constructed generalized Rees ring, whereas problems concerning maximal orders over Krull domains may often be reduced to problems about reflexive Azumaya algebras plus the descent of certain properties of a strongly (or divisorially) graded ring to properties of its part of degree zero. In the context of Krull domains, orders over Krull domains, reflexive modules, the notion of a divisorially graded ring is a very natural one. It provides us with many interesting new examples (e.g., of tame orders) but its main feature is that the construction of divisorially graded rings allows to kill-off given parts of the central class group whereas the construction of strongly graded rings only allows to kill-off given parts of the Picard group. This technique has a general relative counterpart in the construction of $\kappa$-graded rings but this has not been developed into full depth in this book.

The celebrated Skolem-Noether theorem related $\mathrm{Aut}_R(\Lambda)$ to the $\Lambda$-normalizing elements of $\Sigma$. If we go one step further in the search for invariants which neglect the part of $\mathrm{Pic}(\Lambda,\kappa)$ stemming from $\mathrm{Aut}_R(\Lambda)$, then it makes sense to introduce the normalizing class group of $\Lambda$ as being the quotient of the central class group by the classes of principal (fractional) ideals generated by $\Lambda$-normalizing elements. Although this class group seems to have some very meaningful properties, it is not clear what its vanishing will mean for the ring (vanishing should be interpreted as : vanishing modulo the class group of the center). Results obtained so far about the normalizing class group are still rather premature, (e.g., the interpretation of central and normalizing class groups in terms of Weil and Cartier divisors on the non-commutative structure sheaf in L. Le Bruyn's thesis) so we just included some basic properties and notions.

All invariants considered in this book are two-sided invariants in the sense that they are defined in terms of two-sided modules and ideals. This has obvious advantages, but also has the draw back, perhaps unsuspected, that it makes calculations more difficult; this is particularly evident in case one tries to calculate $\mathrm{Pic}(\Lambda,\kappa)$. One-sided invariants do exist, e.g., Fröhlich, Reiner and Ullom investigated the LFP-groups in [ 53 ] . It is plausible that one can proceed along similar lines of thought for orders over Krull domains if one is prepared to undertake a thorough investigation of reflexive left modules or suitable generalizations of these. We did not go into these matters here, nor did we dwell upon the K-theory of orders which makes for another interesting field (see R. Swan, G. Evans [ 141 ] ). Indeed, one of the main aims of the book is to study invariants "measuring" how an order differs from being in the well-understood class of relative Azumaya algebras and to provide both a structure theory of relative Azumaya algebras as well as constructive methods allowing to reduce problems about a certain order to problems about some relative Azumaya algebras combined with descent problems of graded natvre. This aim seems to be well-served by the two-sided invariants and the constructions of generalized Rees rings.

It may be evident from this introduction that the nature of this book is that it presents new methods to infer information about the structure of a ring from knowledge of the structure of certain groups called relative invariants of the ring. Therefore it may be said to be "natural" that certain parts of the book tend to be technical and the exposition may reach a high level of abstraction there. On the other hand the constructive parts make up for this and we also obtain methods for constructing very concrete new examples of arithmetical orders. Obviously, application of these new methods is at a very initial stage; some along-the-way applications have been included in the text and we did include in the References some recent papers contaning applications of the construction of generalized Rees rings. We hope the readers will appreciate the open-ended character of our presentation and also that they will be attracted to some of the problems which remain to be solved.

F. Van Oystaeyen

A. Verschoren

# ACKNOWLEDGMENT

We thank

- The University of Antwerp, UIA and the NFWO, the Foundation for Scientific Research for support and a job, respectively , a job and support.

- A. Fröhlich,for an inspiring conversation and his lecture at the meeting of the Flemish algebraists in Oostende.

- H. Marubayashi, for some joint work and drinking "Elephant" with us.

- M. Coppens for seminar talks, and not being mad when we did not listen.

- S. Caenepeel, E. Nauwelaerts, M. Van den Bergh and L. Le Bruyn, for some joint work and stimulating conversations.

- R. Hirschfeld, for providing some ancient references (but still being young at heart).

- Danielle and Lin, for living with mathematicians at work.

- Parents, relatives and friends for being there.

- Francine, for making our dirty drafts into an impeccably typed book.

We also thank the subdepartment of computer science for making us even more pure mathematicians.

# CONTENTS

# A PRELIMINARIES AND BASIC RESULTS

## A.1. Krull Domains.

In this section we recall some basic facts about Krull domains and lattices; for a complete treatment of the theory we refer to R.M. Fossum's book [ 51] .

A (commutative) domain R with field of fractions K is said to be a Krull domain provided there exists a family $\{\mathcal{O}_v ; v \in V\}$ of discrete valuation rings of K such that

(i) $R = \bigcap_{v \in V} \mathcal{O}_v$ ;

(ii) if $x \in R$, then x is a unit in all but a finite number of the rings $\mathcal{O}_v$.

In the following proposition we summarize some elementary properties of Krull domains.

1.1. Proposition. Let R be a Krull domain with field of fractions K then :

a. R is integrally closed ;

b. if k is a subfield of K, then $R \cap k$ is a Krull domain ;

c. if L is a finite field extension of K, then the integral closure of L is a Krull domain ;

d. if $\{\mathcal{O}_v ; v \in V\}$ is the family of discrete valuation rings defining R, then a subintersection $S = \bigcap_{v \in W} \mathcal{O}_v$ is a Krull domain for any subset $W$ of $V$ ;

e. any polynomial ring over R is a Krull domain ;

f. any formal power series ring over R is a Krull domain ; (Note : there are several ways of defining rings of formal power series in *infinitely* many indeterminates!)

g. if S is a multiplicatively closed subset of R, then $S^{-1}R$ is a Krull domain; this result generalizes to localizations at arbitrary idempotent kernel functors in the sense of O. Goldman [ 60 ]. □

Write $X^{(1)}(R)$ for the set of prime ideals of R having height one. These prime ideals may be used to characterize Krull domains as follows.

1.2. Proposition. A domain R is a Krull domain if and only if the following conditions are satisfied :

a. for each $p \in X^{(1)}(R)$, the local ring $R_p$ is a discrete valuation ring;

b. $R = \bigcap_{p \in X^{(1)}(R)} R_p$;

c. for any $x \neq 0$ in R, the number of prime ideals in $X^{(1)}(R)$ containing x is finite. □

For proofs and full detail the reader may consult N. Bourbaki's [ 22 ].

Let R be a domain with field of fractions K and consider a finite dimensional vector space V over K. An R-submodule M of V is said to be an R-lattice in V if M contains a K-basis for V and M is contained in a finitely generated R-submodule of V.

1.3. Proposition. Let M be an R-submodule of the K-vectorspace V; the following assertions are then equivalent :

a. M is an R-lattice;

b. there exist free R-modules $F_1$ and $F_2$ of rank $\dim_K V$ such that $F_1 \subset M \subset F_2 \subset V$;

c. there exists a free R-submodule F of V with rank equal to $\dim_K V$ and a nonzero $x \in R$ such that $xF \subset M \subset F$;

d. there exist R-lattices $N_1, N_2$ in V such that $N_1 \subset M \subset N_2$. □

Several operations on lattices may be considered, the ones most frequently used are being summed up in :

1.4. Proposition. Let R be a domain with field of fractions K and let U, V, $V_1,\ldots,V_m$, W be finite dimensional K-vectorspaces.

a. If M and N are R-lattices in V, then so are M + N and $M \cap N$.

b. If $U \subset V$ and M is an R-lattice in V, then $M \cap U$ is an R-lattice in U.

c. Consider R-lattices $M_1,\ldots,M_m$ in $V_1,\ldots,V_m$ respectively and let $\mu : V_1 \times \ldots \times V_m \to U$ be a multilinear form, then the R-module generated by $\mu(M_1 \times \ldots \times M_m)$ is an R-lattice in the subspace of U spanned by $\mu(V_1 \times \ldots \times V_m)$.

d. Let M be an R-lattice in V and N an R-lattice in W; define $N:M = \{\alpha \in \mathrm{Hom}_K(V,W);\ \alpha(M) \subset N\}$, then N:M is an R-lattice in $\mathrm{Hom}_K(V,W)$.

e. If $S \supset R$ is a domain with field of fractions L and M an R-lattice in V, then the image of $S \underset{R}{\otimes} M$ in $L \underset{K}{\otimes} V$ is an S-lattice.

Proof. cf. R.M. Fossum [51 ] or N. Bourbaki [ 22]. □

It is easy to see that the R-lattice N:M defined in Proposition 1.4. may be identified with $\mathrm{Hom}_R(M,N)$. Moreover, we may identify V and $\mathrm{Hom}_K(\mathrm{Hom}_K(V,K),K)$ canonically. With this identification M may be viewed as an R-submodule of R:(R:M) and clearly if M is a free R-lattice, then equality holds. For any R-lattice M, we have that R:M = R:(R:(R:M)). This leads to the following definition : an R-lattice M is said to be divisorial if and only if M = R:(R:M). If R is any R-lattice and M is a divisorial R-lattice, then M:N is a divisorial R-lattice (cf. Proposition 2.6 in R.M. Fossum [5 ]).

The terminology introduced above may be extended to arbitrary torsion free R-modules. We say that an R-module M is divisorial if it is torsion free and if within $K \underset{R}{\otimes} M$ we have

$M = \cap\{M_p = R_p \otimes_R M;\ p \in X^{(1)}(R)\}$.

The <u>rank</u> of M is the dimension of the K-vectorspace $K \otimes_R M$. For any torsion free R-module M we define

$R:M = \{f \in \mathrm{Hom}_R(K \otimes_R M, K);\ f(M) \subset R\}$

and we write $M^\star$ for $\mathrm{Hom}_R(M,R)$. Via the canonical isomorphism $K \otimes_R M \cong (K \otimes_R M)^{\star\star}$ (here the bidual K-module is considered), we may view $R:(R:M)$ as an R-submodule of $K \otimes_R M$ in a natural way. If M is an R-lattice, then R:M is isomorphic to $M^\star$ in a natural way and therefore M is divisorial if and only if $M \cong M^{\star\star}$, i.e., if and only if M is a *reflexive* R-module or if and only if $M = \cap\{M_p; p \in X^{(1)}(R)\}$. All of these claims are easily checked.

We maynow apply the theory expounded above to R-lattices in K. An R-lattice in K is called a <u>fractional ideal</u> of R. It follows that R:I is a fractional ideal whenever I is. Moreover in this case :

$R:(R:I) = \cap\{Rx;\ x \in K \text{ and } I \subset Rx\}$

and therefore I is divisorial if and only if it is the intersection of principal fractional ideals. For any three fractional ideals I,J and H of R we have $I:JH = (I:J):H$ and $(I:J)H \subset I:(J:H)$.

Let us now focus on Krull domains. Recall from R.M. Fossum [ 50, § 3 ] that a domain R is <u>completely integrally closed</u> if and only if $R = I:I$ for every fractional ideal I of R.

Let $\mathcal{D}(R)$ be the set of divisorial (fractional) ideals of R. On $\mathcal{D}(R)$ we may define an operation ★ by putting $I \star J = R:(R:IJ)$ and it is easily verified that $\mathcal{D}(R)$ is a commutative monoid with respect to ★. From Proposition 3.4. in R.M. Fossum [ 51 ], we retain that $\mathcal{D}(R)$ is a *group* exactly when R is completely integrally closed. It is well known, cf. loc. cit. Theorem 3.6., that a Krull domain is completely integrally closed and that the set of divisorial ideals satisfies the ascending chain condition. The converse follows from the following useful theorem :

1.5. Theorem. Let R be a completely integrally closed domain such that its set of divisorial ideals satisfies the ascending chain condition, then,

a. $p \in X^{(1)}(R)$ if and only if p is a maximal divisorial prime ideal of R;

b. if $p \in X^{(1)}(R)$, then $R_p$ is a discrete valuation ring;

c. $R = \cap \{R_p ; p \in X^{(1)}(R)\}$;

d. a fractional ideal I of R is divisorial if and only if $I = R:(R:(p_1 . \dots .p_m))$ for certain $p_i \in X^{(1)}(R)$; in this case the appearing prime ideals are determined by I;

e. any divisorial ideal is contained in only finitely many prime ideals in $X^{(1)}(R)$. □

1.6. Corollary Under the above hypotheses;

a. R is a Krull domain;

b. $\mathcal{D}(R)$ is the free group generated by the prime ideals in $X^{(1)}(R)$.

Proof. cf. Theorem 3.12, Corollary 3.13 and Corollary 3.14 in [51 ]. □

For torsion free R-modules M and N, consider $V = K \otimes_R M$ and $W = K \otimes_R N$. Following S. Yuan [ 177], we introduce the notion of "modified tensor-product" as follows. Let M.N be the canonical image of $M \otimes_K N$ in $V \otimes_K W$. Define $M \perp_R N = \cap\{(M.N)_p ; p \in X^{(1)}(R)\}$.In case M and N are R-lattices, there is a natural morphism $M \otimes_R N \to M \perp_R N$, such that $(M \otimes_R N)^{**} \cong M \perp_R N$. It is advantageous to use a localization-theoretic description of $\perp_R$ ; this has been proven useful in the commutative theory of relative invariants of rings, as exposed in [ 162] . Some generalities on "abstract" localization have been included in section A.5. for full detail on kernel functors and localization (of noncommutative rings) we refer to O. Goldman [ 60 ] or J. Golan [ 59 ]. By $\kappa_{R-p}$ we denote the kernel functor associated to the prime ideal p of R, i.e. $\kappa_{R-p}M = \{m \in M;$ $sm = 0$ for some $s \in R-p\}$ for any R-module M. Put $\sigma_1 = \inf\{\kappa_{R-p}; p \in X^{(1)}(R)\}$ and write $Q_{\sigma_1}$ for the localization functor in R-mod associated to $\sigma_1$. For any R-lattices M and N we thus have $Q_{\sigma_1}(M \otimes_R N) = M \perp_R N$. The multiplication law in $\mathcal{D}(R)$ may therefore be described alternatively by $I \star J = Q_{\sigma_1}(IJ)$ for any $I,J \in \mathcal{D}(R)$.

An R-module P is said to be invertible, provided there exists an R-module Q together with an isomorphism $P \otimes_R Q \cong R$. It is clear that the invertible R-modules are exactly the projective rank one R-modules. Let $\mathcal{I}(R)$ be the group of invertible R-submodules of K with respect to the tensor product. This group $\mathcal{I}(R)$ is a subgroup of $\mathcal{D}(R)$, with for any $I,J \in \mathcal{I}(R)$

$$Q_{\sigma_1}(IJ) = I\star J = IJ \cong I \otimes_R J$$

Let $\mathcal{P}(R)$ be the subgroup of $\mathcal{I}(R)$ consisting of *principal* invertible ideals of R. We then define the class group of R as $Cl(R) = \mathcal{D}(R)/\mathcal{P}(R)$ and the Picard group of R to be $Pic(R) = \mathcal{I}(R)/\mathcal{P}(R)$. It is well-known that the Picard group may also be viewed as the group isomorphism classes of invertible R-modules with respect to the tensorproduct. The use of the class group in the study of Krull domains and factorial rings is illustrated by the following elementary result, cf. R.M. Fossum [ 51 ]. Proposition 6.1.

1.7. Proposition. For any domain R, the following assertions are equivalent :

a. R is factorial;

b. every nonzero prime ideal of R contains a nonzero principal prime ideal;

c. R is a Krull domain with $Cl(R) = 1$;

d. R is a Krull domain such that each $P \in X^{(1)}(R)$ is principal;

e. R satisfies the ascending chain condition for principal ideals and two principal ideals intersect to a principal ideal;

f. R satisfies the ascending chain condition for principal ideals and each principal ideal which is maximal in the set of principal ideals is a prime ideal of R. □

Consider Krull domains $R \subset S$ with $Q(R) = K \subset Q(S) = L$ (here Q(R) denotes the field of fractions of R, etc). If $Q \in X^{(1)}(S)$ and $q = Q \cap R$ is in $X^{(1)}(R)$, then we may define the ramification index $e = e(Q/q)$ to be the unique integer e such that $q\, S_Q = Q^e S_Q$. To each fractional ideal I of a Krull domain A say, we associate the <u>divisor</u> div I,

$$\text{div } I : X^{(1)}(A) \to \mathbb{Z}$$

$$p \mapsto v_p(I),$$

where $v_p$ is the valuation associated to $A_p$. In this way we obtain an order preserving isomorphism

$$\text{div} : \mathcal{D}(A) \to \mathbb{Z}^{(X^{(1)}(A))}$$

The monomorphism $i : R \to S$ now defines a group morphism

$$\text{Div}(i) : \mathcal{D}(R) \to \mathcal{D}(S)$$

by sending q to $e(Q/q)Q$, (in this notation we have identified $\mathcal{D}(R)$ with $\mathbb{Z}^{(X^{(1)}(R))}$ and $\mathcal{D}(S)$ with $\mathbb{Z}^{(X^{(1)}(S))}$). The morphism Div(i) induces a morphism $Cl(R) \to Cl(S)$ if and only if i satisfies condition "PDE" (pas d' éclatement) :

(PDE) : if $Q \in X^{(1)}(S)$, then $ht(Q \cap R) \leqslant 1$.

Note that PDE is equivalent to Div(i) mapping $\mathcal{P}(R)$ to $\mathcal{P}(S)$ or to $S_Q$ being a flat R-module, for any $Q \in X^{(1)}(S)$.

1.8. <u>Proposition</u>. If $R \subset S$ are Krull domains, then PDE holds in each of the following situations :

a. S is a flat R-module;

b. S is an integral extension of R;

c. S is a subintersection of R.

Proof : cf. Proposition 6.4. in R.M. Fossum [51 ]. □

Let us conclude this section with some helpful remarks on the classgroup. First, it is clear that every flat extension of a Krull domain R within its field of fractions K is a subintersection of R. Conversely, every subintersection of a Krull domain R is a ring of fractions if and only if Cl(R) is a torsion group. A necessary and sufficient condition for Cl(R) to be a torsion group is that for each $p \in X^{(1)}(R)$, we have that $p^n$ is principal for some $n > 0$, or equivalently, that each nonzero $x \in R$ which is *not* a unit of R has the property that some power $x^n$ of it is a product of primary elements (cf. Proposition 6.8. in [51 ]). If $R \to S$ is a faithfully flat extension of Krull domains, then $\mathrm{Ker}(\mathrm{Cl}(R) \to \mathrm{Cl}(S)) = \mathrm{Ker}(\mathrm{Pic}(R) \to \mathrm{Pic}(S))$.

1.9. Theorem. (Nagata) Let R be a Krull domain and let S be a subintersection $\cap\{R_p ; p \in T\}$ for some $T \subset X^{(1)}(R)$. We then have :

a. the morphism $\mathrm{Cl}(R) \to \mathrm{Cl}(S)$ is surjective;

b. the group $\mathrm{Ker}(\mathrm{Cl}(R) \to \mathrm{Cl}(S))$ is generated by the classes of prime ideals in $X^{(1)}(R)-T$. □

1.10. Corollary. Let S be a multiplicatively closed subset of R generated by prime elements of R, then

a. the morphism $\mathrm{Cl}(R) \to \mathrm{Cl}(S^{-1}R)$ is surjective;

b. the kernel of $\mathrm{Cl}(R) \to \mathrm{Cl}(S^{-1}R)$ is generated by the classes of prime ideals of R that meet S. □

Let us also mention that whereas for any Krull domain R, the canonical

map $Cl(R) \to Cl(R[X])$ is also an isomorphism, this need not be the case for $Cl(R) \to Cl(R[[X]])$, which is only *monomorphic* in general.
For some results on graded Krull domains and gr-Dedekind rings, we refer to F. Van Oystaeyen, A. Verschoren [162].

## A.2. Maximal Orders and Tame Orders over Krull Domains.

This section is an adaptation of a part of R.M. Fossum's paper [50]. The theory of maximal orders over Dedekind domains is well-documented in the literature, e.g. in I. Reiner's book [122]. In the sequel, results available for noetherian integrally closed domains are extended to a theory over arbitrary Krull domains. The parts of R.M. Fossum's paper [50] dealing with a certain class group of an order have not been included in this section. In other chapters we will introduce relative Picard groups and certain class groups of orders over Krull domains.
Throughout this section, R will be a Krull domain with field of fractions K and A will be a central simple algebra over K. An intermediate ring, $R \subset \Lambda \subset A$ such that $K\Lambda = A$ and each element of $\Lambda$ is integral over R is called an R-<u>order</u> of A. The reduced trace $Tr: A \to K$ induces a K-vector-space isomorphism $t : A \to Hom_K(A,K)$ given by $t(a)(b) = Tr(ab)$, for $a,b \in A$. If $a \in A$ is integral over R, then $Tr(a) \in R$. For any K-basis $\{a_1,\dots,a_n\}$ of A we may find $a_1^\star,\dots,a_n^\star \in A$ such that $Tr(a_i^\star a_j) = \delta_{ij}$ and if we write $F = Ra_1 + \dots + Ra_n$ and $F^o = Ra_1^\star + \dots + Ra_n^\star$ then t restricts to an isomorphism

$$t: F^o \to F^\star = R:F$$

2.1. <u>Proposition.</u> a. Consider R-orders $\Lambda \subset \Gamma$ of A and suppose that $\Lambda$ contains a free R-module F, then $F^o \supset \Gamma$.
b. The R-orders of A are R-lattices in A.

Proof. a. It is clear that $\Gamma F \subset \Gamma$. Since $Tr(\Gamma) \subset R$, it follows that $Tr(\Gamma F) \subset R$, i.e. $\Gamma \subset F^{o}$.

b. By definition, each R-order $\Lambda$ contains a free R-module and from a. it follows that $\Lambda$ is contained in a free R-lattice in A; consequently $\Lambda$ is an R-lattice. □

An R-order is said to be maximal if it is not contained properly in another R-order. Every R-order of A is contained in a maximal R-order of A (Theorem 1.4. in [ 50 ], p. 323). A local-global characterization for maximal orders may be derived following the lines of proof used in the classical result over Dedekind domains, as in the basic paper [ 15 ] by M. Auslander and O. Goldman.

2.2. Proposition. An R-order $\Lambda$ of A is a maximal R-order of A if and only if the following conditions are satisfied :

1. $\Lambda$ is a divisorial R-lattice in A;
2. $\Lambda_p$ is a maximal $R_p$-order of A for each $p \in X^{(1)}(R)$. □

To the central simple K-algebra A there corresponds a skewfield $\Delta$ and a finite dimensional right $\Delta$-vectorspace V such that $A \cong End(V_\Delta) \cong M_r(\Delta)$. With these notations we have :

2.3. Theorem : An R-order $\Lambda$ of A is maximal if and only if $\Lambda \cong End(E_\Gamma)$ for some maximal R-order $\Gamma$ of $\Lambda$ and a right $\Gamma$-submodule E of V which is a divisorial R-lattice.

Proof. (cf. Theorem 1.7. in [ 50 ]). In the regular representation of $\Delta$ in $End(V_K)$ we have that $A^{opp} = \{f \in End(V_K);\ fd = df \text{ for all } d \in \Delta\}$, where $A^{opp}$ is the opposite algebra of A. We identify $\Delta$ and $A^{opp}$ with subalgebras of End $(V_K)$. Also, it is known that V is a simple left A-module. Pick $v \neq 0$ in V and consider $\Lambda v$ for some maximal R-order $\Lambda$ of A. It is clear that $\Lambda v$ is an R-lattice in V and $E = \cap \{(\Lambda v)_p;\ p \in X^{(1)}(R)\}$

is a $\Lambda$-module. Put $\Gamma = \Delta \cap (E:E)$, then $\Gamma$ is a divisorial R-order in $\Delta$ and for any $p \in X^{(1)}(R)$, we have that $\Gamma_p = \Delta \cap (E_p:E_p)$ is isomorphic to $\mathrm{Hom}_{\Lambda_p}(E_p, E_p)$, hence $\Gamma_p$ is a maximal $R_p$-order of $\Delta$. On the other hand, maximality of $\Lambda^{opp}$ in $A^{opp}$ entails that $\Lambda^{opp} = A^{opp} \cap (E:E)$ and clearly $A^{opp} \cap (E:E) \cong \mathrm{End}(E_\Gamma)^{opp}$. For the other implication, suppose that $\Gamma$ is a maximal R-order in $\Delta$ and let E be a divisorial R-lattice in V, then $\mathrm{End}(V_\Delta)^{opp} \cap (E:E)$ in $\mathrm{End}(V_K)$ is a divisorial R-order in $\mathrm{End}(V_D)^{opp} = A^{opp}$ Arguing as before yields that it is actually a maximal R-order. □

So far we managed in building a theory of orders over Krull domains, even if they need not be finitely generated as modules over the center. Nevertheless, we want to point out a result in PI-theory, due to A. Braun, which presents a criterion for orders over Krull domains to be finitely generated modules. For the general theory of PI-rings, we refer to C. Procesi [116], N. Jacobson [ 72 ] or P. Cohn [ 38 ].

2.4. Proposition. If $\Lambda$ is an R-order of A such that $\Lambda$ is an affine PI-algebra, then $\Lambda$ is a finitely generated R-module.

Proof. A special case of results by A. Braun [ 23 ]. Note that $\Lambda$ is already a PI-ring, being a subring of one (A!). □

The concept of maximal R-orders of A may be generalized further. A class of divisorial R-orders having received considerable attention in the litterature, cf. [ 50 ], [140], ..., is given by the so-called tame R-orders, which may be viewed as "globalizations" of hereditary orders over discrete valuation rings. Let us point out that in L. Silver's treatment [140] the orders considered are by assumption finite modules over their center, whereas in R.M. Fossum's approach [ 50 ] this is not necessarily the case.

Let R, K and A be as before. An R-order is said to be <u>tame</u>, if it is divisorial as an R-lattice and if $\Lambda_p$ is an hereditary $R_p$-order in A for each $p \in X^{(1)}(R)$. If $\Lambda'$ is a maximal R-order containing the tame order $\Lambda$, then $\Lambda'_p = \Lambda_p$ for almost all $p \in X^{(1)}(R)$ (i.e. all but at most a finite number of exceptions).

Since a PI-algebra with noetherian center is necessarily a finite module over its center, it follows that the above definition reduces to the definition given by L.Silver in [140], whenever R is a *noetherian* Krull domain.

2.5. <u>Lemma</u>. 1. Let E and F be finitely generated $\Lambda$-modules, where $\Lambda$ is an R-algebra and F is reflexive as an R-module, then $\mathrm{Hom}_\Lambda(E,F)$ is reflexive as an R-module.

2. If $\Lambda$ is a reflexive R-algebra, then any finitely generated reflexive $\Lambda$-module is also reflexive as an R-module.

<u>Proof</u>. 1. $\mathrm{Hom}_\Lambda(E,F) = \mathrm{Hom}_\Lambda(E,F^{**}) \cong \mathrm{Hom}_R(F^* \otimes_\Lambda E,R)$, by Cartan-Eilenberg [ 28 ], II.5.2.

2. Put $F = \Lambda$ in 1! □

2.6. <u>Proposition</u> : 1. Let $\Lambda$ be a tame order over the Krull domain R, then every divisorial R-order $\Gamma$ which contains $\Lambda$ is again tame.

2. If we call an ideal I of $\Lambda$ <u>quasi-idempotent</u> if $I = (I^2)^{**}$ (i.e. if $I = Q_{\sigma_1}(I^2)$ in the torsion theoretic wording introduced in the first section), then there is a canonical one-to-one correspondence between divisorial R-orders containing $\Lambda$ and quasi-idempotent ideals of $\Lambda$.

<u>Proof</u>. L. Silver's proof of Theorem 1.8. in [140] may be repeated here without any changes in the slightly more general context of divisorial orders over Krull domains. We include it here for convenience's sake.

1. If $\Gamma$ is a divisorial R-order containing $\Lambda$, then $\Gamma_p \supset \Lambda_p$ for all $p \in X^{(1)}(R)$ entails that $\Gamma_p$ is a hereditary (noetherian) $R_p$-order, hence $\Gamma$ is tame.

2. The conductor $\Lambda : \Gamma = \mathrm{Hom}_\Lambda(\Gamma_\Lambda, {}_\Lambda\Lambda)$ is reflexive as an R-module, and it is an ideal of $\Lambda$, hence an R-lattice. For an arbitrary ideal I of $\Lambda$, we have that $(I^2)_p = I_p^2$, hence in particular $((\Lambda:\Gamma)^2)_p = (\Lambda_p:\Gamma_p)^2$. However, $(\Lambda_p:\Gamma_p)^2 = \Lambda_p:\Gamma_p$, because both $R_p$-orders are hereditary. Since $(\Lambda:\Gamma)^{**} = \cap\{\Lambda_p:\Gamma_p; p \in X^{(1)}(R)\}$ and similarly for $((\Lambda:\Gamma)^2)^{**}$, it follows from the local information that $\Lambda:\Gamma$ is quasi-idempotent. Furthermore, it is clear that $\mathrm{End}((\Lambda:\Gamma)_\Lambda) = \Gamma$. Conversely, if I is a quasi-idempotent ideal of $\Lambda$, then $\Gamma = \mathrm{End}(I_\Lambda)$ is a tame order containing $\Lambda$, since it is divisorial as an R-lattice, while $\Gamma_p$ is a hereditary $R_p$-order containing $\Lambda_p$ at each $p \in X^{(1)}(R)$. By localizing at all $p \in X^{(1)}(R)$, one also establishes that $\Lambda:\Gamma = I$. □

2.7. Corollary. If $\Lambda$ is a tame R-order, then the number of divisorial R-orders containing $\Lambda$ is finite. For each $p \in X^{(1)}(R)$, let e(p) be the number of hereditary $R_p$-orders containing $\Lambda_p$, then there are exactly $e = \sum_{p \in X^{(1)}(R)} e(p)$ tame orders containing $\Lambda$.

Proof. In Theorem 1.7. of M. Harada [ 65 ], it has been established that e(p) is finite. On the other hand, $\Lambda_p$ is a maximal $R_p$-order for almost all $p \in X^{(1)}(R)$. Consequently, e is finite. Any quasi-idempotent ideal I of $\Lambda$ is a divisorial R-lattice by definition, whence the first statement and the fact that e is an upper bound for the number of tame orders containing $\Lambda$. Consider for each $p \in X^{(1)}(R)$ an idempotent ideal I(p) of $\Lambda_p$ and note that, since $\Lambda_p$ is a maximal $R_p$-order for almost all $p \in X^{(1)}(R)$, it follows that $I(p) = \Lambda_p$ for almost all such p. Putting $I = \cap\{I(p); p \in X^{(1)}(R)\}$ yields an R-divisorial ideal of $\Lambda$, which is obviously quasi-idempotent. The fact that the number of tame orders containing $\Lambda$ is exactly e now follows directly. □

An R-lattice M in A, which is a two-sided $\Lambda$-module is called a $\Lambda$-divisor if M is a divisorial R-lattice and $M_p$ is an invertible $\Lambda_p$-module for each $p \in X^{(1)}(R)$. For generalities on Picard groups of orders we refer to the following section. Since $\Lambda_p$ is hereditary, $M_p$ is invertible exactly when $M_p M_p^{-1} = \Lambda_p = M_p^{-1} M_p$ for each $p \in X^{(1)}(R)$.

If E is a right module over any ring $\Gamma$, then we define the (right) trace ideal $t_\Gamma(E) = t_\Gamma^r(E)$ to be the image of $t : E \otimes_\Gamma \mathrm{Hom}_\Gamma(E,\Gamma_\Gamma) \to \Gamma$ given by $t(e \otimes f) = f(e)$. The left trace ideal $t_\Gamma^l(E)$ of a left $\Gamma$-module is defined similarly. Clearly, the trace ideal is always an ideal of $\Gamma$.

2.8. Proposition. Let $\Lambda$ be a tame R-order in A and let M be a twosided $\Lambda$-module in A, which is also a divisorial R-lattice, then the following statements are equivalent :

1. M is a $\Lambda$-divisor;
2. $\mathrm{End}(M_\Lambda) = \Lambda$;
3. $(t_\Lambda^l(M))^{**} = \Lambda$;
4. $(t_\Lambda^r(M))^{**} = \Lambda$;
5. $\mathrm{End}({}_\Lambda M) = \Lambda$.

Proof. It suffices to prove the equivalence of the statements locally at any $p \in X^{(1)}(R)$, i.e. for hereditary orders. Therefore, the proof of 2 ⇔ 3 ⇔ 4 ⇔ 5 is formally the same as that of proposition 2.1. in [ 65 ]. Since $M_p$ is an invertible $\Lambda_p$-module for each $p \in X^{(1)}(R)$, the equivalence of 1 and 2 (or 5) is evident, see also the next section! □

Let $D(\Lambda)$ be the set of $\Lambda$-divisors. In the foregoing it has been shown that the use of divisorial R-lattices, a mild strengthening of the concept of reflexive R-modules, but more general than the finitely generated reflexive R-modules used by L. Silver in [140], allows us to extend many results of [140] to the case where R is merely a Krull domain. We may proceed in the same way and establish :

2.9. Theorem. 1. $\mathcal{D}(\Lambda)$ is an abelian group. In fact, $\mathcal{D}(\Lambda)$ is the free abelian group generated by the ideals $J(\Lambda_p) \cap \Lambda$, for all $p \in X^{(1)}(R)$, where $J(\Lambda_p)$ is the Jacobson radical of $\Lambda_p$.

2. The $\Lambda$-divisor M is invertible if and only if $MM^{-1} = \Lambda = M^{-1}M$.

3. If M and N are invertible divisors, then $M^{-1}$ and $M \star N = (MN)^{\star\star} = Q_{\sigma_1}(MN)$ are invertible.

The invertible $\Lambda$-divisors form a subgroup of $\mathcal{D}(\Lambda)$.

Proof : 1. The product in $\mathcal{D}(\Lambda)$ is defined by $M \star N = (MN)^{\star\star}$ for any $M,N \in \mathcal{D}(\Lambda)$. The proof of the statement is completely similar to the proof of Theorem 2.3. of L. Silver [140].

2. cf. loc. cit. Proposition 3.1.

3. cf. loc. cit. Proposition 3.2. □

Let us also mention :

2.10. Proposition. Let R be a Krull domain with field of fractions K, let A be a central simple algebra over K, and let $\Lambda_1$, $\Lambda_2$ be maximal R-orders in A. The conductor $\Lambda_1 : \Lambda_2$ induces an isomorphism $d(\Lambda_2, \Lambda_1) : \mathcal{D}(\Lambda_1) \to \mathcal{D}(\Lambda_2)$. If $\Lambda_3$ is another maximal R-order in A, then $d(\Lambda_3, \Lambda_2) d(\Lambda_2, \Lambda_1) = d(\Lambda_3, \Lambda_1)$.

Proof. The conductor $\Lambda_1 : \Lambda_2$ is a divisorial R-lattice and it is a $\Lambda_2 - \Lambda_1$-bimodule. It is easily checked that $[(\Lambda_2 : \Lambda_3)(\Lambda_1 : \Lambda_2)]^{\star\star} = \Lambda_1 : \Lambda_3$. Define $d(\Lambda_2, \Lambda_1)(M) = [(\Lambda_1 : \Lambda_2) M (\Lambda_2 : \Lambda_1)]^{\star\star}$, for any $M \in \mathcal{D}(\Lambda_1)$. The statements of the proposition are easily checked. □

2.11. Remark. The foregoing proposition is Theorem 2.1. in R.M. Fossum [ 50 ]. Note, however, that both L. Silver's and R.M. Fossum's treatment of $\mathcal{D}(\Lambda)$ are just a generalization of a divisor group introduced by O. Goldman in [ 61 ].

2.12. Proposition. Let $\Lambda$ be a tame R-order in A and let $Y \subset X^{(1)}(R)$ be arbitrary. If $S = \cap\{R_p ; p \in Y\}$, then $T = \cap\{\Lambda_p ; p \in Y\}$ is a tame S-order in A; moreover, if $\Lambda$ is a maximal R-order, then T is a maximal S-order.

Proof. It is obvious that S is a Krull domain and that T is a divisorial S-order. By localization at each $p \in Y$, the result follows. □

2.13. Corollary. Let S be a multiplicatively closed subset of R. If $\Lambda$ is a tame (resp. maximal) R-order, then $S^{-1}\Lambda$ is a tame (resp. maximal) $S^{-1}R$-order. □

It is also fairly easy to prove that a polynomial ring $\Lambda[X]$ over a tame R-order $\Lambda$ in A is a tame $R[X]$-order in $A \otimes_K K(X)$. Further examples of maximal and tame orders over Krull domains may be constructed by graded techniques; we refer to chapter D for further results.

## A.3. Picard Groups of Noncommutative Rings.

In the first part of this book on relative invariants of rings, we have introduced and studied Picard groups and relative Picard groups of commutative rings. Here, in the treatment of the noncommutative theory, we follow A. Fröhlich's development of this material, cf. [ 52 ]. The reader may also consult H. Bass's book [ 18 ]. Although we do not intend to go into the theory of one-sided class groups of orders, we should point out that such "class groups" exist and have been extensively studied in A. Fröhlich, I. Reiner, S. Ullom [ 53 ], and I. Reiner, S. Ullom [ 123 ]

Consider arbitrary rings A and B. An A-B-bimodule ${}_AM_B$ is said to be <u>invertible,</u> if the following equivalent conditions are satisfied.

(I.1) $M_B$ is a finitely generated projective right B-module and $\mathrm{End}(M_B) \cong A$ via the action of A on ${}_AM_B$, plus the similar statement obtained by interchanging the roles of A and B;

(I.2) there exists a B-A-bimodule ${}_BN_A$, together with isomorphisms $f: N \otimes_A M \to B$ resp. $g : M \otimes_B N \to A$ of B-bimodules resp. A-bimodules, such that the following diagrams commute :

$$\begin{array}{ccc} N \otimes_A M \otimes_B N & \xrightarrow{f \otimes N} & B \otimes_B N \\ \downarrow{\scriptstyle N \otimes g} & & \downarrow{\scriptstyle \cong} \\ N \otimes_A A & \xrightarrow[\cong]{} & N \end{array}$$

$$\begin{array}{ccc} M \otimes_B N \otimes_A M & \xrightarrow{g \otimes M} & A \otimes M \\ \downarrow{\scriptstyle M \otimes f} & & \downarrow{\scriptstyle \cong} \\ M \otimes_B B & \xrightarrow[\cong]{} & M \end{array}$$

If A and B are R-algebras over some commutative ring R, then we say that ${}_AM_B$ is an A-B-bimodule <u>over R</u> if the R-module structures on M induced by A and B coincide. The isomorphism classes [ M ] of invertible A-bimodules equiped with the product $[M][N] = [M \otimes_A N]$ form the <u>Picard group</u> Pic(A) of A. The subgroup $\mathrm{Pic}_R(A)$, where R is a commutative ring over which A is an algebra, given by the isomorphism classes of invertible A-bimodules over R is the <u>Picard group of</u> A <u>over</u> R. Of course, $\mathrm{Pic}(A) = \mathrm{Pic}_{\mathbb{Z}}(A)$!

A <u>Morita equivalence</u> $B \sim A$ over R may be defined to be the isomorphism class [ H ] of an invertible A-B-bimodule M over R. The functor $\mathrm{Pic}_R(-)$

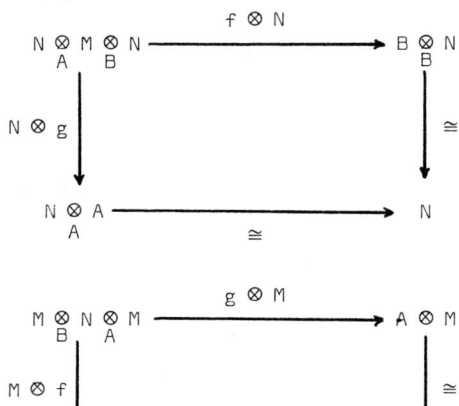

may be viewed as a functor from the category of Morita equivalences of R-algebras to groups and isomorphisms.

If $\mathbb{P}(A)$ denotes the set of isomorphism classes of finitely generated faithfully projective left A-modules, then we obviously have a map $\gamma : \mathrm{Pic}(A) \to \mathbb{P}(A)$, defined by forgetting the right A-module structure. Furthermore, we can define an action of Pic(A) on $\mathbb{P}(A)$ as follows : for $[M] \in \mathrm{Pic}(A)$ and $[P] \in \mathbb{P}(A)$, put $[M].[P] = [M \otimes_A P]$.

Probably the most evident problem concerning $\mathrm{Pic}_R$ is to determine how it relates to the well-known (?) functor $\mathrm{Aut}_R(-)$. The explication of relations between these functors benefits from the use of certain limit groups, based on the observation that the class of Azumaya algebras over R may be regarded as fairly trivial, in terms of Morita equivalences over R.

Indeed, if F denotes one of the functors $\mathrm{Pic}_R(-)$ or $\mathrm{Aut}_R(-)$, and if $E_1$ and $E_2$ are R-Azumaya algebras, then we obtain a morphism $F(A \otimes_R E_1) \to F(A \otimes_R E_1 \otimes_R E_2)$. The same is valid for F = Inn(-), the functor of "inner automorphisms". One verifies easily that these morphisms define directed systems of groups, so we may introduce the limit groups $F_R^\infty(A) = \varinjlim F(A \otimes_R E)$ where E runs through all R-Azumaya algebras.

The obvious morphism $\mathrm{Pic}_R(A_1) \times \mathrm{Pic}_R(A_2) \to \mathrm{Pic}_R(A_1 \otimes_R A_2)$ for R-algebras $A_1$ and $A_2$, is compatible with Morita equivalence, so the same holds for the derived morphism $\mathrm{Pic}_R(A_1) \to \mathrm{Pic}_R(A_1 \otimes_R A_2)$ which sends $[M] \in \mathrm{Pic}_R(A_1)$ to $[M \otimes A_2] \in \mathrm{Pic}_R(A_1 \otimes_R A_2)$. If $A_2$ is an Azumaya algebra, then the latter morphism is an isomorphism. From this it follows that for $F = \mathrm{Pic}_R(-)$ the morphisms in the directed system $\{F(A \otimes_R E)\}$ introduced above are bijective. Consequently, we may identify $\varinjlim \mathrm{Pic}_R(A \otimes_R E) = \mathrm{Pic}_R^\infty(A)$ and $\mathrm{Pic}_R(A)$.

For any $\alpha, \beta \in \mathrm{Aut}_R(A)$, define ${}_\alpha A_\beta$ to be the A-bimodule with underlying additive group A and with twisted A-action given by $a.b.a' = \alpha(a)b\beta(a')$

for any a, a', b $\in$ A. We then have :

3.1. Theorem. 1. The following sequence is exact :

$$1 \to \mathrm{Inn}(A) \to \mathrm{Aut}_R(A) \overset{\tau}{\to} \mathrm{Pic}_R(A),$$

where $\tau$ is defined by putting $\tau(\alpha) = {}_1A_\alpha$ for any $\alpha \in \mathrm{Aut}_R(A)$. The cokernel of $\tau$ corresponds bijectively to the fibres of $\gamma : \mathrm{Pic}_R(A) \to \mathbb{P}(A)$. The orbits of $\mathbb{P}(A)$ under the action of $\mathrm{Pic}_R(A)$ are the $[P] \in \mathbb{P}(A)$, which yield isomorphic R-algebras $\mathrm{End}_A(P)$.

2. For $[M]$, $[N] \in \mathrm{Pic}_R(A)$ and $P \in \mathbb{P}(R)$ we have that $M \otimes_R \mathrm{End}_R(P) \cong N \otimes_R \mathrm{End}_R(P)$ in $A \otimes_R \mathrm{End}_R(P)$-mod if and only if $M \otimes_R P^\star \cong N \otimes_R P^\star$ in A-mod.

3. The exact sequence in 1 yields an exact "limit sequence".

$$1 \to \mathrm{Inn}_R^\infty(A) \to \mathrm{Aut}_R^\infty(A) \overset{\overline{\tau}}{\to} \mathrm{Pic}_R(A)$$

4. For $[M]$, $[N] \in \mathrm{Pic}_R(A)$, the following statements are equivalent :

a. $[M] = [N] . [U]$ for some $[U] \in \mathrm{Im}\,\overline{\tau}$;

b. there is an Azumaya R-algebra E such that $M \otimes_R E \cong N \otimes_R E$ in $A \otimes_R E$-mod.

c. There is a $P \in \mathbb{P}(R)$ such that $M \otimes_R P \cong N \otimes_R P$ in A-mod;

d. for some positive integer n we have $M^{(n)} \cong N^{(n)}$.

Proof. This is Theorem 1 in [52]. □

We define the central Picard group Picent (A) of A to be $\mathrm{Pic}_{Z(A)}(A)$, where Z(A) is the center of A. Similarly, $\mathrm{Autcent}(A) = \mathrm{Aut}_{Z(A)}(A)$!

3.2. Corollary. The following sequences are exact :

$$1 \to \mathrm{Inn}(A) \to \mathrm{Autcent}(A) \overset{\tau}{\to} \mathrm{Picent}(A)$$

resp.

$$1 \to \varinjlim \mathrm{Inn}(A \underset{Z(A)}{\otimes} E) \to \varinjlim \mathrm{Autcent}(A \underset{Z(A)}{\otimes} E) \overset{\overline{\tau}}{\to} \mathrm{Picent}(A). \quad \square$$

3.3. Proposition. Let M be an invertible A-B-bimodule over R; there is a unique R-algebra isomorphism $f: Z(B) \to Z(A)$ such that $f(c)m = mc$

for all $m \in M$, $c \in Z(B)$. From this, one deduces the following properties:

a. If A is commutative, then the following sequence is exact and split :

$1 \to \mathrm{Picent}(A) \to \mathrm{Pic}_R(A) \to \mathrm{Aut}_R(A) \to 1$;

b. If B is Morita equivalent to A over R, then there is an isomorphism :

$\mathrm{Picent}(B) \cong \mathrm{Picent}(A)$;

c. Picent(A) is a normal subgroup of Pic(A) and Autcent(A) is a normal subgroup of Aut(A);

d. if $\pi : S \to R$ is an epimorphism in the category of rings, then $\mathrm{Pic}_R(A) = \mathrm{Pic}_S(A)$ and $\mathrm{Aut}_R(A) = \mathrm{Aut}_S(A)$.

Proof : cf. Theorem 2 and its corollaries in [ 52 ]. □

From our point of view, we will be mainly interested in prime algebras (orders in central simple algebras over Krull domains). For completeness' sake we include A. Fröhlich's result, which reduces problems in the semi-prime case essentially to the prime case. Let $A = \prod_{i \in I} A_i$ be a product of R-algebras, where I need not be finite. Let $\Sigma$ be the subgroup of permutations $\pi$ of I with the property that $A_i$ and $A_{\pi(i)}$ are Morita equivalent over R for any $i \in I$. Finally, let $e_i$ be the central idempotent of A for which $e_iA = Ae_i = A_i$. We then have :

3.4. Theorem. 1. If $\mathrm{Pic}^{\star}_R(A)$ denotes the subgroup of $\mathrm{Pic}_R(A)$ consisting of the classes of bimodules M such that $Me_i = e_iM$ for all $i \in I$, then the maps $[M] \to [Me_i]$ define an injective morphism $\mu : \mathrm{Pic}^{\star}_R(A) \to \prod_{i\in I} \mathrm{Pic}_R(A_i)$ which is bijective if I is finite. A similar property holds upon replacing $\mathrm{Pic}_R(-)$ (and $\mathrm{Pic}^{\star}_R(-)$!) by Picent (-), $\mathrm{Aut}_R(-)$ or Autcent (-).

2. Let each $A_i$ be an indecomposable R-algebra. To each $[M] \in \mathrm{Pic}_R(A)$ there corresponds a unique $\pi \in \Sigma$ such that for all $m \in M$ and all $i \in I$, we have $me_i = e_{\pi(i)}m$. The map $[M] \to \pi$ defines a homomorphism $\mathrm{Pic}_R(A) \to \Sigma$,

with kernel $\mathrm{Pic}_R^{\star}(A)$. If I is finite, this yields an exact sequence

$$1 \to \prod_{i \in I} \mathrm{Pic}_R(A_i) \to \mathrm{Pic}_R(A) \to \Sigma \to 1$$

Proof: cf. Theorem 3 in [52 ]. □

At this point in our survey of the results stemming from the elementary part of [ 52], we omit the results that deal specifically with base change. We do include the following result of [ 52 ], which is of special interest to us. Actually this point is rather "critical" in A. Fröhlich's paper, in the sense that here, for the first time, the assumption that A is a *finitely generated* R-module is used :

3.5. Proposition. Let R be a domain with field of fractions K and let A be an R-algebra, which is a torsionfree, finitely generated R-module such that $A \otimes_R K$ is semisimple. Denote by $I(A)$ the group of invertible fractional ideals of A in $A \otimes_R K$ and by $N(A)$ the normalizer of A in $A \otimes_R K$ (i.e. the group of units $u \in U(A \otimes_R K)$ for which $uAu^{-1} = A$), then the following sequences are exact :

a. $1 \to U(Z(A \otimes_R K))/U(Z(A)) \to N(A)/U(A) \to \mathrm{Outcent}(A) \to 1$, where $\mathrm{Outcent}(A) = \mathrm{Autcent}(A)/\mathrm{Inn}(A)$;

b. $1 \to U(Z(A \otimes_R K)) \to N(A) \to \mathrm{Autcent}(A) \to 1$;

c. $U(Z(A \otimes_R K)) \to N(A) \to \mathrm{Picent}(A) \to 1$;

d. $1 \to \mathrm{Picent}(A) \to \mathrm{Pic}_R(A) \to \mathrm{Pic}_K(A \otimes_R K)$;

e. $1 \to U(A)/U(Z(A)) \to N(A)/U(Z(A \otimes_R K)) \to \mathrm{Picent}(A)$.

Proof. cf. Corollary 2 of Theorem 3 in [ 52 ]. Note that an invertible fractional ideal of A in $A \otimes_R K$ is an A-subbimodule V of $A \otimes_R K$ for which there is a W such that VW = WV = A; in the situation of the proposition, this is compatible with the definition used in the foregoing section. □

This result will be strengthened considerably in the next chapter. Since we aim to study (maximal) orders over Krull domains, taking into account the definition of orders, divisorial and tame orders, given in section A.2., it is clear that we will be facing R-algebras A, which are <u>not</u> necessarily finitely generated R-modules. Looking at the proof of the foregoing proposition, as given in [52], it becomes clear that one only uses the fact that $Z(A) \underset{R}{\otimes} K = Z(A \underset{R}{\otimes} K)$. Now, if we drop the assumption that A is finitely generated, then it still follows from the semisimplicity of $A \underset{R}{\otimes} K$ that A is a PI-algebra over R, so taking $R = Z(A)$, it follows from Posner's theorem [116] that $Z(A \underset{R}{\otimes} K) = Z(A) \underset{R}{\otimes} K = K$. Consequently, at the cost of trivializing part d. in Proposition 3.5., we still have the same results over $R = Z(A)$, and even :

3.6. <u>Corollary</u> : Let R be a domain with field of fractions K and let A be a prime R-algebra, which is a faithful R-module with the property that $A \underset{R}{\otimes} K$ is a central simple algebra, then the results of Proposition 3.5. remain valid.

<u>Proof</u>. Identify A with its image in $A \underset{R}{\otimes} K$. If $x \in Z(A \underset{R}{\otimes} K)$, then $rx \in Z(A)$ for some $0 \neq r \in R$. As $r x \neq 0$, we have $x \in Z(A)r^{-1} \subset Z(A) \underset{R}{\otimes} K$. The rest of the proof is as in [52]. □

We conclude this survey by giving some properties of the "localization sequence". From this point on, results are no longer true for orders over arbitrary Krull domains, and in order to obtain similar results it will be necessary to switch over to the relative theory. Actually, the failure of the "absolute" techniques at this point was one of the main motivations to develop further the theory of relative Picard groups of orders over Krull domains !

If $p$ is a prime ideal of $R$, then $R \to R_p$ and $A \to A_p$ are ringepimorphisms, hence the localization maps $\text{Pic}_R(A) \to \text{Pic}_{R_p}(A_p)$ commute with the isomorphisms induced by Morita-equivalence as well as with tensorproducts of R-algebras. Moreover, the canonical morphism

$$T : \text{Picent}(Z(A)) \to \text{Picent}(A) : [M] \to [M \otimes_{Z(A)} A]$$

commutes with Morita equivalence over $Z(A)$, as well. If $E$ is an Azumaya algebra over $Z(A)$, then we have a commutative diagram

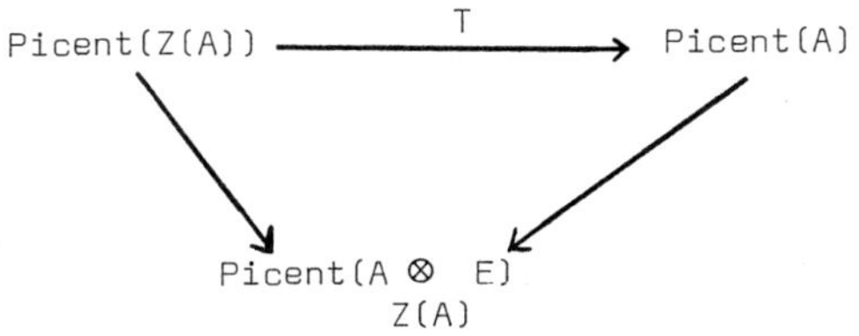


For any A-bimodule $N$, write $N^A = \{n \in N;\ na = an \text{ for all } a \in A\}$; clearly $N^A$ is a $Z(A)$-module.

3.7. <u>Theorem</u> 1. The morphism $T$ is injective and $\text{Im}\, T$ is a central subgroup in $\text{Picent}(A)$.

2. $\text{Ker}(\text{Pic}_R(A) \to \prod_p \text{Pic}_R(A_p)) \subset \text{Picent}(A)$;

3. if $Z(A)_m$ is semilocal for every maximal ideal $m$ of $R$, then $\text{Im}\, T \subset \text{Ker}(\text{Picent}(A) \to \prod_p \text{Pic}_R(A_p))$;

4. if $M^A$ is a finitely generated $Z(A)$-module for every invertible A-bimodule $M$ and if for all maximal ideals $m$ of $R$, the induced map $(M^A)_m \to (M_m)^{A_m}$ is bijective, then $\text{Ker}(\text{Picent}(A) \to \prod_p \text{Pic}_R(A_p)) \subset \text{Im}\, T$;

5. if $Z(A)_m$ is semilocal for every maximal ideal $m$ of $R$ and $A$ is a finitely generated module over its noetherian center $Z(A)$, then :

a. for any homomorphism $R \to S$ of commutative rings such that $A$ is also an S-algebra, we have an exact sequence

$$1 \to \text{Picent}(Z(A)) \to \text{Pic}_S(A) \to \prod_p \text{Pic}_S(A_p)\ ;$$

b. the following sequence is exact :

$$1 \to \mathrm{Picent}(Z(A)) \to \mathrm{Picent}(A) \to \prod_{p} \mathrm{Picent}(A_p);$$

(all products over a subset of Spec(R) containing the maximal ideals of R!). □

3.8. Note : The proof of this theorem, as given in [52 ], p. 20, relies on two lemmas, one of which depends on a finite presentation argument. The statements 1, 2, 3, 4 of the above theorem do not depend on that lemma, so they are still true in the situations we aim to study. However, exactly the localization sequences given in 5.a and 5.b depend on the finiteness conditions and similar results over non Noetherian Krull domains require an approach through relative Picard groups.
Since the lines of attack separate here, we leave to the interested reader the further investigation of the very interesting properties of Pic(-) and Picent(-) of orders in the sense of [ 19 ] and [ 52 ]. Let us just mention two more results of a rather explicit nature.

3.9. Theorem Let R be a Dedekind domain an A an R-order such that $A \otimes_R K$ is semisimple, then :

1. $I(A) = \bigsqcup_m I(A_m)$, where $m$ runs through the maximal ideals of R;
2. $\mathrm{Picent}(A_m) = 1$ for almost all $m$;
3. The following sequence is exact :

$$1 \to \mathrm{Picent}(Z(A)) \to \mathrm{Picent}(A) \to \bigsqcup_m \mathrm{Picent}(A_m) \to 1.$$

Proof. cf. [ 52 ], Theorem 6, p. 22. □

If A is a maximal order over a Dedekind domain R in a finite dimensional separable K-algebra $\Sigma$, then Picent(A) is independent of the actual choice of A in $\Sigma$, any other maximal R-order in $\Sigma$ being Morita equivalent to it cf. [122]. Moreover, all invertible A-bimodules are locally free in A-mod, and $I(A)$ is a free abelian group generated by the maximal ideals of A, hence Picent(A) is *abelian*.

Equally explicit results may be obtained for congruence orders over a Dedekind domain R. If J is an ideal of R, then we define for every positive integer n a congruence order $A = R+JM_n(R) \subset M_n(K)$. For a maximal ideal $m$ of R, let $m^j$ be the precise power of $m$ dividing J, then $A_m = R_m + m^j M_n(R_m)$.

3.10. Theorem. Let A be a congruence order over R, then :

1. every invertible A-bimodule is locally free;
2. $N(A) = N(M_n(R))$, hence $Gl_n(R) \subset N(A)$; if R is a principal ideal domain then the resulting morphism $Gl_n(R) \to \text{Outcent}(A)$ is surjective;
3. $\text{Picent}(A_m) = PGl(R_m/m^j)$, and hence $\prod_m \text{Picent}(A_m) = PGl_n(R/J)$;
4. the localization sequence $1 \to \text{Picent}(R) \to \text{Picent}(A) \to PGl_n(R/J) \to 1$ is split;
5. the map $M \to M\,M_n(R)$ defines a morphism $I(A) \to I(M_n(R))$; the map $[M] \mapsto [M \otimes_A M_n(R)]$ defines a morphism $\text{Picent}(A) \to \text{Picent}(M_n(R))$, such that the following diagram commutes :

$$\begin{array}{ccc} \text{Picent}(R) & \longrightarrow & \text{Picent}(A) \\ & \searrow \quad \swarrow & \\ & \text{Picent}(M_n(R)) & \end{array}$$

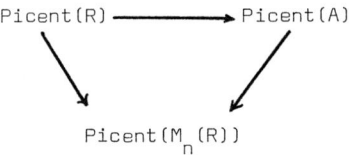

A.4. Group Graded Rings.

The constructive methods used in chapter D are based upon graded techniques. In this section we review some basic results, but we refer to C. Năstăsescu and F. Van Oystaeyen in [109] for full detail. In the commutative theory developed in [162], Chapter VI, we already introduced some elementary facts about commutative $\mathbb{Z}$-graded rings. Here, we consider gradations by arbitrary groups. Let G be a multiplicative group with identity element e. An associative ring R, with unit as always, is a graded ring of type G if $R = \bigoplus_{\sigma \in G} R_\sigma$ for additive subgroups $R_\sigma$ of R, such that $R_\sigma R_\tau \subset R_{\sigma\tau}$ for all $\sigma, \tau \in G$. The elements of R in $R_\sigma$ for some $\sigma \in G$ is said to be homogeneous of degree $\sigma$. Obviously, $R_e$ is a subring of R and $1 \in R_e$. If H is a subgroup of G, then we define the graded ring $R^{(H)}$ of type H by $(R^{(H)})_\sigma = R_\sigma$ for any $\sigma \in H$. A left R-module M is said to be a graded left R-module of type G if $M = \bigoplus_{\sigma \in G} M_\sigma$ for some family of additive subgroups $M_\sigma \subset M$ such that $R_\tau M_\sigma \subset M_{\tau\sigma}$ for all $\tau, \sigma \in G$. Unless otherwise mentioned, "R-module" means "left R-module"; we write R-gr for the category of graded R-modules.

A submodule N of M is a graded submodule if for any $x \in N$, the homogeneous components of x are again in N. An R-linear map $f : M \to N$ with $M, N \in$ R-gr is said to be a graded morphism of degree $\tau \in G$, if $f(M_\sigma) \subset M_{\sigma\tau}$ for all $\sigma \in G$. These morphisms build into an additive subgroup $HOM_R(M,N)_\tau$ of $Hom_R(M,N)$ and we define $HOM_R(M,N) = \bigoplus_{\tau \in G} HOM_R(M,N)_\tau$. It is clear that $HOM_R(M,N)$ is a graded abelian group of type G; we write $Hom_{R\text{-}gr}(M,N)$ for $HOM_R(M,N)_e$ and these are considered to be the morphisms in the category R-gr. All good properties of this category may be summarized into : R-gr is a *Grothendieck category*.

For $M \in$ R-gr and $\sigma \in G$, we define the $\sigma$-shift functor $T_\sigma$:R-gr $\to$ R-gr by $T_\sigma(M) = M(\sigma)$, where $M(\sigma)$ is the graded R-module with underlying set M and graded structure given by $M(\sigma)_\tau = M_{\tau\sigma}$ for all $\tau \in G$. The functor $T_\sigma$ is an equivalence of categories for each $\sigma \in G$, and one may prove that $\{R(\sigma);\ \sigma \in G\}$ is a set of generators for R-gr. A module $F \in$ R-gr is said to be gr-free if it has a basis of homogeneous elements. The functor R-gr $\to$ R-mod, which just forgets gradation is denoted $\underline{\ \ }$, i.e. if $M \in$ R-gr, then $\underline{M}$ is the ungraded module underlying M.

4.1. Lemma 1. A graded R-module P is a projective object in R-gr if and only if $\underline{P}$ is a projective R-module;
2. Let N be a graded R-submodule of $M \in$ R-gr; if N is an essential subobject of M in R-gr, then $\underline{N}$ is essential in $\underline{M}$;
3. if G is finite or if $\underline{M}$ is finitely generated, then we have $\underline{HOM_R(M,N)}$ $Hom_R(\underline{M},\underline{N})$.

Proof. 1. cf. Corollary I.2.2. in [109];
2. cf. Lemma I.2.8. in [109];
3. cf. Corollary I.2.11 in [109]. □

It is clear what we mean by a graded left ideal of R or a graded ideal of R. We use the notation gr-R to denote the category of graded right R-modules (and R-gr-R for graded R-bimodules) with corresponding graded morphisms of degree e.

If $M \in$ gr-R and $N \in$ R-gr, let $(M \otimes_R N)_\sigma$ be the additive subgroup of $\underline{M} \otimes_R \underline{N}$ generated by the $x_\alpha \otimes y_\beta$ with $x_\alpha \in M_\alpha$ and $y_\beta \in N_\beta$ such that $\alpha\beta = \sigma$. The graded tensorproduct of M and N is then just $M \otimes_R N = \bigoplus_{\sigma \in G} (M \otimes_R N)_\sigma$; this graded tensorproduct satisfies the usual properties with respect to the HOM-functor. It follows easily that an $M \in$ R-gr is gr-flat (i.e.

flat in R-gr) if and only if $\underline{M}$ is flat in R-mod.

We now introduce a notion which has particular importance, in view of the results in Chapter D. The graded ring R of type G is said to be strongly graded if $R_\sigma R_\tau = R_{\sigma\tau}$ for all $\sigma, \tau \in G$; a graded R-module M is strongly graded if $R_\tau M_\sigma = M_{\tau\sigma}$ for all $\tau, \sigma \in G$.

4.2. Theorem. The following statements are equivalent for a graded ring R of type G :

1. R is strongly graded;
2. every $M \in$ R-gr is strongly graded;
3. the functors $(-)_e$ : R-gr $\to$ $R_e$-mod and $R \otimes_{R_e} -$ : $R_e$-mod $\to$ R-gr define inverse equivalences between the Grothendieck categories R-gr and $R_e$-mod (the left-right symmetric statements of 2. and 3. are equivalent to each of the foregoing as well).

Proof. cf. Theorem I.3.4 in [109]. □

If R is a strongly graded ring of type G, then for every $\sigma, \tau \in G$ the canonical morphism $R_\sigma \otimes_{R_e} R_\tau \to R_{\sigma\tau}$ is an isomorphism of $R_e$-bimodules. It is easy to verify that for each $\sigma \in G$ we have $[R_\sigma] \in \mathrm{Pic}(R_e)$. In the commutative theory we already noted some results in the $\mathbb{Z}$-graded case, cf. Lemma VI.1. in [162], but as it turns out, strongly graded rings are particularly useful in case G is a finite group. This is mainly due to the generalized crossed product structure, which is available for any strongly graded ring of type G. Let G be an arbitrary group and consider a group homomorphism $\varphi : G \to \mathrm{Pic}(A)$ for some (noncommutative) ring A. Write $\varphi[\sigma] = [P_\sigma]$, for any $\sigma \in G$. A factor set associated to $\varphi$ is a family $f = \{f_{\sigma,\tau}; \sigma, \tau \in G\}$ of bimodule isomorphisms $f_{\sigma,\tau} : P_\sigma \otimes_A P_\tau \to P_{\sigma\tau}$ such that the following diagrams commute for every $\rho, \sigma, \tau \in G$, where

$\alpha : P_e \to A$ is a fixed A-bimodule isomorphism

$$\begin{array}{ccc} P_\rho \underset{A}{\otimes} P_\sigma \otimes P_\tau & \xrightarrow{\;P_\sigma \otimes f_{\sigma,\tau}\;} & P_\rho \otimes P_{\sigma\tau} \\ \Big\downarrow {\scriptstyle f_{\rho,\sigma} \otimes P_\tau} & & \Big\downarrow {\scriptstyle f_{\rho,\sigma\tau}} \\ P_{\rho\sigma} \otimes P_\tau & \xrightarrow[\;f_{\rho\sigma,\tau}\;]{} & P_{\rho\sigma\tau} \end{array}$$

$$\begin{array}{ccc} P_\sigma \underset{A}{\otimes} P_e & \xrightarrow{\;P_\sigma \otimes \alpha\;} & P_\sigma \underset{A}{\otimes} A \\ {\scriptstyle f_{\sigma,e}} \searrow & & \swarrow {\scriptstyle \cong} \\ & P_\sigma & \end{array}$$

$$\begin{array}{ccc} P_e \underset{A}{\otimes} P_\sigma & \xrightarrow{\;\alpha \otimes P_\sigma\;} & A \underset{A}{\otimes} P_\sigma \\ {\scriptstyle f_{e,\sigma}} \searrow & & \swarrow {\scriptstyle \cong} \\ & P_\sigma & \end{array}$$

Let $F(\varphi)$ be the collection of factor sets associated to $\varphi$. For every $f \in F(\varphi)$, we write $A < f, \varphi, G >$ for the ring obtained by defining multiplication on $\underset{\sigma \in G}{\oplus} P_\sigma$ by $x_\sigma y_\tau = f_{\sigma,\tau}(x_\sigma \otimes y_\tau)$ for any $x_\sigma \in P_\tau$.

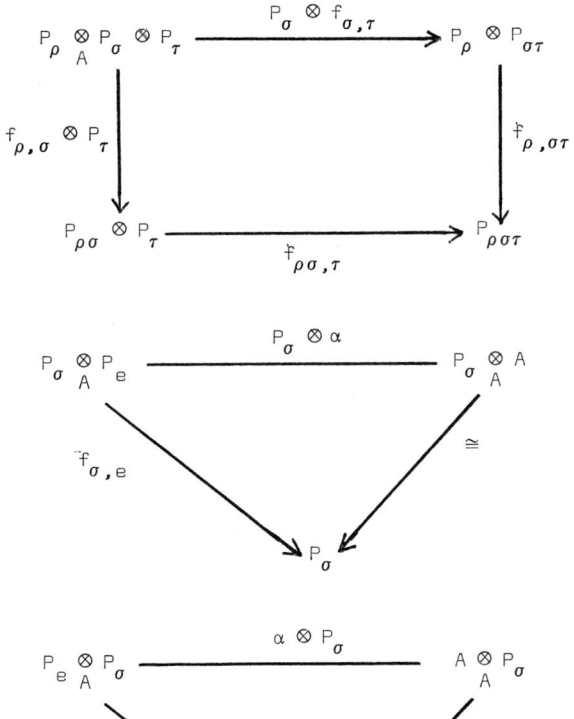

4.3. <u>The generalized crossed product theorem.</u> The ring $R = A < f,\varphi,G >$ constructed above is strongly graded of type G and it has an identity. Furthermore, R contains a subring isomorphism to A, actually $R_e = P_e \cong A$. Upto graded isomorphism $A < f,\varphi,G >$ is independent of the choice of the family $\{P_\sigma;\sigma \in G\}$ representing the $\{\varphi(\sigma);\sigma \in G\}$. Conversely, if R is a strongly graded ring of type G, then there exists a group homomorphism $\varphi:G \to \mathrm{Pic}(R_e)$ and a factor set f associated to $\varphi$ such that $R \cong R_e <f,\varphi,G>$ as graded rings.

Proof. Proposition I.3.13 and Remark I.3.15 in [109]. □

A rather straightforward verification shows that two factor sets $f,g \in F(\varphi)$ necessarily differ by a 2-cocycle, i.e. $g=qf$ for some $q \in Z^2(G,U(Z(A)))$ and consequently the elements of $H^2(G,U(Z(A)))$ correspond bijectively to the graded A-isomorphism classes of strongly graded rings $A < f,\varphi,G >$. Moreover, a generalized crossed product $A < f,\varphi,G >$ is a *genuine* crossed product whenever $\mathrm{Im}\,\varphi = [A]$, in particular, if Pic(A) = 1!

If R is strongly graded of type G, then we write $\mathrm{Mod}(R_e,R)$ for the class of all $R_e$-subbimodules of R. If $N \in \mathrm{Mod}(R_e,R)$ and $\sigma \in G$, then $N^\sigma = R_{\sigma^{-1}}NR_\sigma$ is called the $\sigma$-conjugate of N.

4.4. Proposition. 1. The map $\varphi_\sigma : \mathrm{Mod}(R_e,R) \to \mathrm{Mod}(R_e,R)$ defined by $\varphi_\sigma(N) = N^\sigma$ for any $N \in \mathrm{Mod}(R_e,R)$ is bijective and preserves inclusions, sums, intersections and products in R.

2. The correspondence $I \to I_e$ defines a bijection between the set of graded ideals of R and the set of G-invariant ideals of $R_e$.

3. If M is a simple left $R_e$-module, then for every $\sigma \in G$, the left $R_e$-module $R_\sigma \otimes_{R_e} M$ is simple and if $I = \mathrm{Ann}_{R_e}(M)$, then $I = (\mathrm{Ann}_{R_e}(R_\sigma \otimes_{R_e} M))^\sigma$.

4. If P is a prime ideal of $R_e$, then $P^\sigma$ is a prime ideal of $R_e$ for all $\sigma \in G$.

5. The prime radical $\mathrm{rad}(R_e)$ and the Jacobson radical $J(R_e)$ are G-invariant.

Proof. p. 34, 35, 36 in [109]. □

In conclusion let us mention :

4.5. Theorem 1. If R is graded of type $\mathbb{Z}$, then $M \in$ R-gr is gr-Noetherian if and only if $\underline{M}$ is noetherian in R-mod.

2. If R is strongly graded of type G and $M \in$ R-gr, then M is gr-noetherian if and only if $M_e$ is noetherian in $R_e$-mod.

3. If R is graded by a finite group, then $M \in$ R-gr is gr-noetherian (resp. gr-artinian) if and only if $\underline{M}$ is noetherian (resp. artinian) in R-mod.

4. If R is left gr-noetherian of type G and H is a subgroup of finite index in G, then $R^{(H)}$ is left gr-noetherian.

Proof. cf. Theorem II.3.5., Lemma II.3.6. and Corollary II.3.15 in [109] . □

4.6. Comment : Some recent results about strongly graded rings have appeared in E. Dade [ 40 ], T. Kanzaki [ 79 ], C. Năstăsescu and F. Van Oystaeyen [ 107,109 ].

## A.5. Localization and Sheaves.

Terminology and preliminaries concerning localization in the setting of idempotent kernel functors have already been introduced in "The Commutative Theory" [162], section I.1., where we did not restrict to commutative rings. For a general treatment we may also refer to O. Goldman [ 60 ], J. Golan [ 59 ] or F. Van Oystaeyen, A. Verschoren [ 159 ] . For the theory of sheaves and presheaves we refer to R. Godement's classic [ 58 ] or F. Van Oystaeyen, A. Verschoren [ 159 ].

In this book we only need very rudimentary notions of sheaf theory, anyway and only at a few particular places. So we restrict ourselves here to pointing out some facts, usually in connection with the non-commutativity of the ground ring, which have been neglected in the commutative theory [ 162] .

If $\kappa$ is an idempotent kernel functor in R-mod, then the quotient category $(R,\kappa)$-mod is the full subcategory of R-mod consisting of the $\kappa$-closed left R-modules M, i.e. those $M \in$ R-mod, for which the localizing morphism $j_\kappa : M \to Q_\kappa(M)$ is an isomorphism. Here $Q_\kappa(-)$ is the localization functor associated to $\kappa$.

Theorem I.1.3. in [162 ] provides a good number of equivalent characterizations of T-functors, in particular $\kappa$ is a T-functor if and only if for every left ideal $L \in \mathcal{L}(\kappa)$ we have $Q_\kappa(R)j_\kappa(L) = Q_\kappa(R)$. Note that we will also say : $\kappa$ induces a perfect localization.

We now focus on somewhat weaker finiteness conditions on kernel functors We call $\kappa$ (or $\mathcal{L}(\kappa)$!) <u>of finite type</u> if every $L \in \mathcal{L}(\kappa)$ contains a finitely generated $L' \subset L$ such that $L' \in \mathcal{L}(\kappa)$. We say that $\kappa$ is <u>noetherian</u> if any ascending chain of left ideals $L_1 \subset \ldots \subset L_\alpha \subset \ldots$ such that $\bigcup_\alpha L_\alpha \in \mathcal{L}(\kappa)$ has the property that $L_\kappa \in \mathcal{L}(\kappa)$ for some $\kappa$. We say that R is $\kappa$-<u>noetherian</u> if $Q_\kappa(R)$ satisfies the ascending chain condition for $\kappa$-closed left ideals. We will come back to this notion in Chapter B, where it will be used throughout. The implications between the conditions "$\kappa$ is of finite type" and "$\kappa$ is noetherian" are easily explicited. On the other hand, we have

5.2. <u>Lemma</u> : Let $\kappa$ be an idempotent kernel functor in R-mod, then

1. If $\kappa$ is a T-functor, then $\kappa$ has finite type;
2. If R is $\kappa$-noetherian, then $\kappa$ has finite type.

<u>Proof</u>. 1. If $L \in \mathcal{L}(\kappa)$, then $Q_\kappa(R)j_\kappa(L) = Q_\kappa(R)$ yields that $1 = \sum_{i=1}^n q_i\overline{l}_i$, where $q_i \in Q_\kappa(R)$ and $\overline{l}_i \in j_\kappa(L)$. Choose $l_i$ representing $\overline{l}_i$ in L and

consider $L' = Rl_1 + \dots + Rl_n \subset L$. From $Q_\kappa(R)j_\kappa(L') = Q_\kappa(R)$ one easily deduces that $R/L'$ is $\kappa$-torsion and thus $L' \in \mathcal{L}(\kappa)$.

2. Consider a left ideal $L \in \mathcal{L}(\kappa)$. Pick $l_1 \in L$ and put $L_1 = Rl_1$. If there is $l_2 \in L - Rl_1$, consider $L_2 = Rl_1 + Rl_2$; ... . Finally, one arrives at $L = Rl_1 + \dots + Rl_n$ for some n, or else one obtains an ascending chain of $\kappa$-closed left ideals $Q_\kappa(L_1) \subset Q_\kappa(L_2) \subset \dots \subset Q_\kappa(L) = Q_\kappa(R)$. Since R is $\kappa$-noetherian, $Q_\kappa(L_m) = Q_\kappa(R)$ holds for some $m \in \mathbb{N}$, hence $R/L_m$ is $\kappa$-torsion i.e. $L_m \in \mathcal{L}(\kappa)$. □

In section I.1. of [162] we introduced the kernel functor $\kappa_I$ associated to an ideal I of R, by letting $\mathcal{L}(\kappa_I)$ consist of all left R-ideals containing an R-ideal J such that $I \subset \text{rad}(J)$. On the other hand, to any prime ideal P of R we associated a kernel functor $\kappa_{R-P}$ by letting $\mathcal{L}(\kappa_{R-P})$ consist of all left R-ideals containing an ideal J of R such that $J \not\subset P$. In general an *idempotent* kernel functor $\kappa$ is said to be <u>symmetric</u> if $\mathcal{L}(\kappa)$ has a cofinal set consisting of ideals of R. It should be noted that in general $\kappa_I$ or $\kappa_{R-P}$ are not necessarily idempotent. However, we have the following result :

5.2. <u>Lemma</u> : 1. If I is an ideal of R such that $\kappa_I$ has finite type, then $\kappa_I$ is an idempotent kernel functor.

2. If P is a prime ideal of R such that $\kappa_{R-P}$ has finite type, then $\kappa_{R-P}$ is an idempotent kernel functor.

<u>Proof.</u> First note that the definition of "having finite type" makes sense for not necessarily idempotent kernel functors. Both statements follow from the following more general fact : if $\mathcal{L}(\kappa)$ is multiplicatively closed and has a cofinal system of ideals of R, then $\kappa$ is idempotent whenever it has finite type. In order to verify idempotency of $\kappa$, it suffices to show that any left ideal L contained in a $J \in \mathcal{L}(\kappa)$ such that

J/L is $\kappa$-torsion is itself $\kappa$-torsion. Now, let $J_1 \subset J$ be a finitely be a finitely generated left ideal in $L(\kappa)$, say $J_1 = Rx_1 + \dots + Rx_n$. For each $x_i$ $(i = 1,\dots,n)$, there is an ideal of R, say $I_i$, such that $I_i \in L(\kappa)$ and $I_i x_i \subset L$. Therefore $I = I_1 \cap \dots \cap I_n \in L(\kappa)$ and $IJ_1 \subset L$ implies that $L \in L(\kappa)$, since $L(\kappa)$ is multiplicatively closed. □

If $\kappa_I$ or $\kappa_{R-P}$ is idempotent (e.g. if R is left noetherian, by the foregoing), then we denote by $Q_I(-)$ resp. $Q_{R-P}(-)$ the corresponding localization functors. If R is commutative these functors are well-known. Indeed, $Q_{R-P}(-)$ is then just the usual localization at the prime ideal P. On the other hand, assume for simplicity's sake that I is a finitely generated R-ideal (then $\kappa_I$ is of finite type, hence idempotent, by the above lemma) and let $X(I) \subset \mathrm{Spec}(R)$ be the open subset of Spec(R), consisting all prime ideals P of R with $I \not\subset P$, then for any $M \in R$-mod, we have $Q_I(M) = \Gamma(X(I), \widetilde{M})$. Here $\widetilde{M}$ is the canonical sheaf of Spec(R) associated to M - see also below.

It is well-known (cf. [59], [60], ...) that the set of idempotent kernel functors in R-mod is a lattice and we may define, for any family $\{\kappa_i; i \in I\}$ the idempotent kernel functors $\kappa' = \bigwedge_{i \in I} \kappa_i = \inf\{\kappa_i; i \in I\}$, by putting $L(\kappa') = \bigcap_{i\in I} L(\kappa_i)$ and $\kappa'' = \bigvee_{i\in I} \kappa_i = \sup\{\kappa_i; i \in I\}$, by taking for $L(\kappa'')$ the idempotent filter generated by $\bigcup_{i\in I} L(\kappa_i)$. Note that the supremum of symmetric kernel functors need not be symmetric in general - this is true, however in special cases, e.g. if R is left noetherian !

If the symmetric kernel functor $\kappa$ has finite type, then we let $C(\kappa)$ consist of the ideals of R which are maximal with respect to the property of not being contained in $L(\kappa)$. It is easy to see that $C(\kappa)$ consists of prime ideals of R. As in [162] we denote by $X(\kappa) \subset \mathrm{Spec}(R)$ the set of prime ideals of R not contained in $L(\kappa)$ , hence $C(\kappa) \subset X(\kappa)$!

5.3. Lemma If the symmetric kernel functor $\kappa$ has finite type, then $\mathcal{L}(\kappa) = \cap\{\mathcal{L}(\kappa_{R-P});\ P \in \mathcal{C}(\kappa)\}$.

Proof. That $\mathcal{L}(\kappa) \subset \mathcal{L}(\kappa_{R-P})$ for all $P \in \mathcal{C}(\kappa)$ is clear. Now, assume that an ideal I of R is not in $\mathcal{L}(\kappa)$. By the hypothesis that $\kappa$ has finite type, each ideal in $\mathcal{L}(\kappa)$ is contained in a maximal one not in $\mathcal{L}(\kappa)$, say $I \subset P$. As it follows that $I \notin \mathcal{L}(\kappa_{R-P})$, we find that $\mathcal{L}(\kappa)$ and $\cap\{\mathcal{L}(\kappa_{R-P});\ P \in \mathcal{C}(\kappa)\}$ contain the same ideals of R. Moreover, any left ideal L in the latter filter contains, for every $P \in \mathcal{C}(\kappa)$, an ideal $J_P$ of R such that $J_P \not\subset P$. But then L contains $J = \sum_P J_P$, an ideal of R which has the property that $J \not\subset P$ for all $P \in \mathcal{C}(\kappa)$. By the first part, $J \in \mathcal{L}(\kappa)$, hence $L \in \mathcal{L}(\kappa)$. □

5.4. Proposition. Let R be $\kappa$-noetherian for some symmetric kernel functor $\kappa$, then

1. $\kappa_{R-P}$ is symmetric for every $P \in \mathcal{C}(\kappa)$ ;
2. we have $\kappa = \Lambda\{\kappa_{R-P};P \in \mathcal{C}(\kappa)\}$.

Proof. Since a $\kappa_{R-P}$-closed ideal of R is necessarily $\kappa$-closed too, it follows that R is $\kappa_{R-P}$-noetherian for each $P \in \mathcal{C}(\kappa)$. The first statement is therefore an immediate consequence of Lemma 5.1.2. and Lemma 5.2.2.. In view of 1., it is clear that 2. follows from Lemma 5.3. □

5.5. Comment. In many applications R will be commutative and $\kappa$ noetherian, but we will also study R-algebras A which will then be localized only at central kernel functors. The main examples are obtained by considering maximal or tame orders over a Krull domain R, where we take for $\kappa$ the idempotent kernel functor $\sigma_1 = \Lambda\{\kappa_{R-p};p \in X^{(1)}(R)\}$, in which case R is certainly $\sigma_1$-noetherian !

If $R \to S$ is a ring morphism and $\kappa$ a kernel functor in R-mod, then we may define a kernel functor $\overline{\kappa}$ in S-mod, by taking for the class of torsion modules for $\overline{\kappa}$ exactly those S-modules that are $\kappa$-torsion, when considered as an R-module. If $R \to S$ is an arbitrary morphism, then nothing can be said about the relation between $C(\kappa)$ and $C(\overline{\kappa})$. Only in very special cases, e.g. if S is a Galois extension of a commutative ring R, or if S is an Azumaya algebra over R, there is an evident relation between $C(\kappa)$ and $C(\overline{\kappa})$.

In the sequel of this book, we will not use sheaf theory as a technical tool. Indeed, whereas sheaves appeared in a natural way in the commutative theory [162], mainly because of the connection between relative Brauer groups and Brauergroups of ringed space, the use of this theory is not so natural when dealing with noncommutative rings. We just introduce some notations and terminology, plus a few really basic facts, so that we may occasionally use these as a language, mainly with an eye to economizing on notation.

The set of prime ideals of a (not necessarily commutative) ring R, denoted by Spec(R), may be endowed with its Zariski topology by taking for the open sets of X = Spec(R), the sets $X(I) = \{P \in X; P \not\supset I\}$ associated to the ideals I of R. Note that $X(I) = X(\kappa_I)$. To an open set X(I), determining I upto radical, we associate the kernel functor $\kappa_I$ as defined before. Note that $\kappa_I$ also depends only upon rad(I)!
Assume for simplicity's sake R to be left noetherian, then assigning $Q_I(R)$ to the open set X(I) defines a presheaf on Spec(R), in the obvious way. One of the main advantages of symmetric localization, is that this sheaf theoretic machinery works :

5.6. Theorem (D. Murdoch, F. Van Oystaeyen). If R is a prime, left noetherian ring, then the above construction actually yields a sheaf $\mathcal{O}_R$ on Spec(R). The stalk of $\mathcal{O}_R$ at $P \in \mathrm{Spec}(R)$ is given by $\mathcal{O}_{R,P} = Q_{R-P}(R)$.

Proof. cf. [103]. □

If R is left noetherian but not prime, then we may use the sheafification functor applied to the above construction in order to derive a structure sheaf $\mathcal{O}_R$ on Spec(R). The statement about the stalks of $\mathcal{O}_R$ then remains valid. In many situations a similar result holds if R is not necessarily left noetherian, e.g. if R is a prime P.I. ring or if R is a prime Goldie ring, cf. A. Verschoren [113] or F. Van Oystaeyen, A. Verschoren [161]. In these cases, one describes the rings assigned to X(I) directly within the simple artinian ring of quotients Q(R) of R. Since the left noetherian hypothesis is only needed to assure the idempotency of the occuring kernel functors, it will be superfluous in those particular cases.

If M is an R-module, then we define a sheaf of R-modules over Spec(R) by sheafification of the presheaf obtained by assigning $Q_I(M)$ to X(I); this sheaf will be denoted by $\tilde{M}$ or $\mathcal{O}_M$. For an attempt to construct a noncommutative algebraic geometry by means of these and other sheaves we refer to F. Van Oystaeyen, A. Verschoren [161]. A projective counterpart of the above constructions may be carried out over Proj(R) for a positively graded ring $R = \bigoplus_n R_n$, but so far some success has been achieved only in the case of graded PI rings, cf. [161]. We do not go into any details, but we conclude by pointing out a special case of the sheaf of modules construction, which may be useful in the sequel.

Lat A be an R-algebra. Since R is central in A, it is easy to verify that $\tilde{A}$ is actually a sheaf of R-algebras over Spec(R). In most applications R will be actually the center of A; we then refer to $\tilde{A}$ as the central structure sheaf of A and denote it by $\widetilde{A}_R$ in order to avoid confusion with the structure sheaf of A on Spec(A). The central structure sheaf is especially useful in the study of orders over R, as a good deal of information about the order may be inferred from "central" data. For example, a maximal order over a Dedekind domain in a central simple algebra is a Zariski central ring in the sense of [150], meaning that the central structure sheaf and the "noncommutative" stucture sheaf of the order coincide. A "divisorial" version of such a result holds for maximal orders over Krull domains, we will come back to this in Chapter C.3.

5.7. Comment : It is clear that from Sections 1 and 2, or in fact from the commutative theory in [162], that the subset $X^{(1)}(R)$ of X = Spec(R) is most important. We have seen (after 1.6.) that the operation $(-)^{**}$ may be given as $Q_{\sigma_1}(-)$, where $\sigma_1 = \Lambda\{\kappa_{R-p}; p \in X^{(1)}(R)\}$, at least when applied to R-lattices. For a finitely generated R-module M, being reflexive is a property which is evident in the behaviour of $\widetilde{M}$ over the set $X^{(1)}(R)$. Of course, $X^{(1)}(R)$ is not open, but still a geometrically stable subset of Spec(R) in the sense of [162,166], The latter fact accounts for some theoretic applications concerning quasi coherent sheaves and reflexive modules in [162], chapter IV; Section 3 and 4. If R is a noncommutative version of a Krull domain, e.g. a maximal order over a Krull domain or an $\Omega$-Krull ring in the sense of [176] (which we will not consider in the sequel because in the PI case $\Omega$-Krull rings are just maximal orders over Krull rings), then both $X^{(1)}(R)$ and $X^{(1)}(Z(R))$ combined with their respective sheaves $\widetilde{R}$ and $\widetilde{R}_{Z(R)}$ play a similar role. However, the noncommutative sheaftheory in connection with reflexive modules or quasicoherent sheaves over R resp. over Z(R) will not be expounded here, because it moves very quickly outside of the scope of this book.

# B RELATIVE PICARD GROUPS

## B.1. Invertible Bimodules.

Throughout A and B are are arbitrary rings and $\lambda$ and $\mu$ are idempotent kernel functors in A-mod resp B-mod. If we want to stress the fact that a certain module P is an A-B-bimodule, then we sometimes denote this by ${}_AP_B$. If we only wish to consider the left action, say, then we denote this by ${}_AP$, etc. We will write $\mathrm{Hom}_{A\text{-mod}}(P,Q)$ resp. $\mathrm{Hom}_{\text{mod-}B}(P,Q)$ as ${}_A[P,Q]$ resp. $[P,Q]_B$ and $\mathrm{Hom}_{A\text{-mod-}B}(P,Q)$ as ${}_A[P,Q]_B$, if P and Q are in the appropriate categories. We need the following :

1.1. Lemma Let $N \in$ A-mod, $M \in$ B-mod and $P \in$ A-mod-B, then

1. ${}_A[P,N] \in$ B-mod;
2. there is a canonical bijection $\theta : {}_A[P \otimes_B M, N] \to {}_B[M, {}_A[P,N]]$.

Proof : Let $p \in P$, $b, b' \in B$ and $\varphi \in {}_A[P,N]$, then we define the left B-module structure on ${}_A[P,N]$ by $(b.\varphi)(p) = \varphi(pb)$. Clearly, $b.\varphi \in {}_A[P,N]$. Moreover, $(b.(b'.\varphi))(p) = (b'.\varphi)(pb) = \varphi(pbb') = ((bb').\varphi)(p)$, so this yields a B-module structure, indeed. Now, if $\varphi \in {}_A[P \underset{B}{\otimes} M, N]$, then $\theta(\varphi)(m)$ is defined for any $m \in M$ by $\theta(\varphi)(m)(p) = \varphi(p \otimes m) \in N$, for $p \in P$. Clearly, $\theta(\varphi)$ is left B-linear, since for $b \in B$, we have

$$\theta(\varphi)(bm)(p) = \varphi(p \otimes bm) = \varphi(pb \otimes m) = \theta(\varphi)(m)(pb) = (b.\theta(\varphi)(m))(p).$$

Similarly, $\theta(\varphi)(m)$ is left A-linear, since for $a \in A$ we have

$$\theta(\varphi)(m)(ap) = \varphi(ap \otimes m) = a\varphi(p \otimes m) = a\theta(\varphi)(m)(p) =$$

The rest is verification. □

1.2. Lemma Let P be an A-B-bimodule, then for any idempotent kernel functor $\kappa$ in A-mod, the module of quotients $Q_\kappa(P)$ may be canonically endowed with an A-B-bimodule structure extending that of P and with this structure, if $j_\kappa : P \to Q_\kappa(P)$ denotes the localizing morphism, then $j_\kappa \in {}_A[P, Q_\kappa(P)]_B$.

<u>Proof</u>. It is clear that $\kappa P$ is an A-B-bimodule. Indeed, if $p \in \kappa P$, then $Ip = 0$ for some $I \in \mathcal{L}(\kappa)$, so for all $b \in B$ we have $Ipb = 0$ too, hence $pb \in \kappa P$. It follows that $\overline{P} = P/\kappa P$ is an A-B-bimodule, hence in what follows, we may assume P to be $\kappa$-torsion free, i.e. $P \hookrightarrow Q_\kappa(P)$. We already know that $Q_\kappa(P)$ has a unique left A-module structure extending that of P. Let us endow $Q_\kappa(P)$ with a right B-module structure as follows. Denote by $\tau_b : P \to P : p \mapsto pb$ the right multiplication by b, for any $b \in B$, then since P is an A-B-bimodule, we have for all $p \in A$ that $a\tau_b(m) = a(mb) = (am)b = \tau_b(am)$, i.e. $\tau_b$ is left A-linear. It follows that $\tau_b$ extends uniquely to $\hat{\tau}_b : Q_\kappa(P) \to Q_\kappa(P)$ :

$$\begin{array}{ccc} P & \xrightarrow{\tau_b} & P \\ {\scriptstyle j_\kappa}\downarrow & & \downarrow{\scriptstyle j_\kappa} \\ Q_\kappa(P) & \xrightarrow[\hat{\tau}_b]{} & Q_\kappa(P) \end{array}$$

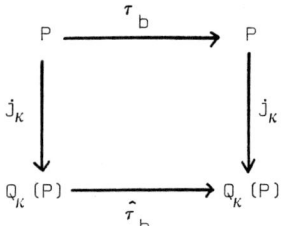

Pick another $c \in B$ and denote by $\tau_c : P \to P$ the right multiplication by c and let $\hat{\tau}_c$ be the unique extension of $\tau_c$ to $Q_\kappa(P)$. Similarly, let $\hat{\tau}_{cb} : Q_\kappa(P) \to Q_\kappa(P)$ denote the unique A-linear extension of $\tau_{cb}$ to $Q_\kappa(P)$. We obviously have $\tau_b\tau_c = \tau_{cb}$, hence a uniqueness argument yields $\hat{\tau}_{cb} = \hat{\tau}_b\hat{\tau}_c$. Let us define right multiplication on $Q_\kappa(P)$ by putting $\pi b = \hat{\tau}_b(\pi)$ for all $\pi \in Q_\kappa(P)$ and $b \in B$. The foregoing then yields that $\pi(cb) = (\pi c)b$ and the A-linearity of $\hat{\tau}_b$ for all $b \in B$ implies that this canonically endows $Q_\kappa(P)$ with a right B-module structure. Since $\hat{\tau}_b$ is left B-linear by construction, for all $a \in A$ we have $a(\pi b) = a\hat{\tau}_b(\pi) = \hat{\tau}_b(a\pi) = (a\pi)b$ for any $\pi \in Q_\kappa(P)$, hence the left and right actions are compatible. It is now clear that this is the only twosided A-B action on $Q_\kappa(P)$ extending the one on P. Indeed, assume that $Q_\kappa(P)$ possesses another compatible right B-action and for any $b \in B$, denote by $\sigma_b : Q_\kappa(P) \to Q_\kappa(P)$ the right multiplication by b for this action, then by assumption $\sigma_b$ extends $\tau_b : P \to P$ and on the other hand, it is left A-linear, as it is compatible with the left A-action. But, $\hat{\tau}_b$ is the unique A-linear extension of $\tau_b$, hence $\hat{\tau}_b = \sigma_b$ and the two actions of B on the right coincide, as $b \in B$ is arbitrary. Finally, from the commutativity of the above diagram, it follows that $j_\kappa \tau_b = \hat{\tau}_b j_\kappa$ for all $b \in B$, hence for all $p \in P$, we have $j_\kappa(pb) = j_\kappa(\tau_b(p)) = \hat{\tau}_b j_\kappa(p) = j_\kappa(p)b$, i.e. $j_\kappa \in {}_A[P, Q_\kappa(P)]_B$. This finishes the proof. □

Recall from [162] the following lemma.

1.3. Lemma Consider an exact diagram in A-mod of the form

$$\begin{array}{ccccccc}
M_1 & \xrightarrow{u} & M_2 & \xrightarrow{p} & P & \longrightarrow & 0 \\
\downarrow{\scriptstyle \alpha_1} & & \downarrow{\scriptstyle \alpha_2} & & \downarrow{\scriptstyle \gamma} & & \\
N_1 & \xrightarrow[v]{} & N_2 & \xrightarrow[q]{} & Q & \longrightarrow & 0
\end{array}$$

If $\kappa$ is an idempotent kernel functor and $\mathrm{Ker}(\alpha_2)$, $\mathrm{Coker}\,(\alpha_1)$ and $\mathrm{Coker}\,(\alpha_2)$ are $\kappa$-torsion, then so are $\mathrm{Ker}(\gamma)$ and $\mathrm{Coker}(\gamma)$.

Proof Let $x \in \mathrm{Ker}(\gamma)$; as p is surjective, there is some $y \in M_2$ with $p(y) = x$ and then $0 = \gamma(x) = \gamma(p(y)) = q(\alpha_2(y))$, hence $\alpha_2(y) = v(z)$ for some $z \in N_1$. Let $\bar{z}$ be the image of z in $\mathrm{Coker}(\alpha_1)$, then, since $\mathrm{Coker}(\alpha_1)$ is $\kappa$-torsion by assumption, $J\bar{z} = 0$ for some $J \in \mathcal{L}(\kappa)$, hence $Jz \subset \mathrm{Im}(\alpha_1)$, so $J\alpha_2(y) = Jv(z) \subset \mathrm{Im}(v\alpha_1) = \mathrm{Im}(\alpha_2 u)$. Pick $j \in J$, then $j\alpha_2(y) = \alpha_2 u(t)$ for some $t \in M_1$, so $\alpha_2(jy - u(t)) = 0$ and $jy - u(t) \in \mathrm{Ker}(\alpha_2)$. But $\mathrm{Ker}(\alpha_2)$ is $\kappa$-torsion, so for some $I \in \mathcal{L}(\kappa)$, we have $I(jy - u(t)) = 0$ and it follows that $0 = Ip(jy - u(t)) = Ijp(y)$, i.e. $jp(y) \in \kappa P$. Since this holds for all $j \in J$, we have $Jp(y) \subset \kappa P_2$ and so $p(y) \in \kappa P$, i.e. $\mathrm{Ker}(\gamma)$ is $\kappa$-torsion. On the other hand, $\mathrm{Coker}(\alpha_2)$ maps onto $\mathrm{Coker}(\gamma)$, so $\mathrm{Coker}(\gamma)$ is $\kappa$-torsion, as $\mathrm{Coker}(\alpha_2)$ is. □

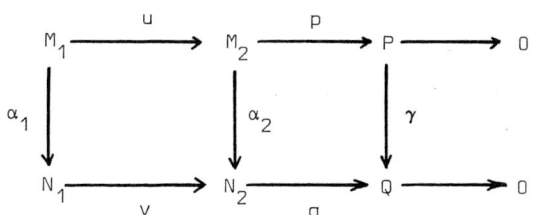

1.4. Note The full subcategory of A-mod, consisting of all $\kappa$-closed left A-modules and denoted by $(A,\kappa)$-mod is a Grothendieck category, as we pointed out before. A morphism $\varphi : M \to N$ between $\kappa$-closed A-modules is an isomorphism (resp. an epimorphism) in $(A,\sigma)$-mod if and only if $\mathrm{Ker}(\varphi)$ and $\mathrm{Coker}(\varphi)$ are $\kappa$-torsion in A-mod (hence $\mathrm{Ker}(\varphi) = 0$!) resp. if $\mathrm{Coker}\ (\varphi)$ is $\kappa$-torsion in A-mod. The assumptions in the foregoing lemma then imply that we have an exact diagram in $(A,\kappa)$-mod

$$
\begin{array}{ccccccc}
 & & 0 & & & & \\
 & & \downarrow & & & & \\
Q_\kappa(M_1) & \longrightarrow & Q_\kappa(M_2) & \longrightarrow & Q_\kappa(P) & \longrightarrow & 0 \\
{\scriptstyle Q_\kappa(\alpha_1)}\downarrow & & \downarrow{\scriptstyle Q_\kappa(\alpha_2)} & & \downarrow{\scriptstyle Q_\kappa(\gamma)} & & \\
Q_\kappa(N_1) & \longrightarrow & Q_\kappa(N_2) & \longrightarrow & Q_\kappa(Q) & \longrightarrow & 0 \\
\downarrow & & \downarrow & & & & \\
0 & & 0 & & & &
\end{array}
$$

An easy diagram-chase, then yields the conclusion that $Q_\kappa(\gamma)$ is an isomorphism in $(A,\kappa)$-mod. From this it follows immediately that $\gamma$ has $\kappa$-torsion kernel and cokernel !

In a similar way one proves :

1.5. Lemma Consider an exact diagram in A-mod of the form

$$
\begin{array}{ccccccc}
0 & \longrightarrow & P & \longrightarrow & M_1 & \longrightarrow & M_2 \\
 & & {\scriptstyle \gamma}\downarrow & & \downarrow{\scriptstyle \alpha_1} & & \downarrow{\scriptstyle \alpha_2} \\
0 & \longrightarrow & Q & \longrightarrow & N_1 & \longrightarrow & N_2
\end{array}
$$

If $\mathrm{Ker}(\alpha_1)$, $\mathrm{Ker}(\alpha_2)$ and $\mathrm{Coker}(\alpha_1)$ are torsion for some idempotent kernel functor $\kappa$ in A-mod, then so are $\mathrm{Ker}(\gamma)$ and $\mathrm{Coker}(\gamma)$.

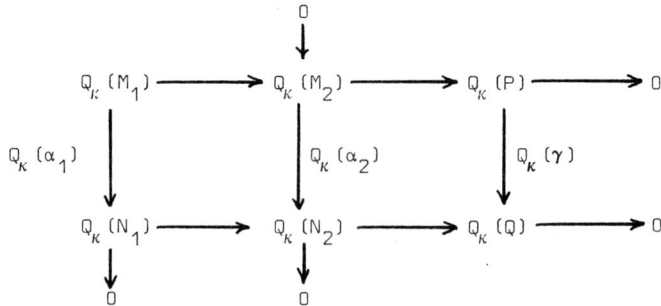

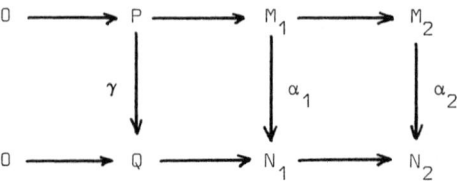

Let $\lambda$ resp. $\mu$ be an idempotent kernel functor in A-mod resp. B-mod. An A-B-bimodule P is said to be <u>$(\lambda,\mu)$-flat</u> if it has the following property: if $\varphi : M \to N$ is a morphism of left B-modules with the property that $\mathrm{Ker}(\varphi)$ is $\mu$-torsion, then $\mathrm{Ker}(P \otimes_B \varphi)$, the kernel of the induced map $P \otimes_B M \to P \otimes_B N$ is $\lambda$-torsion.

1.6. <u>Proposition</u>. Let P be an A-B-bimodule, then

1. ${}_AP_B$ is $(\lambda,\mu)$-flat if and only if the following two conditions hold:
   a. if $M'' \in$ B-mod is $\mu$-torsion, then $P \otimes_B M'' \in$ A-mod is $\lambda$-torsion;
   b. if $i : M' \to M$ is a monomorphism in B-mod, then $\mathrm{Ker}(P \otimes_B i) \in$ A-mod is $\lambda$-torsion;

2. ${}_AP_B$ is $(\lambda,\mu)$-flat if and only if ${}_AQ_\lambda(P)_B$ is $(\lambda,\mu)$-flat.

<u>Proof</u>. One implication in 1. is obvious, indeed, it suffices to apply the definition of $(\lambda,\mu)$-flatness to a monomorphism $i : M' \to M$ (with $\mu$-torsion kernel $\mathrm{Ker}(i) = 0$ !) and the canonical map $\pi : M'' \to 0$ (with $\mu$-torsion kernel $\mathrm{Ker}(\pi) = M''$), when $M''$ is $\mu$-torsion. Conversely, let $\alpha$ be an arbitrary surjective B-linear map with kernel $\mathrm{Ker}(\alpha) = K$ which is $\mu$-torsion, say

$$0 \longrightarrow K \longrightarrow M \longrightarrow M'' \longrightarrow 0$$

Then we obtain a commutative exact diagram

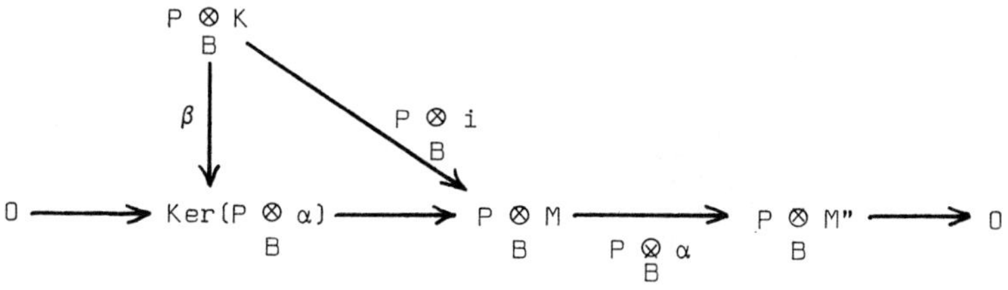

Here, the map $\beta$ is surjective and $P \otimes_B K$ is $\lambda$-torsion by assumption, hence $\mathrm{Ker}(P \otimes_B \alpha)$ is $\lambda$-torsion too. We thus have verified the characteristic property of $(\lambda,\mu)$-flatness for surjective $\alpha$; the injective case is part of the assumptions. Now, for the general case, consider an arbitrary morphism $\alpha : M \to N$ and factorize it as follows :

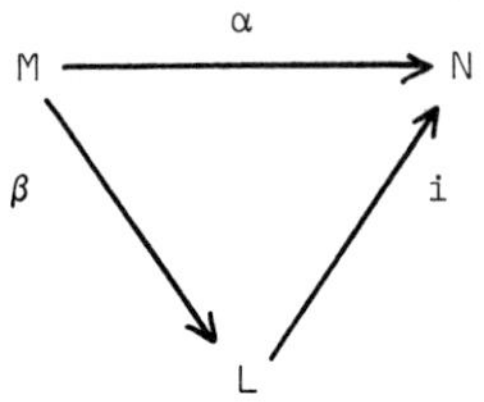


where $\beta$ is the canonical morphism $M \to L = \mathrm{Im}(\alpha) = M/\mathrm{Ker}(\alpha)$ and $i : L \to N$ is the canonical inclusion. We then obtain an exact commutative diagram

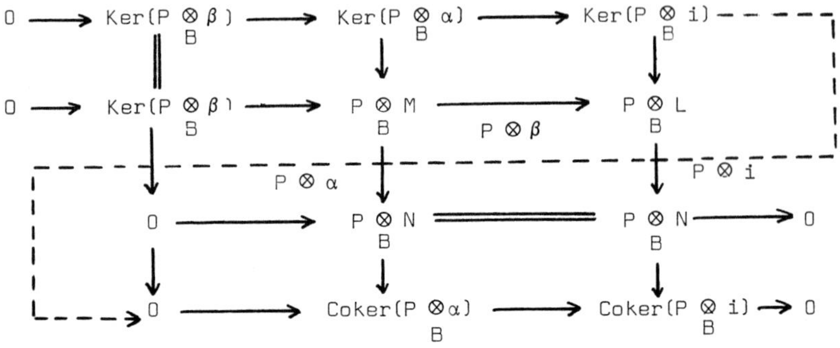


Here $\mathrm{Ker}(P \otimes_B i)$ is $\lambda$-torsion, since $i : L \to N$ is a monomorphism and $P$ is $(\lambda,\mu)$-flat, while $\mathrm{Ker}(\alpha) = \mathrm{Ker}(\beta)$ implies that $\mathrm{Ker}(P \otimes \beta)$ is $\lambda$-torsion too, $\beta : M \to L$ being surjective. But then the snake-lemma yields an exact sequence

$$0 \longrightarrow \mathrm{Ker}(P \underset{B}{\otimes} \beta) \longrightarrow \mathrm{Ker}(P \underset{B}{\otimes} \alpha) \longrightarrow \mathrm{Ker}(P \underset{B}{\otimes} i) \longrightarrow 0$$

and since both extremes are $\lambda$-torsion, so is the middle term. This proves 1.

In order to prove 2, note that we already know that $Q_\lambda(P)$ is an A-B-bimodule and that $j_\lambda \in {}_A[P, Q_\lambda(B)]_B$. Let $\varphi : M \to N$ be a morphism of left B-modules such that $\mathrm{Ker}(\varphi)$ is $\mu$-torsion and consider the following exact diagram in A-mod:

$$\begin{array}{ccccccc}
0 \to & \mathrm{Ker}(P \underset{B}{\otimes} \varphi) & \longrightarrow & P \underset{B}{\otimes} M & \longrightarrow & P \underset{B}{\otimes} N \\
 & \downarrow \theta & & \downarrow j_\lambda \otimes M & & \downarrow j_\lambda \otimes N \\
0 \to & \mathrm{Ker}(Q_\lambda(P) \underset{B}{\otimes} \varphi) & \longrightarrow & Q_\lambda(P) \underset{B}{\otimes} M & \longrightarrow & Q_\lambda(P) \underset{B}{\otimes} N
\end{array}$$

We want to show that both Ker $(j_\lambda \underset{B}{\otimes} Q)$ and Coker$(j_\lambda \underset{B}{\otimes} Q)$ are $\lambda$-torsion for any $Q \in$ B-mod, for in that case Ker$(\theta)$ and Coker$(\theta)$ are $\lambda$-torsion by Lemma 1.5., hence Ker$(P \underset{B}{\otimes} \varphi)$ is $\lambda$-torsion if and only if Ker$(Q_\lambda(P) \underset{B}{\otimes} \varphi)$ is $\lambda$-torsion, proving that ${}_AP_B$ is $(\lambda,\mu)$-flat if and only if ${}_AQ_\lambda(P)_B$ is $(\lambda,\mu)$-flat. Now, in order to prove this, one proceeds as follows : view $Q \in$ B-mod as the cokernel of a left B-linear morphism $B^{(I)} \to B^{(J)}$, then we obtain an exact commutative diagram in A-mod

$$\begin{array}{ccccccc}
P \underset{B}{\otimes} B^{(I)} & \longrightarrow & P \underset{B}{\otimes} B^{(J)} & \longrightarrow & P \underset{B}{\otimes} Q & \longrightarrow & 0 \\
\downarrow \alpha_1 & & \downarrow \alpha_2 & & \downarrow \gamma & & \\
Q_\lambda(P) \underset{B}{\otimes} B^{(I)} & \longrightarrow & Q_\lambda(P) \underset{B}{\otimes} B^{(J)} & \longrightarrow & Q_\lambda(P) \underset{B}{\otimes} Q & \longrightarrow & 0
\end{array}$$

Now, $P \underset{B}{\otimes} B$ is left A-isomorphic to P and $Q_\lambda(P) \underset{B}{\otimes} B$ to $Q_\lambda(P)$, hence Ker$(\alpha_i)$ and Coker$(\alpha_i)$ are $\lambda$-torsion for $i = 1,2$, so applying 1.3. yields the assertion. □

1.7. <u>Note</u>. In [166], where R is a commutative ring, the definition of $\kappa$-flatness (= $(\kappa,\kappa)$-flatness) of an R-module P only included property b. above. However, both definitions coincide here, since for R-commutative and $P \in$ R-mod, $P \underset{R}{\otimes} M''$ is $\kappa$-torsion, whenever $M''$ is $\kappa$-torsion. Indeed, if $\sum_{i=1}^{n} p_i \otimes m_i \in P \underset{R}{\otimes} M''$ for some $p_i \in P$ resp. $m_i \in M''$, then if $M''$ is $\kappa$-torsion we may find $I \in \mathcal{L}(\kappa)$ such that $I m_i = 0$ for all $i = 1,\dots,n$ hence $I(\sum_{i=1}^{n} p_i \otimes m_i) = 0$, as for $j \in J$ we have $j(p_i \otimes m_i) = (jp_i) \otimes m_i = (p_i j) \otimes m_i =$

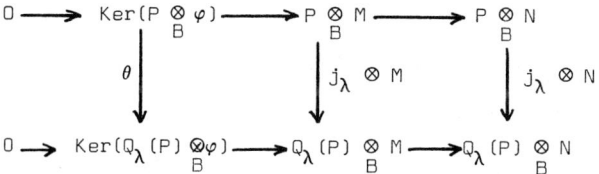

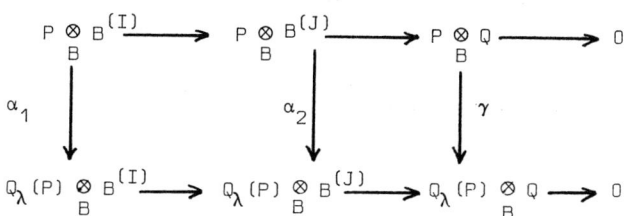

$p_i \otimes jm_i = 0$. Note that we used the fact that the left and right actions on P coincide. We will see below, that if A is noncommutative and an algebra over some commutative subring R on which $\kappa$ is defined (an exact definition will be given below), then, if P is defined over R, it follows that the same argument applies.

1.8. <u>Proposition</u>. Let P be an A-B-bimodule and M a left B-module, then

1. ${}_AQ_\lambda(Q_\lambda(P) \underset{B}{\otimes} M) = {}_AQ_\lambda(P \underset{B}{\otimes} M)$.
2. if P is a $(\lambda,\mu)$-flat A-B-bimodule, then ${}_AQ_\lambda(P \underset{B}{\otimes} Q_\mu(M)) = {}_AQ_\lambda(P \underset{B}{\otimes} M)$.

<u>Proof</u>. First note that 1. has essentially been proved in (1.6.2.). Now, in order to prove 2., consider the following exact sequence

$$0 \longrightarrow \mu M \longrightarrow M \xrightarrow{j_\mu} Q_\mu(M) \longrightarrow T \longrightarrow 0$$

where $T = \mathrm{Coker}(j_\mu)$ is $\mu$-torsion. It splits into two parts :

$$0 \longrightarrow \mu M \xrightarrow{i} M \xrightarrow{\pi} \overline{M} \longrightarrow 0$$

resp.

$$0 \longrightarrow \overline{M} \xrightarrow{j} Q_\mu(M) \longrightarrow T \longrightarrow 0$$

where $\overline{M} = M/\mu M$, so we obtain two exact sequences :

$$0 \longrightarrow \mathrm{Ker}(P \underset{B}{\otimes} i) \longrightarrow P \underset{B}{\otimes} \mu M \xrightarrow{P \otimes i} P \underset{B}{\otimes} M \xrightarrow{P \otimes \pi} P \underset{B}{\otimes} \overline{M} \longrightarrow 0$$

resp.

$$0 \longrightarrow \mathrm{Ker}(P \underset{B}{\otimes} j) \longrightarrow P \underset{B}{\otimes} \overline{M} \xrightarrow{P \otimes j} P \underset{B}{\otimes} Q_\mu(M) \longrightarrow P \underset{B}{\otimes} T \longrightarrow 0$$

Now, since P is $(\lambda,\mu)$-flat, both $\mathrm{Ker}(P \underset{B}{\otimes} j)$ and $P \underset{B}{\otimes} T$ are $\lambda$-torsion, in the latter sequence, so $P \underset{B}{\otimes} j$ induces an isomorphism $Q_\lambda(P \underset{B}{\otimes} \overline{M}) = Q_\lambda(P \underset{B}{\otimes} Q_\mu(M))$. In the former sequence, $P \underset{B}{\otimes} \mu M$ is $\lambda$-torsion, hence so is $\mathrm{Ker}(P \underset{B}{\otimes} \pi)$ and $P \underset{B}{\otimes} \pi$ induces an isomorphism $Q_\lambda(P \underset{B}{\otimes} M) = Q_\lambda(P \underset{B}{\otimes} \overline{M})$. □

1.9. <u>Lemma</u>. If $\{P_i; i \in I\}$ is an arbitrary family of A-B-bimodules, then $\underset{i \in I}{\oplus} P_i$ is $(\lambda,\mu)$-flat if and only if for each $i \in I$ the A-B-bimodule $P_i$ is $(\lambda,\mu)$-flat. □

1.10. Lemma. An inductive limit of $(\lambda,\mu)$-flat A-B-bimodules is $(\lambda,\mu)$-flat.

Proof. This follows immediately from the fact that for each inductive family $(P_i, i \in I)$ and each left B-module M, we have $(\varinjlim_I P_i) \otimes_B M = \varinjlim_I (P_i \otimes_B M)$. □

Let P be an A-B-bimodule, then we call it $(\lambda,\mu)$-compatible, if it has the property that for all $\mu$-torsion left B-modules T the left A-module $P \otimes_B T$ is $\lambda$-torsion. Call ${}_AP_B$ a flat A-B-bimodule, if it has the property that for any injective $i : M' \to M$ in B-mod, the induced $P \otimes_B i : P \otimes_B M' \to P \otimes_B M$ is injective in A-mod. If $A = B$, $\lambda = \mu$, then we speak of a $\lambda$-flat, $\lambda$-compatible or flat A-bimodule.

1.11. Corollary Any $(\lambda,\mu)$-compatible flat A-B-bimodule ${}_AP_B$ is $(\lambda,\mu)$-flat.

Proof. If suffices to check (1.6.1.a),as (1.6.1.b) holds by assumption But, if $i = M' \to M$ is an injective map of left B-modules, then $\mathrm{Ker}(P \otimes_B i) = 0$ is certainly $\lambda$-torsion ! □

1.12. Corollary. Any $\lambda$-compatible right projective A-bimodule is $\lambda$-flat. □

1.13. Lemma. Let ${}_AP_B$ be an A-B-bimodule and N a left A-module, then

1. ${}_A[P,N] \in$ B-mod;

2. if ${}_AP_B$ is $(\lambda,\mu)$-flat and ${}_AN$ is $\lambda$-injective (resp. $\lambda$-closed), then ${}_A[P,N]$ is $\mu$-injective (resp. $\mu$-closed).

Proof. 1. Let $p \in P$, $b \in B$ and $\varphi \in {}_A[P,N]$, then we define $b.\varphi \in {}_A[P,N]$ by $(b.\varphi)(p) = \varphi(pb)$; this is easily seen to yield a B-module structure on ${}_A[P,N]$.

2. There is a canonical bijection $\theta : {}_A[P \otimes_B M, N] \to {}_B[M, {}_A[P,N]]$ for any ${}_AN$, ${}_BM$ and ${}_AP_B$, where ${}_A[P,N]$ is given the left B-module structure defined in 1. It is given by sending $\varphi \in {}_A[P \otimes_B M, N]$ to $\theta(\varphi) \in {}_B[M, {}_A[P,N]]$ defined by

$(\theta(\varphi)(m))(p) = \varphi(p \otimes m) \in N$ for any $m \in M$ and $p \in P$. Consider an exact sequence of left B-modules

$$0 \longrightarrow I \longrightarrow B \longrightarrow B/I \longrightarrow 0$$

where $B/I$ is $\mu$-torsion, i.e. $I \in L(\mu)$, then this induces an exact sequence in A-mod of the form

$$0 \longrightarrow K \longrightarrow P \underset{B}{\otimes} I \longrightarrow P \underset{B}{\otimes} B \longrightarrow P \underset{B}{\otimes}(B/I) \longrightarrow 0$$

where K and $P \underset{B}{\otimes}(B/I)$ are $\lambda$-torsion, as P is $(\lambda,\mu)$-flat. As N is $\lambda$-injective (resp. $\lambda$-closed), the canonical map

$${}_A[P \underset{B}{\otimes} B, N] \to {}_A[P \underset{B}{\otimes} I, N]$$

is surjective (resp. bijective). But, applying the isomorphism $\theta$, it then follows that the canonical map

$${}_B[B, {}_A[P,N]] \to {}_B[I, {}_A[P,N]]$$

is surjective (resp. bijective), i.e. ${}_A[P,N]$ is $\mu$-injective (resp. $\mu$-closed). □

1.14. <u>Corollary</u>. Let $\kappa$ be an idempotent kernel functor in A-mod and let L be a left finitely generated $\kappa$-flat A-bimodule, then for any $\kappa$-torsionfree $M \in$ A-mod, we have ${}_A[Q_\kappa(L), Q_\kappa(M)] = Q_\kappa({}_A[L,M])$.

<u>Proof</u>. First note that $Q_\kappa(L)$ is $\kappa$-flat, since L is, by (1.6.2.), hence the foregoing lemma implies that ${}_A[Q_\kappa(L), Q_\kappa(M)] = {}_A[L, Q_\kappa(M)]$ is also $\kappa$-closed. Hence the assertion will be established, if we are able to prove that in the exact sequence

$$0 \longrightarrow K \longrightarrow [L,M] \xrightarrow{f} [L, Q_\kappa(M)] \longrightarrow T \longrightarrow 0$$

both $K = \mathrm{Ker}(f)$ and $T = \mathrm{Coker}(f)$ are $\kappa$-torsion. First, suppose $\varphi : L \to M$ is in K and let $j : M \to Q_\kappa(M)$ be the localization morphism, then $\varphi \in K$ yields that $j\varphi = 0$, i.e. $\varphi(L) \subset \mathrm{Ker}(j) = \kappa M$. Since L is finitely generated, we may find $I \in \mathcal{L}(\kappa)$ such that $I\varphi(L) = 0$, i.e. $I\varphi = 0$ and $\varphi \in \kappa([L,M])$ Therefore, K has to be $\kappa$-torsion. Next, in order to prove that T is $\kappa$-torsion, it suffices to verify that for any $\varphi : L \to Q_\kappa(M)$, there is $L_1 \subset L$ with $L/L_1$ being $\kappa$-torsion and a morphism $\varphi_1 : L_1 \to M$ such that the following square is commutative :

$$\begin{array}{ccc} L_1 & \longrightarrow & L \\ {\scriptstyle \varphi_1}\downarrow & & \downarrow{\scriptstyle \varphi} \\ M & \longrightarrow & Q_\kappa(M) \end{array}$$

Now, if $L' = \varphi(L) \subset Q_\kappa(M)$, then as L' is finitely generated, we may find $I \in \mathcal{L}(\kappa)$ such that $IL' \subset M$. Put $L_1 = IL$ and $\varphi_1 = \varphi|L_1$, then $(L_1, \varphi_1)$ satisfies our requirements. □

Let A resp. B and $\lambda$ resp. $\mu$ be as before and let C be a ring endowed with an idempotent kernel functor $\nu$ in C-mod. If ${}_AP_B$ resp. ${}_BQ_C$ is $(\lambda,\mu)$-flat resp. $(\mu,\nu)$-flat, then it is easy to see that ${}_A(P \underset{B}{\otimes} Q)_C$ is $(\lambda,\nu)$-flat. Now, let ${}_AP_B$ be an A-B-bimodule and M a left B-module, then we will write $P \underset{\lambda}{\perp} M$ for the left A-module $Q_\lambda(P \underset{B}{\otimes} M)$. Let $(A,\lambda)$, $(B,\mu)$, $(C,\nu)$, $(E,\pi)$ be couples consisting of a ring and an idempotent kernel functor in the category of left modules over that ring; let ${}_AP_B$ resp. ${}_BQ_C$ resp. ${}_CT_E$ be $(\lambda,\mu)$-flat resp. $(\mu,\nu)$-flat resp. $(\nu,\pi)$-flat, then it is clear from the foregoing that the A-D-bimodules $P \underset{\lambda}{\perp} (Q \underset{\mu}{\perp} T)$ resp. $(P \underset{\lambda}{\perp} Q) \underset{\lambda}{\perp} T$ are canonically isomorphic. When no ambiguity arises, we will simply write $P \perp Q \perp T$ for this bimodule.

We may now state the following definition :

Let P be an A-B-bimodule, then ${}_AP_B$ is <u>$(\lambda,\mu)$-invertible</u> if the following conditions are satisfied :

a. ${}_AP_B$ is $(\lambda,\mu)$-flat;

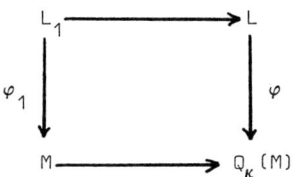

b. there exists a B-A-bimodule ${}_BQ_A$ which is $(\mu,\lambda)$-flat and isomorphisms $\varphi : P \underset{\lambda}{\perp} Q \to Q_\lambda(A)$ resp. $\psi : Q \underset{\mu}{\perp} P \to Q_\mu(B)$ in A-mod-A resp. B-mod-B.

1.15. Proposition. Let $(A,\lambda)$, $(B,\mu)$ and $(C,\nu)$ be couples consisting of a ring and an idempotent kernel functor in the category of left modules over that ring. Let ${}_AP_B$ be a $(\lambda,\mu)$-invertible A-B-bimodule and ${}_BQ_C$ a $(\mu,\nu)$-invertible B-C-bimodule, then $P \underset{\lambda}{\perp} Q$ is $(\lambda,\nu)$-invertible.

Proof. By the preceding remarks, $P \underset{B}{\otimes} Q$ is $(\lambda,\nu)$-flat, hence so is $P \underset{\lambda}{\perp} Q$ by (1.6.2.). Now, if ${}_AP_B$ has "inverse" ${}_BP'_A$ and ${}_BQ_C$ has "inverse" ${}_CQ'_B$, then there are isomorphisms $((P \underset{\lambda}{\perp} Q) \underset{\lambda}{\perp} (Q' \underset{\nu}{\perp} P')) \xrightarrow{\sim} Q_\lambda(A)$ and $((Q' \underset{\nu}{\perp} P') \underset{\nu}{\perp} (P \underset{\lambda}{\perp} Q)) \xrightarrow{\sim} Q_\nu(C)$ by (1.8.1.) and (1.8.2.). □

For any idempotent kernel functor $\kappa$ in A-mod, let us say that an A-bimodule P is $\kappa$-invertible if it is $(\kappa,\kappa)$-invertible, i.e. P is $\kappa$-flat and there exists another $\kappa$-flat A-bimodule Q, an "inverse" for P together with isomorphisms $\varphi : P \underset{\kappa}{\perp} Q \xrightarrow{\sim} Q_\kappa(A)$ resp. $\psi : Q \underset{\kappa}{\perp} P \xrightarrow{\sim} Q_\kappa(A)$ of A-bimodules. Let us prove some generalities about these modules.

1.16. Proposition. Let P be a $\kappa$-closed, $\kappa$-invertible A-bimodule, then

1. ${}_A[P,P] = Q_\kappa(A)^{opp}$;
2. ${}_A[P,P]_A = Z(Q_\kappa(A))$.

Proof. 1. Define $\theta : Q_\kappa(A)^{opp} \to {}_A[P,P]$ by $\theta(a)(p) = pa$ for any $p \in P$. If $\theta(a) = 0$, then $Pa = 0$, so $(Q \underset{A}{\otimes} P)a = 0$ for any right A-module Q. Since any left A-linear map $Q \underset{A}{\otimes} P \to Q \underset{A}{\otimes} P$ extends uniquely to an A-linear map $Q_\kappa(Q \underset{A}{\otimes} P) \to Q_\kappa(Q \underset{A}{\otimes} P)$, whenever Q is two-sided, it suffices to choose Q such that $Q_\kappa(Q \underset{A}{\otimes} P) \cong Q_\kappa(P \underset{A}{\otimes} Q) \cong Q_\kappa(A)$, i.e. choose Q to be an "inverse" for P, to deduce that multiplication by a yields the zero map $Q_\kappa(A) \to Q_\kappa(A)$, i.e. a = 0 ! It follows that $\theta$ is injective.

Now, let $f \in {}_A[P,P]$, then $Q \underset{A}{\otimes} f : Q \underset{A}{\otimes} P \to Q \underset{A}{\otimes} P$ extends in a unique way to a left A-linear map $g = Q_\kappa(Q \underset{A}{\otimes} f) : Q_\kappa(A) \to Q_\kappa(A)$, which makes the following diagram commutative :

$$\begin{array}{ccc} Q \underset{A}{\otimes} P & \xrightarrow{Q \underset{A}{\otimes} f} & Q \underset{A}{\otimes} P \\ \downarrow & & \downarrow \\ Q_\kappa(A) & \xrightarrow[g]{} & Q_\kappa(A) \end{array}$$

If $g(1) = a$, then g is right multiplication by a. Define $h : P \to P$ by $p \to pa$, then we have $Q_\kappa(Q \underset{A}{\otimes} f) = Q_\kappa(Q \underset{A}{\otimes} h)$. Now $Im(f-h) \subset P$, is thus a $\kappa$-torsion free left A-module, while on the other hand $Q_\kappa(Q \otimes Im(f-h)) = 0$ i.e. $f = h$.

2. If f is an A-bimodule endomorphism of P, then it is given by right multiplication by some $a \in Q_\kappa(A)$. Right linearity then implies that $P(xa - ax) = 0$ for all $x \in A$ ; as before it follows that $ax - xa = 0$ for all $x \in A$, i.e. $a \in Q_\kappa(A)^A = Z(Q_\kappa(A))$. □

1.17. Proposition. If P is isomorphic to $Q_\kappa(A)$ in A-mod, then there is an $\alpha \in Aut(Q_\kappa(A))$ such that P is isomorphic to $Q_\kappa(A)_\alpha$ in A-mod-A.

Proof. Let $f : P \cong Q_\kappa(A)$ be an isomorphism in A-mod and let $a \in Q_\kappa(A)$. Define $g : P \to P$ by $g(p) = f^{-1}(f(p)a)$. Since $Q_\kappa(A)^{opp} = {}_A[P,P]$, there is $\alpha(a) \in Q_\kappa(A)$ such that $g(p) = p\,\alpha(a)$, i.e. $f(p\,\alpha(a)) = f(p)a$, for all $p \in P$. Evidently, $\alpha \in Aut(Q_\kappa(A))$ and then f may be considered as an A-bimodule isomorphism $P \cong {}_1Q_\kappa(A)_\alpha$. □

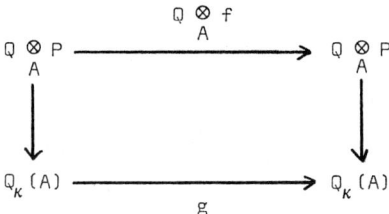

## B.2. The Picard Group of a Grothendieck Category.

Let R be a commutative ring, then an abelian category $\mathcal{A}$ is said to be an R-category if it is endowed with a ringhomomorphism $R \to Z(\mathcal{A})$, the center of the category $\mathcal{A}$, which by definition is just the ring $End(1_{\mathcal{A}})$. Thus, saying that $\mathcal{A}$ is an R-category amounts to endow each set $\mathrm{Hom}_{\mathcal{A}}(A',A'')$ with an R-module structure. In particular, if A is an R-algebra with structure morphism $R \to A$, then the fact that $Z(A\text{-mod}) = Z(A)$, the center of A. (cf. Lemma 2.1. below !); shows that each R-algebra structure on A endows A-mod with an R-category structure and conversely.

An R-functor between R-categories $\mathcal{A}_1$ and $\mathcal{A}_2$ is a functor (★) $F : \mathcal{A}_1 \to \mathcal{A}_2$ inducing for each couple of objects $(A',A'')$ of $\mathcal{A}_1$ an R-linear map $\mathrm{Hom}_{\mathcal{A}_1}(A',A'') \to \mathrm{Hom}_{\mathcal{A}_2}(F(A'), F(A''))$. Other concepts relating to R-categories are defined similarly.

Let $\mathcal{A}$ be an R-category, then we define the R-Picard group $\mathrm{Pic}_R(\mathcal{A})$ of $\mathcal{A}$, to the group of isomorphism classes $[F]$ of R-autoequivalences $F : \mathcal{A} \to \mathcal{A}$. If $R = \mathbb{Z}$, then we will simply $\mathrm{Pic}(\mathcal{A})$ for $\mathrm{Pic}_{\mathbb{Z}}(\mathcal{A})$ and call it *the* Picard group of $\mathcal{A}$. On the other hand, for any R-algebra A, the group $\mathrm{Pic}_R(A)$ was defined (in chapter A) to consist of all isomorphism classes of invertible left $A \underset{R}{\otimes} A^{opp}$-modules, i.e. of all classes of A-modules P defined over R and such that for a similar Q there exist compatible bimodule isomorphisms $P \underset{A}{\otimes} Q \cong A$ resp. $Q \underset{A}{\otimes} P \cong A$. It is well-known, cf. Bass [18] for example, that the map

$$\mathrm{Pic}_R(A) \to \mathrm{Pic}_R(A\text{-mod}): [P] \to [P \underset{A}{\otimes} -]$$

defines an isomorphism of groups, with inverse given by sending $[F] \in \mathrm{Pic}_R(A\text{-mod})$ to $[F(A)] \in \mathrm{Pic}_R(A)$ for any R-autoequivalence F of A-mod. If is the purpose of this section to derive a similar result for $(A,\kappa)$-mod.

(★) All functors of this text will be additive at least.

2.1. Lemma. For every ring A and every idempotent kernel functor $\kappa$ in A-mod, the rings $Z((A,\kappa)\text{-mod})$ and $Z(Q_\kappa(A))$ are isomorphic.

Proof. Define a map

$$\begin{array}{rcl} h : Z(Q_\kappa(A)) & \longrightarrow & End(1_{(A,\kappa)\text{-mod}}) \\ c & \longmapsto & h(c) \end{array}$$

where h(c) maps x to cx "locally". Since one easily verifies that h is a well-defined ringhomomorphism, it suffices to check that h is bijective. Now, the injectivity of h is clear, since one has $h(c)_R(1) = c.1 = c$. On the other hand, let us take $\varphi \in End(1_{(A,\kappa)\text{-mod}})$, then $\varphi_{Q_\kappa(A)} : Q_\kappa(A) \to Q_\kappa(A)$ is defined by $\varphi_{Q_\kappa(A)}(1) = c$. If $M \in (A,\kappa)$-mod, then a morphism $f : Q_\kappa(A) \to M$ is completely determined by $f(1) = m \in M$, since ${}_A[Q_\kappa(A), M] = {}_A[A, M]$. The following diagram is commutative :

$$\begin{array}{ccc} Q_\kappa(A) & \xrightarrow{\ \psi\ } & Q_\kappa(A) \\ f \downarrow & & \downarrow f \\ M & \xrightarrow[\ \varphi_M\ ]{} & M \end{array}$$

Hence $\varphi_M(m) = \varphi_M f(1) = f\,\psi(1) = f(c) = cm = h(c)_M(m)$. Since m is arbitrary this implies that $\varphi_M = h(c)_M$ for all M, i.e. $\varphi = h(c)$, proving that h is surjective too, modulo the verification that c is central in $Q_\kappa(A)$. Now, this last fact may be verified as follows : in the above commutative diagram, let $M = Q_\kappa(A)$ and let $f : Q_\kappa(A) \to Q_\kappa(A)$ be defined by right multiplication by some $a \in Q_\kappa(A)$. Since $\varphi$ is natural this yields that $\psi f = f\psi$, hence $ac = \psi f(1) = f\psi(1) = ca$, and as $a \in Q_\kappa(A)$ is arbitrary, this proves the assertion. □

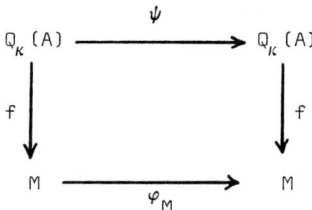

2.2. Note. The foregoing proof actually shows that if $\mathcal{A}$ is a full subcategory of A-mod, containing A, then $Z(\mathcal{A}) = Z(A)$. The reader should also note that we have used the following fact in the above proof : if $f : M \to N$ is an A-linear map between $M, N \in (A,\kappa)$-mod, then f is $Q_\kappa(A)$-linear. This may be seen as follows. First note that f obviously "extends" to an $\overline{A} = A/\kappa A$-linear map, as M and N are $Q_\kappa(A)$-modules, hence also $\overline{A}$-modules in the obvious way. We may thus assume $A = \overline{A}$, i.e. A is $\kappa$-torsion free. Let $a \in Q_\kappa(A)$, then we may find $I \in \mathcal{L}(\kappa)$ such that $I a \subset A$, so for all $i \in I$ we have $if(am) = f(iam) = iaf(m)$ for any $m \in M$, hence $I(f(am) - af(m)) = 0$, so $f(am) - af(m) \in \kappa N = 0$. It follows that f is $Q_\kappa(A)$-linear indeed !

Let $\kappa$ be an idempotent kernel functor in A-mod, then it is well-known that $(A,\kappa)$-mod is a Grothendieck category and that the (left exact !) endofunctor $Q_\kappa(-)$ in A-mod induces an *exact* functor $\underline{\underline{a}}_\kappa$ : A-mod $\to (A,\kappa)$-mod, which is left adjoint to the canonical inclusion $i_\kappa : (A,\kappa)$-mod $\to$ A-mod. The Gabriel-Popescu theorem states essentially that every Grothendieck category $\mathcal{C}$ is of this form. More precisely, let $\mathcal{C}$ be a Grothendieck category and G an object in $\mathcal{C}$, then we define the representable functor $q_G$ by

$$q_G : \mathrm{Hom}_{\mathcal{C}}(G,-) : \mathcal{C} \to \text{mod-}q_G(G).$$

This functor possesses a left adjoint

$$T_G : \text{mod-}q_G(G) \to \mathcal{C}.$$

Recall that G is said to be a <u>generator</u> for $\mathcal{C}$ if for each morphism $\alpha : B \to C$ in $\mathcal{C}$, there exists a morphism $\beta : G \to B$ such that $\alpha\beta \neq 0$. The Gabriel-Popescu theorem [ 56 ] then says that for $T = T_G$, $q = q_G$ and $\varphi : Tq \to 1$ the adjunction arrow, the following statements are equivalent :

a. G is a generator for $\mathcal{C}$;

b. $\varphi$ is fully faithful;

c. $\varphi$ is a functorial isomorphism and T is exact;

d. T is exact and induces an equivalence of categories between $\mathcal{C}$ and mod(q(G), Ker T).

(This last category is the quotient category of mod-q(G) relative to the localizing subcategory Ker T, which consists of all $M \in$ mod-$q_G(G)$ such that $T(M) = 0$, see also P. Gabriel [55] or B. Stenström [142].)

In order to study equivalences, and in particular antiequivalences, between Grothendieck categories, it thus suffices to restrict attention to equivalences between categories $(A,\lambda)$-mod and $(B,\mu)$-mod, say. Note also that we have :

2.3. <u>Corollary</u>. For any Grothendieck category $\mathcal{C}$ with generator G, the center $Z(\mathcal{C})$ of $\mathcal{C}$ is isomorphic to $Z(\mathrm{End}_{\mathcal{C}}(G))$, the center of the $\mathcal{C}$-endomorphism ring of G. □

2.4. <u>Proposition</u>. Let $S : (B,\mu)\text{-mod} \to (A,\lambda)\text{-mod}$ possess a right adjoint, then S is isomorphic to a functor $\underline{\underline{a}}_\lambda(P \otimes_B -)$, for some $\lambda$-closed A-B-bimodule.

<u>Proof</u>. It suffices to mimick the demonstration of a similar result for ordinary left module categories. Since S has a right adjoint, it is right exact and commutes with direct sums. Let $P = S(Q_\mu(B))$, then S induces a ring morphism

$$B \longrightarrow Q_\mu(B) = {}_B[Q_\mu(B), Q_\mu(B)]^{\mathrm{opp}} = \mathrm{End}_{(B,\mu)\text{-mod}}(Q_\mu(B))^{\mathrm{opp}} \longrightarrow$$

$$\longrightarrow \mathrm{End}_{(A,\lambda)\text{-mod}}(P)^{\mathrm{opp}} = {}_A[P,P]^{\mathrm{opp}},$$

endowing P with an A-B-bimodule structure. Take $M \in (B,\mu)$-mod, then we may find an exact sequence of left B-modules $B^{(I)} \to B^{(J)} \to M \to 0$ for some index sets I and J. Let us write $B^{(I,\mu)} = \underline{\underline{a}}_\mu(B^{(I)})$ for the direct sum of I copies of $Q_\mu(B)$ in $(B,\mu)$-mod, then applying the functors $\underline{\underline{a}}_\lambda(P \otimes_B -)$ and S yields exact sequences

$$\underline{\underline{a}}_\lambda(P \otimes_B B^{(I)}) \longrightarrow \underline{\underline{a}}_\lambda(P \otimes_B B^{(J)}) \longrightarrow \underline{\underline{a}}_\lambda(P \otimes_B M) \longrightarrow 0$$

resp.

$$S(B^{(I,\mu)}) \to S(B^{(J,\mu)}) \to S(M) \to 0$$

in $(A,\lambda)$-mod, which combine into a commutative diagram

$$\begin{array}{ccccccc}
\underline{\underline{a}}_\lambda(P^{(I)}) & \longrightarrow & \underline{\underline{a}}_\lambda(P^{(J)}) & \longrightarrow & \underline{\underline{a}}_\lambda(P \underset{B}{\otimes} M) & \longrightarrow & 0 \\
\| & & \| & & \| & & \\
S(B^{(I,\mu)}) & \longrightarrow & S(B^{(J,\mu)}) & \longrightarrow & S(M) & \longrightarrow & 0
\end{array}$$

for indeed, we have $\underline{\underline{a}}_\lambda(P^{(I)}) = P^{(I,\lambda)} = (S(Q_\mu(B)))^{(I,\lambda)} = S(B^{(I,\mu)})$ and similarly for the index set J. But then, both rows being exact, we obtain that $\underline{\underline{a}}_\lambda(P \underset{B}{\otimes} M)$ and $S(M)$ are isomorphic. An easy verification then yields that this isomorphism is actually functorial. □

2.5. Proposition. Let P be a $\mu$-closed A-B-bimodules arising from an equivalence $S : (B,\mu)\text{-mod} \to (A,\lambda)\text{-mod}$ as in (2.4), then P is $(\lambda,\mu)$-flat.

Proof. Recall that if $M,N \in (B,\mu)$-mod, then $M \xrightarrow{f} N \to 0$ is exact in $(B,\mu)$-mod if and only if Coker(f) is $\mu$-torsion, where Coker(f) is taken in B-mod or $Q_\mu(B)$-mod. Let M be $\mu$-torsion and write M in an exact sequence

$$(1)\quad B^{(I)} \xrightarrow{u} B^{(J)} \longrightarrow M \longrightarrow 0$$

Tensoring by P yields an exact sequence of left A-modules

$$(2)\quad P \underset{B}{\otimes} B^{(I)} \longrightarrow P \underset{B}{\otimes} B^{(J)} \longrightarrow P \underset{B}{\otimes} M \longrightarrow 0$$

where we want to verify that $P \underset{B}{\otimes} M$ is $\lambda$-torsion. Now, the exact sequence (1) yields an exact sequence

$$(3)\quad B^{(I,\mu)} \xrightarrow{Q_\mu(u)} B^{(J,\mu)} \longrightarrow N \longrightarrow 0$$

where $B^{(I,\mu)} = Q_\mu(B^{(I)})$ and where $N = \mathrm{Coker}(Q_\mu(u))$. Then, tensoring by P gives

$$(4)\quad P \underset{B}{\otimes} B^{(I,\mu)} \longrightarrow P \underset{B}{\otimes} B^{(J,\mu)} \longrightarrow P \underset{B}{\otimes} N \longrightarrow 0$$

The sequences (1) and (3) combine into an exact, commutative diagram of left B-modules

$$\begin{array}{ccccccc}
B^{(I)} & \longrightarrow & B^{(J)} & \longrightarrow & M & \longrightarrow & 0 \\
\alpha \downarrow & & \downarrow \beta & & \downarrow f & & \\
B^{(I,\mu)} & \longrightarrow & B^{(J,\mu)} & \longrightarrow & N & \longrightarrow & N
\end{array}$$

Here $\alpha$ and $\beta$ have $\mu$-torsion kernel and cokernel by definition, hence by (1.3.), so has f. As M is $\mu$-torsion by assumption, it follows that N is $\mu$-torsion too.

Now, consider the exact sequences (2) and (4); they combine into an exact commutative diagram of left A-modules

$$\begin{array}{ccccccc}
P \underset{B}{\otimes} B^{(I)} & \longrightarrow & P \underset{B}{\otimes} B^{(J)} & \longrightarrow & P \underset{B}{\otimes} M & \longrightarrow & 0 \\
\gamma \downarrow & & \downarrow \delta & & \downarrow g & & \\
P \underset{B}{\otimes} B^{(I,\mu)} & \longrightarrow & P \underset{B}{\otimes} B^{(J,\mu)} & \longrightarrow & P \underset{B}{\otimes} N & \longrightarrow & 0
\end{array}$$

We have $\underline{\underline{a}}_{\chi}(P \underset{B}{\otimes} B^{(I,\mu)}) = S(B^{(I,\mu)}) = S(Q_{\mu}(B))^{(I,\lambda)} = P^{(I,\lambda)} = Q_{\lambda}(P^{(I)})$

and $\underline{\underline{a}}_{\lambda}(P \underset{B}{\otimes} B^{(I)}) = \underline{\underline{a}}_{\lambda}(P^{(I)}) = Q_{\lambda}(P^{(I)})$, so it easily follows that $Q_{\lambda}(\gamma) : Q_{\lambda}(P \underset{B}{\otimes} B^{(I)}) \to Q_{\lambda}(P \underset{B}{\otimes} B^{(I,\mu)})$ is an isomorphism. A similar statements holds for the index set J, hence we derive that Ker(g) and Coker(g) are $\lambda$-torsion. In order to prove that $P \underset{B}{\otimes} M$ is $\lambda$-torsion, it thus suffices to show that $P \underset{B}{\otimes} N$ is.

First note that (2) entails that the sequence

$$B^{(I,\mu)} \longrightarrow B^{(J,\mu)} \longrightarrow 0$$

is exact in $(B,\mu)$-mod, so, since $\underline{\underline{a}}_{\lambda}(P \underset{B}{\otimes} -)$ is right exact and commutes with direct sums, we obtain an exact sequence in $(A,\lambda)$-mod of the form

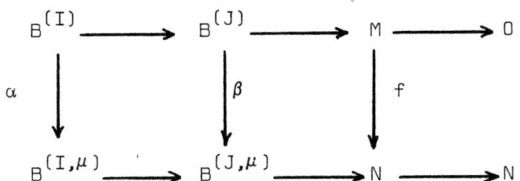

$$Q_\lambda(P \underset{B}{\otimes} B^{(I,\mu)}) \longrightarrow Q_\lambda(P \underset{B}{\otimes} B^{(J,\mu)}) \longrightarrow 0$$

hence $K = \text{Coker}(Q_\lambda(P \underset{B}{\otimes} Q_\mu(u)))$ is $\lambda$-torsion. Therefore, the following is a commutative exact diagram

$$\begin{array}{ccccccc}
P \underset{B}{\otimes} B^{(I,\mu)} & \longrightarrow & P \underset{B}{\otimes} B^{(J,\mu)} & \longrightarrow & P \underset{B}{\otimes} N & \longrightarrow & 0 \\
\downarrow v & & \downarrow w & & \downarrow h & & \\
Q_\lambda(P \underset{B}{\otimes} B^{(I,\mu)}) & \longrightarrow & Q_\lambda(P \underset{B}{\otimes} B^{(J,\mu)}) & \longrightarrow & K & \longrightarrow & 0
\end{array}$$

Here the kernel and cokernel of $v$ and $w$ are $\lambda$-torsion, hence so are Ker(h) and Coker(h) by (1.3.). As K is $\lambda$-torsion, it then follows that $P \underset{B}{\otimes} N$ is $\lambda$-torsion as well.

It remains to show that if $i : M' \to M$ is injective in B-mod, then $\text{Ker}(P \underset{B}{\otimes} i)$ is $\lambda$-torsion in A-mod. This will be done below.

2.6. Lemma. Let $S : (B,\mu)\text{-mod} \to (A,\lambda)\text{-mod}$ be an equivalence and let $T : (A,\lambda)\text{-mod} \to (B,\mu)\text{-mod}$ be an inverse for S, then we may find a $\lambda$-closed A-B-bimodule ${}_AP_B$ and a $\mu$-closed B-A-bimodule ${}_BQ_A$ together with isomorphisms of bimodules $\varphi : Q_\lambda(P \underset{B}{\otimes} Q) \cong Q_\lambda(A)$ and $\psi : Q_\mu(Q \underset{B}{\otimes} P) \cong Q_\mu(B)$ and such that we have isomorphisms $S \cong \underline{\underline{a}}_\lambda(P \underset{B}{\otimes} -)$ and $T \cong \underline{\underline{a}}_\mu(Q \underset{A}{\otimes} -)$.

Proof. It has been pointed out that we may find P and Q with the property that $S \cong \underline{\underline{a}}_\lambda(P \underset{B}{\otimes} -)$ and $T \cong \underline{\underline{a}}_\mu(Q \underset{A}{\otimes} -)$, i.e. take $P = S(Q_\mu(B))$ and $Q = T(Q_\lambda(A))$. In order to derive the isomorphism $\varphi : Q_\lambda(P \underset{B}{\otimes} Q) \cong Q_\lambda(A)$ note that $ST \cong 1_{(A,\lambda)\text{-mod}}$, hence there is a *natural* isomorphism in $(A,\lambda)$-mod

$$\Phi : Q_\lambda(P \underset{B}{\otimes} Q_\mu(Q \underset{A}{\otimes} -)) \cong 1_{(A,\lambda)\text{-mod}},$$

so in particular

$$\Phi_{Q_\lambda(A)} : Q_\lambda(P \underset{B}{\otimes} Q_\mu(Q \underset{A}{\otimes} Q_\lambda(A))) \cong Q_\lambda(A)$$

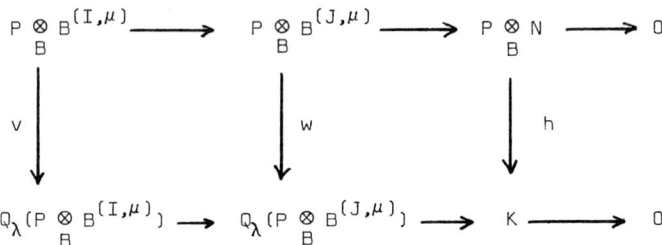

as left A-modules and $Q_\mu(Q \underset{A}{\otimes} Q_\lambda(A)) = T(Q_\lambda(A)) = Q$, i.e. we obtain an isomorphism $\varphi : Q_\lambda(P \underset{B}{\otimes} Q) \cong Q_\lambda(A)$ in A-mod. On the other hand, for any $a \in A$, the right multiplication $\tau_a : Q_\lambda(A) \to Q_\lambda(A) : \alpha \mapsto \alpha a$ is left A-linear, so by the naturality of $\Phi$ the following diagram is commutative :

$$\begin{array}{ccc} Q_\lambda(P \underset{B}{\otimes} Q_\mu(Q \underset{A}{\otimes} Q_\lambda(A))) & \xrightarrow{\Phi_{Q_\lambda(A)}} & Q_\lambda(A) \\ \downarrow{\scriptstyle \tau_a} & & \downarrow{\scriptstyle \tau_a} \\ Q_\lambda(P \underset{B}{\otimes} Q_\mu(Q \underset{A}{\otimes} Q_\lambda(A))) & \xrightarrow[\Phi_{Q_\lambda(A)}]{} & Q_\lambda(A) \end{array}$$

where we also have denoted by $\tau_a$ the induced right multiplication $Q_\lambda(P \underset{B}{\otimes} Q_\mu(Q \underset{A}{\otimes} \tau_a))$. It follows that for any $p \in Q_\lambda(P \underset{B}{\otimes} Q_\mu(Q \underset{A}{\otimes} Q_\lambda(A)))$ $\cong Q_\lambda(P \underset{B}{\otimes} Q)$ we have $\Phi_{Q_\lambda(A)}(p)a = \tau_a \Phi_{Q_\lambda(A)}(p) = \Phi_{Q_\lambda(A)}\tau_a(p) = \Phi_{Q_\lambda(A)}(pa)$, i.e. $\varphi = \Phi_{Q_\lambda(A)}$ is an A-bimodule isomorphism, indeed. □

<u>End of the proof of (2.5.)</u>. Let P and Q be as above. If $i : M' \to M$ is a monomorphism in B-mod, then the exact sequence

$$0 \longrightarrow K \xrightarrow{j} P \underset{B}{\otimes} M' \xrightarrow{P \otimes i} P \underset{B}{\otimes} M$$

with $K = \mathrm{Ker}(P \underset{B}{\otimes} i)$, yields that the composite $Q_\mu(Q \otimes P \otimes i)\, Q_\mu(Q \otimes j)$: $Q_\mu(Q \underset{A}{\otimes} K) \to Q_\mu(Q \underset{A}{\otimes} P \underset{B}{\otimes} M') \to Q_\mu(Q \underset{A}{\otimes} P \underset{B}{\otimes} M)$ is the zero-map. But, since we have $Q_\mu(Q \underset{A}{\otimes} P \underset{B}{\otimes} -) = Q_\mu(Q_\mu(Q \underset{A}{\otimes} P) \underset{B}{\otimes} -) = Q_\mu(Q_\mu(B) \underset{B}{\otimes} -) = Q_\mu(-)$, it follows that $Q_\mu(Q \underset{A}{\otimes} j) : Q_\mu(Q \underset{A}{\otimes} K) \to Q_\mu(Q \underset{A}{\otimes} P \underset{B}{\otimes} M')$ has to be the zero-map. The commutativity of the diagram :

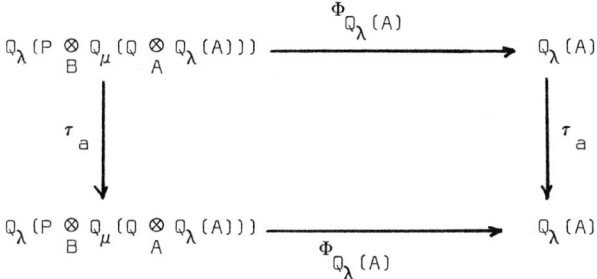

$$\begin{array}{ccc}
Q \underset{A}{\otimes} K & \xrightarrow{\;Q \otimes j\;} & Q \underset{A}{\otimes} P \underset{B}{\otimes} M \\
\big\downarrow & & \big\downarrow \theta \\
Q_{\mu}(Q \underset{A}{\otimes} K) & \xrightarrow[\;\hat{O}\;]{} & Q_{\mu}(Q \underset{A}{\otimes} P \underset{B}{\otimes} M)
\end{array}$$

entails that $T = \mathrm{Im}(Q \otimes j) \subset \mathrm{Ker}(\theta) = \mu(Q \underset{A}{\otimes} P \underset{B}{\otimes} M)$. But, $Q \underset{A}{\otimes} j$ factorizes as $Q \underset{A}{\otimes} K \to T \hookrightarrow Q \underset{A}{\otimes} P \underset{B}{\otimes} M'$, hence $P \underset{B}{\otimes} Q \underset{A}{\otimes} j$ factorizes as $P \underset{A}{\otimes} Q \underset{B}{\otimes} K \to P \underset{A}{\otimes} T \to P \underset{A}{\otimes} Q \underset{B}{\otimes} P \underset{A}{\otimes} M'$. Here $P \underset{A}{\otimes} T$ is $\lambda$-torsion, as $T$ is $\mu$-torsion, so $Q_{\lambda}(P \underset{A}{\otimes} Q \underset{B}{\otimes} K) \to Q_{\lambda}(P \underset{A}{\otimes} Q \underset{B}{\otimes} P \underset{A}{\otimes} M')$ is the zero-map. Now this map reduces to $Q_{\lambda}(K) \to Q_{\lambda}(P \underset{A}{\otimes} M')$ and it follows that $\mathrm{Im}(j) \subset \lambda(P \underset{B}{\otimes} M')$, i.e. $K$ is $\lambda$ -torsion. □

2.7. Corollary. Let $(A,\lambda)$ and $(B,\mu)$ be couples consisting of a ring and an idempotent kernel functor in the category of left modules over that ring, then there is a one-to-one correspondence between

1. The isomorphism classes of equivalences $(B,\mu)$-mod $\to$ $(A,\lambda)$-mod;
2. The isomorphism classes of $\lambda$-closed $(\lambda,\mu)$-invertible A-B-bimodules.

Proof. It is clear from the foregoing that any equivalence $S:(B,\mu)$-mod $\to$ $(A,\lambda)$-mod yields a $\lambda$-closed $(\lambda,\mu)$-invertible A-B-bimodule $P = S(Q_{\mu}(B))$. Conversely, let ${}_{A}P_{B}$ be a $\lambda$-closed $(\lambda,\mu)$-invertible A-B-bimodule with "inverse" ${}_{B}Q_{A}$ and put $S = \underline{\underline{a}}_{\lambda}(P \underset{B}{\otimes} -)$ resp. $T = \underline{\underline{a}}_{\mu}(Q \underset{A}{\otimes} -)$. Let us show that $S$ and $T$ are inverse equivalences between the categories $(B,\mu)$-mod and $(A,\lambda)$-mod. Pick $M \in (A,\lambda)$-mod, then $ST(M) = \underline{\underline{a}}_{\lambda}(P \underset{B}{\otimes} Q_{\mu}(Q \underset{A}{\otimes} M)) = \underline{\underline{a}}_{\lambda}(P \underset{B}{\otimes} Q \underset{A}{\otimes} M) = \underline{\underline{a}}_{\lambda}(Q_{\lambda}(P \underset{B}{\otimes} Q) \underset{A}{\otimes} M) = \underline{\underline{a}}_{\lambda}(Q_{\lambda}(A) \underset{A}{\otimes} M) = \underline{\underline{a}}_{\lambda}(M) = M$, i.e. $ST \cong 1_{(A,\lambda)\text{-mod}}$, since the isomorphisms are natural in $M$. Similarly for $TS \cong 1_{(B,\mu)\text{-mod}}$. □

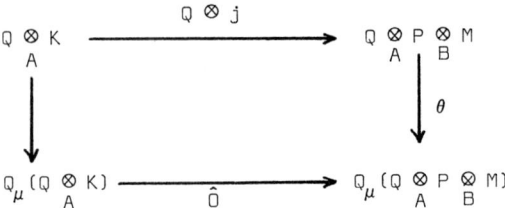

2.8. <u>Proposition.</u> The isomorphisms $\varphi : P \perp_{\lambda} Q \to Q_{\lambda}(A)$ and $\psi : Q \perp_{\mu} P \to Q_{\mu}(B)$ in the definition of a $(\lambda,\mu)$-invertible A-B-bimodule P may be chosen in such a way that the following diagrams are commutative.

1.
$$\begin{array}{ccc} P \perp Q \perp P & \xrightarrow{\varphi \perp P} & A \perp P \\ {\scriptstyle P \perp \psi}\downarrow & & \downarrow \\ P \perp B & \longrightarrow & P \end{array}$$

2.
$$\begin{array}{ccc} Q \perp P \perp Q & \xrightarrow{\psi \perp Q} & B \perp Q \\ {\scriptstyle Q \perp \varphi}\downarrow & & \downarrow \\ Q \perp A & \longrightarrow & Q \end{array}$$

where the unlabeled maps are canonical.

<u>Proof</u>. Denote by $\alpha : A \perp P \to P$ resp. $\beta : P \perp B \to P$ the canonical maps and let S be the equivalence of $(B,\mu)$-mod to $(A,\lambda)$-mod associated to ${}_A P_B$. In particular, $S(Q_{\mu}(B)) = P$ ! It is clear that $\beta(P \perp \psi) = u\,\alpha(\varphi \perp P)$ for some automorphism u of P in A-B-mod.

But $u \in {}_A[P,P] = {}_A[S(Q_{\mu}(B)), S(Q_{\mu}(B))] = {}_B[Q_{\mu}(B), Q_{\mu}(B)] = Q_{\mu}(B)^{opp}$, hence u is right multiplication with some unit of $Q_{\mu}(B)$, which we will also denote by u, and it is clear that u being an A-B-automorphism, we have $u \in Z(Q_{\mu}(B))$, the center of $Q_{\mu}(B)$. Let $v = u^{-1}$, then we claim that $v\beta = \beta(P \perp v)$. Note that on the right hand side v is acting on $Q_{\mu}(B)$ and on the left hand side v is acting on P. However, this does not cause any ambiguity. Let us denote by $j_M : M \to Q_{\lambda}(M)$ the localization morphism for M in A-mod, then from the definition of $P \perp v : P \perp B \to P \perp B$, it follows that $(P \perp v) j_{P \otimes B} = j_{P \otimes B}(P \otimes v)$; so pick $p \in P$ and $b \in B$; then $v\,\beta\, j_{P \otimes B}(p \otimes b) = v(pb)$ (since $\beta$ is induced by the isomorphism

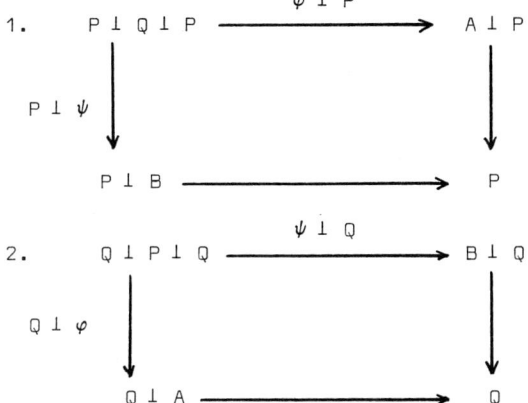

$P \underset{B}{\otimes} B \to P) = pbv$ and $\beta(P \perp v)j_{P \otimes B}(p \otimes b) = \beta j_{P \otimes B}(P \otimes v)(p \otimes b) =$

$= \beta j_{P \otimes B}(p \otimes v(b)) = \beta j_{P \otimes B}(p \otimes bv) = pbv$, so $vpj_{P \otimes B} = \beta(P \perp v)j_{P \otimes B}$ and it follows that $v\beta = \beta(P \perp v)$ indeed. Now, combine this with the identity $v\beta(P \perp \psi) = \alpha(\varphi \perp P)$, then we find $\alpha(\varphi \perp P) = \beta(P \perp v)(P \perp \psi) = \beta(P \perp v\psi)$, so, upon replacing $\psi$ by $v\psi$ (which maps $q \in Q \perp P$ to $\psi(q)v \in Q_\mu(B)$), we get the commutativity of the first diagram.

Assume this has been done, then we will show (★) that the second diagram is automatically commutative too : denote by $\gamma : B \perp Q \to Q$ resp. $\delta : Q \perp A \to Q$ the canonical maps, then we claim that $\gamma(\psi \perp Q) = \delta(Q \perp \varphi)$. For $\tau = \alpha, \ldots$ denote by $\tau'$ the underlying map $A \underset{A}{\otimes} P \to P, \ldots$ . Note that upto isomorphism $\tau$ and $\tau'$ may be identified. We will identify $P \perp (Q \perp \ldots)$ and $(P \perp Q) \perp \ldots$, $P \underset{B}{\otimes}(Q \underset{A}{\otimes} \ldots)$ and $(P \underset{B}{\otimes} Q) \underset{A}{\otimes} \ldots$ etc. We then have, denoting by $m : A \otimes A \to A$ the multiplication map :

$\varphi(P \perp \delta(Q \perp \varphi))j_{P \otimes Q \otimes P \otimes Q} = \varphi j_{P \otimes Q}(P \otimes \delta(Q \perp \varphi))(P \otimes j_{Q \otimes P \otimes Q}) =$

$= \varphi j_{P \otimes Q}(P \otimes \delta(Q \perp \varphi)j_{Q \otimes P \otimes Q}) = \varphi j_{P \otimes Q}(P \otimes \delta(j_{Q \otimes A}(Q \otimes \varphi)(Q \otimes j_{P \otimes Q}))) =$

$= \varphi j_{P \otimes Q}(P \otimes \delta(j_{Q \otimes A}(Q \otimes \varphi j_{P \otimes Q})))) = \varphi j_{P \otimes Q}(P \otimes \delta'(Q \otimes \varphi j_{P \otimes Q})) =$

$m(\varphi j_{P \otimes Q} \otimes \varphi j_{P \otimes Q})$ (right linearity of $\varphi j_{P \otimes Q}$) $= \varphi j_{P \otimes Q}(\alpha'(\varphi j_{P \otimes Q} \otimes P) \otimes Q)$

(left linearity of $\varphi j_{P \otimes Q}$) $= \varphi j_{P \otimes Q}(\alpha j_{A \otimes P}(\varphi j_{P \otimes Q} \otimes P) \otimes Q) =$

$\varphi j_{P \otimes Q}(\alpha(\varphi \perp P)j_{P \otimes Q \otimes P} \otimes Q) = \varphi j_{P \otimes Q}(\beta(P \perp \psi)j_{P \otimes Q \otimes P} \otimes Q) =$

$\varphi j_{P \otimes Q}(\beta'(P \otimes \psi j_{Q \otimes P}) \otimes Q) = \varphi j_{P \otimes Q}(P \otimes \gamma'(\psi j_{Q \otimes P} \otimes Q)) =$

$\varphi j_{P \otimes Q}(P \otimes \gamma(\psi \perp Q)j_{Q \otimes P \otimes Q}) = \varphi j_{P \otimes Q}(P \otimes \gamma(\psi \perp Q))(P \otimes j_{Q \otimes P \otimes Q} =$

$\varphi(P \perp \gamma(\psi \perp Q))j_{P \otimes Q \otimes P \otimes Q}$. It thus follows that $\varphi(P \perp \gamma(\psi \perp Q)) =$ $\varphi(P \perp \delta(Q \perp \varphi))$. Let $\theta = \gamma(\psi \perp Q) - \delta(Q \perp \varphi) : Q \to Q$, then $\theta$ is a B-A-endomorphism of Q. Let S be the equivalence from $(B,\mu)$-mod to $(A,\lambda)$-mod associated to P, then $\varphi(S\theta) = 0$, hence $S\theta = 0$ as $\varphi$ is an isomorphism .

(★) If you trust the authors, skip the calculation that follows : it is long, technical and very uninteresting.

But, S being an equivalence, this yields that $\theta = 0$, i.e. $\gamma(\psi \perp Q) = \delta(Q \perp \varphi)$ and the second square is commutative indeed. □

If A and B are algebras over some commutative ring R, then the above correspondence is easily seen to extend to equivalences between R-categories and $(\lambda,\mu)$-invertible A-B-bimodules defined over R. Let A be an R algebra, then $(A,\kappa)$-mod is an R-category, hence it makes sense to define its R-Picard group $\mathrm{Pic}_R((A,\kappa)$-mod. On the other hand, denote by $\mathrm{Pic}_R(A,\kappa)$ the set of isomorphism classes $[P]$ of $\kappa$-closed, $\kappa$-invertible A-bimodules over R. Since we know that for any A-bimodule M, defined over R, we have that $Q_\kappa(M)$ is defined over M too, it is readily seen that $\mathrm{Pic}_R(A,\kappa)$ may be given a group structure by putting $[P_1].[P_2] = [P_1 \perp P_2]$ for any pair of $\kappa$-closed, $\kappa$-invertible A-bimodules $P_1$, $P_2$ defined over R. The class $[Q_\kappa(A)]$ then plays the role of identity element. If the $\kappa$-closed $\kappa$-invertible A-bimodule P has "inverse" Q (which we may also choose to be $\kappa$-closed since P is $\kappa$-flat by assumption), then $[Q] = [P]^{-1}$ in $\mathrm{Pic}_R(A,\kappa)$. The group $\mathrm{Pic}_R(A,\kappa)$ thus obtained is called the <u>relative Picard group</u> (over R) of A with respect to $\kappa$. If $R = \mathbb{Z}$, then we will simply write $\mathrm{Pic}(A,\kappa)$ for $\mathrm{Pic}_{\mathbb{Z}}(A,\kappa)$. We have proved :

2.9. <u>Theorem</u>. Let A be an R-algebra and let $\kappa$ be an idempotent kernel functor in A-mod, then $\mathrm{Pic}_R((A,\kappa)\text{-mod}) = \mathrm{Pic}_R(A,\kappa)$. □

2.10. <u>Proposition</u>. Let ${}_AP_B$ be a $(\lambda,\mu)$-invertible A-B-bimodule, then ${}_A[P,Q_\lambda(A)]$ is an "inverse" for P.

<u>Proof</u>. We may choose P to be $\lambda$-closed and assume that it is associated to an equivalence $S = \underline{\underline{a}}_\lambda(P \otimes_B -)$: $(B,\mu)$-mod $\to$ $(A,\lambda)$-mod with inverse $T : (A,\lambda)$-mod $\to$ $(B,\mu)$-mod. An "inverse" for P is then $T(Q_\lambda(A)) \in (B,\mu)$-mod, which is canonically endowed with a right A-module structure by the naturality of T. Let $M \in (A,\lambda)$-mod and $N \in (B,\mu)$-mod, then we have

${}_A[S(N),M] = {}_B[N,T(M)]$ , but also (upto canonical morphism) ${}_A[S(N),M] =$ $= {}_A[\underline{\underline{a}}_\lambda(P \otimes_B N), M] = {}_A[P \otimes_B N,M] = {}_B[N,{}_A[P,M]]$ and ${}_A[P,M] \in (B,\mu)$-mod, since ${}_AP_B$ is $(\lambda,\mu)$-flat, cf. (1.13). It follows that the functors $T$ and ${}_A[P,-]$ are naturally isomorphic, being both right adjoint to the same functor S, hence in particular ${}_BT(Q_\lambda(A)) = {}_A[P,Q_\lambda(A)]$ in B-mod. But this is also an isomorphism of *right* A-modules, because the isomorphism $T \xrightarrow{\sim} {}_A[P,-]$ is a natural one. □

## B.3. Morita Contexts.

We have seen in (2.7) that there is a one-to-one correspondence between isomorphism classes of equivalences between $(B,\mu)$-mod and $(A,\lambda)$-mod and isomorphism classes of $\lambda$-closed $(\lambda,\mu)$-invertible A-B-bimodules. Let us now point out how equivalences between $(B,\mu)$-mod and $(A,\lambda)$-mod occur in "nature". We will rely heavily on material covered by B. Mueller in [102] and D. Turnidge in [144].

Recall the following definition. A Morita context consists of two rings A and B, two bimodules ${}_BP_A$ and ${}_AQ_B$ and two bimodule homomorphisms (the "pairings") denoted by $(-,-) : Q \otimes_B P \to A$ and $[-,-] : P \otimes_A Q \to B$ satisfying the associativity relations $q[p,q'] = (q,p)q'$ resp. $p(q,p') = [p,q]p'$ for all $p,p' \in P$ resp. $q,q' \in Q$. The images of these pairings are called the trace ideals of the Morita context; they are (two sided) ideals, and they will be denoted by $J_A$ and $J_B$. We will see below, how any Morita context induces an equivalence between certain quotient categories of A-mod and B-mod. We need the following :

3.1. Lemma. Let I be an ideal of A, then the set $\mathcal{F}_I$, consisting of all

$M \in A$-mod such that $Ann_M(I) = 0$, is the torsionfree class of a hereditary torsion theory associated to some idempotent kernel functor $\kappa^I$. The corresponding idempotent filter $\mathcal{L}(\kappa^I)$ is the set of all left ideals L of A such that $Ann_M(I) = 0$ implies $Ann_M(L) = 0$ for all $M \in A$-mod. The quotient category $(A,\kappa^I)$-mod consists of all $M \in A$-mod such that the canonical map $M = {}_A[A,M] \to {}_A[I,M]$ is bijective.

Proof. It is easy to see that $F_I$ is closed under products, submodules, extensions and injective hulls, hence $F_I$ is the torsionfree class of an idempotent kernel functor $\kappa^I$-indeed. Moreover, $L \in \mathcal{L}(\kappa^I)$ if and only if $A/L$ is $\kappa^I$-torsion, i.e. $0 = [A/L,M] = Ann_M(L)$ for all $M \in F_I$, which yields the description of $\mathcal{L}(\kappa^I)$. Finally, the natural map $M \to {}_A[I,M] : m \mapsto h(m)$, where $h(m)(i) = im$ for all $i \in I$, is a monomorphism, exactly when $Ann_M(I) = 0$ i.e. $M \in F_I$. Moreover, ${}_A[I,M]$ may then be identified with the set of all $e \in E = E(M)$, the injective hull of M in A-mod, with the property that $Ie \subset M$. It follows that the natural map $M \to [I,M]$ is surjective if and only if $Ie \subset M$ implies $e \in M$, i.e. $Ann_{E/M}(I) = 0$ and so $E/M \in F_I$. But it is well-known that for any $M \in A$-mod, we have that M is $\kappa^I$-closed if and only if M and E/M are $\kappa^I$-torsion free.

3.2. Note. If A is left noetherian, then $\kappa^I$ is well-known : it coincides with the symmetric kernel functor $\kappa_I$ in A-mod, whose idempotent filter $\mathcal{L}(\kappa_I)$ is generated by the positive powers $I^n$ of I, cf. (A.5.2.). Indeed, clearly $\kappa_I \geqslant \kappa^I$. Conversely, if $\kappa^I M = 0$ and $m \in \kappa_I M$, then $I^n m = 0$ for some n, hence $Ann_M(I) = 0$ implies $m = 0$, hence $\kappa_I M = 0$. So $\kappa_I = \kappa^I$, indeed. We also have the following :

3.3. Corollary. $\mathcal{L}(\kappa^I)$ consists of all left ideals of A containing I if and only if $I = I^2$.

Proof. If the set of all left A-ideals containing I is a Gabriel topology, then necessarily $I = I^2$. Conversely, if $I = I^2$ then we may consider for any $L \in \mathcal{L}(\kappa^I)$ the left A-module $M = A/(IL:I)$, for which we have $\mathrm{Ann}_M(I) = 0$, hence $\mathrm{Ann}_M(L) = 0$. But, $I\,\bar{1} = \bar{0}$, so $\bar{1} \in \mathrm{Ann}_M(L)$, hence $1 \in (IL:I)$ and $I \subset IL \subset L$. This finishes the proof. □

Warning. For technical reasons, we will now temporarily switch from left to right, i.e. we will work in mod-A, mod(A,$\kappa$), etc. The idempotent kernel functor associated to an ideal I as above will still be denoted by $\kappa^I$ if we define it in mod-A. The reason for this choice is that, in this way we avoid switching to opposite rings occasionally. Moreover, there is a social aspect as well, since it appears that most results involving Morita-theory are traditionally formulated in terms of *right* modules.

3.4. Proposition. Every Morita context between rings A and B induces an equivalence between the quotient categories mod-$(A,\kappa^{J_A})$ and mod-$(B,\kappa^{J_B})$.

Proof. Let us write $\kappa_A$ and $\kappa_B$ for $\kappa^{J_A}$ resp. $\kappa^{J_B}$. We will use notations as above, i.e. the Morita context is described by ${}_BP_A$, ${}_AQ_B$, ... . The bimodules ${}_BP_A$ and ${}_AQ_B$ determine functors $[P,-]_A$ : mod-A → mod-B and $[Q,-]_B$ : mod B → mod-A. There are natural transformations

$$\varphi: 1_{\text{mod-}A} \to [Q \otimes_B P,-] \cong [Q,[P,-]_A]_B$$

(resp.

$$\psi: 1_{\text{mod-}B} \to [P \otimes_A Q,-] \cong [P,[Q,-]_B]_A)$$

given explicitly by $\varphi_M(m)(q)(p) = m(q,p)$ for any $M \in$ mod-A and any $m \in M$, $p \in P$, $q \in Q$. We claim that $\varphi_M$ is an isomorphism exactly when $M \in$ mod-$(A,\kappa_A)$. Of course $\mathrm{Ker}(\varphi_M) = \mathrm{Ann}_M(J_A)$, so $\varphi_M$ is injective if and only if $M \in \mathcal{T}_{J_A}$, the torsion class associated to $\kappa_A$. Under this assumption, any map $\varphi \in [Q \underset{B}{\otimes} P, M]_A$ factors through $(-,-): Q \underset{B}{\otimes} P \to J_A$, since $\sum_i (q_i, p_i) = 0$ implies that $\varphi(\sum_i q_i \otimes p_i)(q,p) = \varphi(\sum_i q_i \otimes p_i(q,p)) = \varphi(\sum_i q_i \otimes [p_i, q] p) = \varphi(\sum_i q_i[p_i, q] \otimes p) = \varphi(\sum_i (q_i, p_i) q \otimes p) = 0$, hence $\varphi(\sum_i q_i \otimes p_i) J_A = 0$, so $\varphi(\sum_i q_i \otimes p_i) = 0$. This yields an isomorphism $[J_A, M]_A = [Q \underset{B}{\otimes} P, M] = [Q, [P, M]_A]_B$, whose composition with the natural map $M \to [J_A, M]_A$ is just $\varphi_M$. So $\varphi_M$ is bijective if and only if $M \to [J_A, M]_A$ is.

Consider the equivalence between the quotient categories mod-$(A, \kappa_A)$ and mod-$(B, \kappa_B)$ of mod-A and mod-B determined by the trace ideals of the context above. If $\mathcal{C}_A$ is a quotient category of mod-$(A, \kappa_A)$, i.e. a quotient category of mod A with respect to an idempotent kernel functor $\lambda$ in mod-A with $\kappa_A \leqslant \lambda$ or also, equivalently, $J_A \in \mathcal{L}(\lambda)$, then, restricting the original equivalence to $\mathcal{C}_A$ yields an equivalence between $\mathcal{C}_A =$ mod-$(A, \lambda)$ and some quotient category $\mathcal{C}_B =$ mod-$(B, \mu)$ of mod-$(B, \kappa_B)$, where $\mu$ is an idempotent kernel functor in mod-B with $J_B \in \mathcal{L}(\mu)$. One easily verifies that for any right B-ideal L we have that $L \in \mathcal{L}(\mu)$ if and only if P/LP is a $\lambda$-torsion right A-module. Indeed, B/L is $\mu$-torsion if and only if $0 = [B/L, [P, M]_A]_B = [B/L \underset{B}{\otimes} P, M]_A = [P/LP, M]_A$ for all M in mod-$(A, \lambda)$, i.e. if and only if the second statement holds. This device yields a bijective correspondence between the idempotent kernel functors $\lambda$ in A-mod with $\kappa_A \leqslant \lambda$ and the idempotent kernel functors $\mu$ in B-mod with $\kappa_B \leqslant \mu$.

For any module $P \in$ mod-A and any ring isomorphism $\sigma : B \to \mathrm{End}(P_A)$, we may construct a Morita context by putting $Q = [P, A]_A$ and by defining $(\alpha, p) = \alpha(p)$ and $[p, \alpha] = \sigma^{-1}(p\alpha)$ for any $\alpha \in Q$ and $p \in P$. This context is called the <u>derived context of</u> P <u>and</u> $\sigma$. If $B = \mathrm{End}(P_A)$ and $\sigma$ is the identity map, then we speak of the <u>derived context of</u> P.

Let us call a Morita context <u>right normalized</u> if the four natural morphisms $P \to Q^{\star}$, $Q \to P^{\star}$, $A \to End(Q_B)$ and $B \to End(P_A)$ are isomorphisms. It is obvious that the derived context of a generator $P_A$ is right normalized. Moreover, we have the following :

3.5. <u>Lemma</u>. The derived context of $P \in$ mod-A is right normalized if and only if the natural maps $A \to [T,A]_A$ and $P \to [T,P]_A$ are bijective for $T = tr(P_A) = J_A$.

<u>Proof</u>. By (3.1.) it is clear that both maps are isomorphisms if and only if A and P are lying in mod-$(A,\kappa_A)$, which is equivalent to $\varphi_A$ and $\varphi_P$ being isomorphic by (3.4.). Since by assumption $Q = [P,A]_A$ and $B = End(P_A)$, we have that $\varphi_A$ and $\varphi_P$ may be identified with $\varphi_A : A \to [Q,[P,A]_A]_B = End(Q_B)$ resp. $\varphi_P : P \to [Q,[P,P]_A]_B = Q^{\star}$, hence $\varphi_A$ and $\varphi_P$ are isomorphisms if and only if the context is right normalized. □

Recall, e.g. from H. Bass [ 18 ], that the Morita theorems establish a one-to-one correspondence between :

1. isomorphism classes of category equivalences between mod-A and mod-B;
2. isomorphism classes of Morita contexts between A and B satisfying $J_A = A$ resp. $J_B = B$;
3. isomorphism classes of B-A-bimodules ${}_BP_A$ for which the natural map $B \to End(P_A)$ is isomorphic and $P_A$ is a progenerator.

The following result generalizes this to essentially all category equivalences between full subcategories of module categories containing the base ring. We will first give a proof of the general result, then specialize to the case of Grothendieck categories.

3.5. <u>Theorem [ 102]</u>. Every (additive) equivalence between full subcategories of mod-A and mod-B containing the modules $A_A$ and $B_B$ has a unique, maximal extension. Domain and range of such "extremal" equivalences are quotient categories determined by an ideal. Moreover, there is a one-to-one correspondence between :

1. isomorphism classes of external equivalences between full subcategories of mod-A and mod-B containing the modules $A_A$ and $B_B$;

2. isomorphism classes of right normalized Morita contexts between A and B;

3. isomorphism classes of B-A-bimodules ${}_BP_A$ for which the maps $B \to \mathrm{End}(P_A)$ $A \to [T,A]_A$ and $P \to [T,P]_A$ are bijective, if $T = \mathrm{tr}(P_A)$.

<u>Proof</u>. The foregoing lemma asserts that the conditions in 3. may be replaced by the requirement that $B \to \mathrm{End}(P_A)$ should be bijective and that the derived context of $P_A$ should be right normalized. We denote this modified statements by 3'. and give a correspondence between 1., 2. and 3'.

a. To any right normalized context we associate the equivalence $[P,-]_A : \text{mod-}(A,\kappa_A) \to \text{mod-}(B,\kappa_B)$. This defines a function U. We have $A \in \text{mod-}(A,\kappa_A)$, since $\varphi_A : A \xrightarrow{\sim} \mathrm{End}(Q_B) \cong [Q,[P,A]_A]_B$ is an isomorphism; similarly, $B \in \text{mod-}(B,\kappa_B)$.

b. We define a function V which associates with any equivalence $F:\mathcal{C}_A \to \mathcal{C}_B$ between full subcategories of mod-A and mod-B containing $A_A$ and $B_B$ a bimodule ${}_BP_A$ satisfying 3'. as follows. Let $G : \mathcal{C}_B \to \mathcal{C}_A$ be an inverse of F, then F is representable by the bimodule $P = G(B)$ and G by $F(A) \cong [P,A]_A = P^\star$. Therefore, $\mathrm{End}(P^\star_B) = [P^\star,[P,A]_A] \cong G(F(A)) \cong A$ resp. $P^{\star\star} \cong [P^\star,[P,P]_A]_B \cong G(F(P)) \cong P$ and $\mathrm{End}(P_A) \cong F(P) = F(G(B)) \cong B$, by natural maps, i.e. the conditions of 3' hold for P.

c. A third function W associates with any bimodule satisfying 3'. the derived context of $P_A$ and $\sigma : S \xrightarrow{\sim} \mathrm{End}(P_A)$, which is right normalized, as

has been pointed out before.
The composition WVU and VUW are identities (upto isomorphism), as one easily verifies. In order to investigate the relationship between F and (UVW)(F) for any equivalence F : $\mathcal{C}_A \to \mathcal{C}_B$ between full subcategories of mod-A and mod-B containing $A_A$ and $B_B$, we choose an inverse $G:\mathcal{C}_B \to \mathcal{C}_A$ of F and an isomorphic natural transformation $\theta : 1_{\mathcal{C}_A} \to GF$. Then V(F) = G(B), the bimodule representing F. Note first that if F' is an extension of F and G' an inverse for F', then G' extends G, hence G'(B) = G(B), proving that V(F) = V(F'). Moreover (WV)(F) is the derived context of G(B) and $\sigma : B \xrightarrow{\sim} \mathrm{End}(G(B)_A)$ and we observe that $G(B)^\star =$ $= [\,G(B),A\,]_A \cong F(A)$, the bimodule which represents G; therefore we obtain :

$$A \xrightarrow{\varphi_A} [\,G(B)^\star,[\,G(B),A\,]_A\,]_B \cong G(F(A)) \xrightarrow{\theta_A^{-1}} A,$$

which is a bimodule isomorphism (by naturality), i.e. given by an invertible central element $c \in Z(A)$. Naturality also implies that $\varphi_M$ is the composition of $\theta_M c$ with the isomorphism $G(F(M)) \to$ $\to [\,G(B)^\star,[\,G(B),M\,]_A\,]_B$, hence itself an isomorphism, for all $M \in \mathcal{C}_A$, proving that $\mathcal{C}_A \subset \text{mod-}(A,\kappa_A)$. This shows that (UWV)(F) is the unique maximal extension of F, which finishes the proof. □

3.7. <u>Corollary</u>. If a right normalized context exists between rings A and B, then their centers are isomorphic.

<u>Proof</u>. As we pointed out in (2.2.), there is an isomorphism between Z(A) and the center $Z(\mathcal{C}_A)$ of any full subcategory of mod-A containing $A_A$. □

It is now easy to give a straightforward description of the results of (3.6.), if we work in the case of Grothendieck categories from the start, i.e. $\mathcal{C}_A = (A,\lambda)\text{-mod}$ and $\mathcal{C}_B = (B,\mu)\text{-mod}$ for some idempotent kernel functors $\lambda$ and $\mu$ in A-mod resp. B-mod. We assume A to be $\lambda$-closed and B to be $\mu$-closed. If $(A,\lambda)\text{-mod} \underset{U}{\overset{V}{\rightleftarrows}} (B,\mu)\text{-mod}$ are inverse equivalences then we know that $U \cong \underline{\underline{a}}_\lambda (P \underset{B}{\otimes} -)$ and $V \cong \underline{\underline{a}}_\mu (Q \underset{A}{\otimes} -)$ for some $(\lambda,\mu)$-invertible ${}_AP_B$ and some $(\mu,\lambda)$-invertible ${}_BQ_A$ for which there are bimodule isomorphisms $\varphi : Q_\lambda(P \underset{B}{\otimes} Q) \cong A$ resp. $\psi : Q_\mu(Q \underset{A}{\otimes} P) \cong B$. Moreover, according to (2.8.) the maps $\varphi$ and $\psi$ may be chosen to make the diagrams 1. and 2. in loc. cit. commutative. Now, if we let $[-,-] = \varphi j_{P \otimes Q} : P \underset{B}{\otimes} Q \to Q_\lambda(P \underset{B}{\otimes} Q) \to A$ resp. $(-,-) = \psi j_{Q \otimes P}: Q \underset{A}{\otimes} P \to Q_\mu(Q \underset{A}{\otimes} P) \to B$, then it is easy to see that we thus obtain an Morita context between A and B. By definition, $J_A$ is the image of $[-,-]$. If we denote by $\overline{P \underset{B}{\otimes} Q}$ the quotient $P \underset{B}{\otimes} Q/\lambda(P \underset{B}{\otimes} Q) \subset Q_\lambda(P \underset{B}{\otimes} Q)$, then $[-,-]$ factorizes through $\overline{P \underset{B}{\otimes} Q}$ and as $Q_\lambda(P \underset{B}{\otimes} Q)/\overline{P \underset{B}{\otimes} Q}$ is $\lambda$-torsion, it follows that $A/J_A$ is $\lambda$-torsion too, hence $J_A \in \mathcal{L}(\lambda)$. Similarly, $J_B \in \mathcal{L}(\mu)$. We now claim that if $J_A \in \mathcal{L}(\lambda)$, then $\kappa_A \leqslant \lambda$ and conversely. Indeed, we have $L \in \mathcal{L}(\kappa_A)$ if and only if $\mathrm{Ann}_M(J_A) = 0$ implies $\mathrm{Ann}_M(L) = 0$ for any $M \in A\text{-mod}$. If $M \in A\text{-mod}$ is $\lambda$-torsion free, then certainly $\mathrm{Ann}_M(J_A) = 0$ (since by assumption $J_A \in \mathcal{L}(\lambda)$), so $\mathrm{Ann}_M(L) = 0$ as well. Applying this to any M of the form $M/\lambda M$, we find that for all $M \in A\text{-mod}$ and $m \in M$, we have that $Lm = 0$ implies $m \in \lambda M$. Finally, applying this remark to $M = A/L$, we find $L\,\overline{1} = \overline{0}$, so $\overline{1} \in \lambda(A/L)$, i.e. $A/L$ is $\lambda$-torsion and $L \in \mathcal{L}(\lambda)$. Conversely, if $\kappa_A \leqslant \lambda$, then $J_A \in \lambda$, then $J_A \in \mathcal{L}(\kappa_A) \subset \mathcal{L}(\lambda)$, which proves the assertion.

It follows in particular that P and Q are $\kappa_A$- resp. $\kappa_B$-closed and that we already have isomorphisms $\alpha : Q_{\kappa_A}(P \underset{B}{\otimes} Q) \cong A$ and $\beta : Q_{\kappa_B}(Q \underset{A}{\otimes} P) \cong B$. As ${}_AP_B$ is $(\kappa_A, \kappa_B)$-flat and ${}_BQ_A$ is $(\kappa_B, \kappa_A)$-flat for obvious reasons, this shows that they are relatively invertible with respect to the appropriate idempotent kernel functors, so the original equivalences U and V extend to equivalences between $(A,\kappa_A)$-mod and $(B,\kappa_B)$-mod. We leave the (easy) verification that this is the one constructed above to the reader.

Let $\sigma$ be an idempotent kernel functor in A-mod and $\tau$ an idempotent kernel functor in B-mod, and assume that $\sigma \leqslant \lambda$, $\tau \leqslant \mu$ and that there exists extensions of U resp. V :

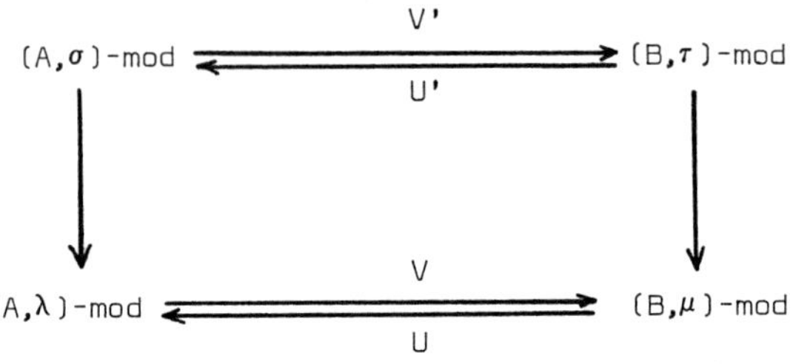


Since any $\lambda$-closed A-module is also $\sigma$-closed and similarly over B, it is clear that $P \in (A,\sigma)$-mod resp. $Q \in (B,\tau)$-mod, hence the fact that U' resp. V' extends U resp. V yields that we obtain isomorphisms $Q_\sigma(P \otimes_B Q) \cong A$ and $Q_\tau(Q \otimes_A P) \cong B$, fitting into commutative diagrams.

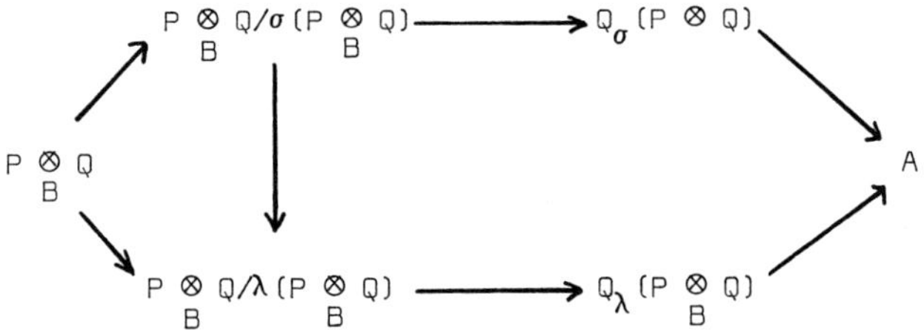


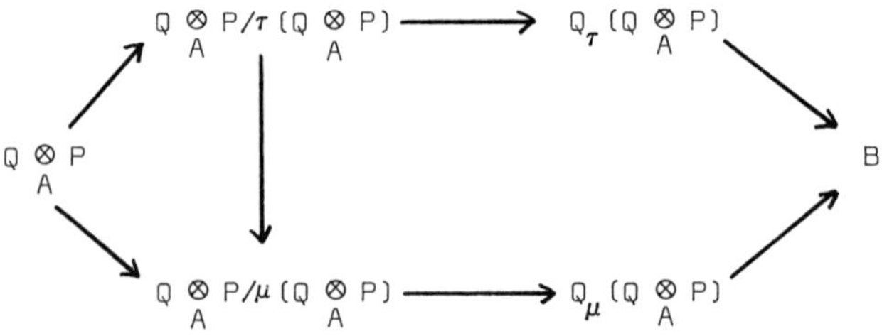


From the commutativity of these diagrams, it follows as before that $J_A \in L(\sigma)$ and $J_B \in L(\tau)$, where $J_A$ = trace ideal of "$\lambda$-context" = trace ideal of "$\sigma$-context", and similarly for $J_B$. By the foregoing we obtain

$\kappa_A \leq \sigma$ and $\kappa_A \leq \tau$, which proves the maximality of the context constructed above.

Let $\lambda$ be an idempotent kernel functor in A-mod and $\mu$ an idempotent kernel functor in B-mod, then we call A and B (left) relatively Morita equivalent with respect to $(\lambda,\mu)$ or simply : $(\lambda,\mu)$-Morita equivalent if there exists an equivalence $(A,\lambda)\text{-mod} \rightleftarrows (B,\mu)\text{-mod}$. We will write this as $A \underset{(\lambda,\mu)}{\frown} B$.

A $(\lambda,\mu)$-Morita equivalence between A and B is just the isomorphism class of a $\mu$-closed $(\mu,\lambda)$-invertible B-A-bimomdule ${}_BP_A$. It is clear that if A and B are $(\lambda,\mu)$-Morita equivalent, then there is a $(\lambda,\mu)$-Morita equivalence between them, and conversely, cf.(2.7.). If P determines a $(\lambda,\mu)$-Morita equivalence between A and B, then we will sometimes write this as $A \underset{(\lambda,\mu)}{\frown} B$. There is a notion of partial composition of relative Morita equivalences given by

$$P \overset{P}{\underset{(\lambda,\mu)}{\frown}} B \overset{Q}{\underset{(\mu,\nu)}{\frown}} C = A \overset{Q_\nu(Q \underset{B}{\otimes} P)}{\underset{(\lambda,\nu)}{\frown}} C.$$

It is natural to ask when a relative Morita equivalence between two rings determines a genuine Morita equivalence. In concluding this section we provide some partial answers to this question. The results below are not used elsewhere in the text, so they may be skipped at first reading. For more details, the reader is referred to D. Turnidge [144] and J. Hutchinson, D. Turnidge [69] .

As before, we will work in the "right" terminology. For any (right) Morita context $(P,Q,[-,-], (-,-))$ between A and B, we denote by ^the localization functor at $\kappa_A$ and $\kappa_B$. It is easy to show that the above constructions yield a Morita context between $\hat{A} \cong [\hat{Q},\hat{Q}]_{\hat{B}} \cong [Q,\hat{Q}]_B$ and $\hat{B} \cong [\hat{P},\hat{P}]_{\hat{A}} \cong [P,\hat{P}]_A$ with $\hat{P} \cong [\hat{Q},\hat{B}]_{\hat{B}} \cong [Q,\hat{B}]_B$ and $\hat{Q} \cong [\hat{P},\hat{A}]_{\hat{A}} \cong [P,\hat{A}]_A$, the pairings being given by the canonical extensions of $[-,-]$ and $(-,-)$. We then have the following :

3.8. Proposition. Assume that $\kappa_B$ has property (T) in mod-B then the following are equivalent :

1. $\kappa_A$ has property (T) in mod-A;
2. $\hat{A}$ and $\hat{B}$ are Morita equivalent via the quotient context given by $\hat{P}$ and $\hat{Q}$;
3. $\hat{Q}$ is a finitely generated projective right $\hat{B}$-module.

Proof. (1.) ⇒ (2.) Since $\kappa_A$ and $\kappa_B$ have Goldman's property (T), we have mod-$(A,\kappa_A)$ = mod-$\hat{A}$ and mod-$(B,\kappa_B)$ = mod-$\hat{B}$. Since the equivalence between mod-$(A,\kappa_A)$ and mod-$(B,\kappa_B)$ is given by $[P,-]_A$ and $[Q,-]_B$, and since for all $M \in$ mod-$(A,\kappa_A)$ resp. $N \in$ mod-$(B,\kappa_B)$, we have $[P,M]_A \cong [\hat{P},M]_{\hat{A}}$ resp. $[Q,N]_B \cong [\hat{Q},N]_{\hat{B}}$, it follows that mod-$\hat{A}$ and mod-$\hat{B}$ are equivalent via the functors $[\hat{P},-]_{\hat{A}}$ and $[\hat{Q},-]_{\hat{B}}$.

(2.) ⇒ (3.) This is a well-known basic result in Morita-theory, cf. H. Bass [ 18].

(3.) ⇒ (1.) It suffices to show that $\hat{M} = M \otimes_A \hat{A}$ for all $M \in$ mod-A. Pick such an M, then $M \otimes_A \hat{Q} \in$ mod-$\hat{B}$ = mod-$(B,\kappa_B)$ as $\kappa_B$ has property (T). Now, $\hat{Q}$ being a finitely generated projective $\hat{B}$-module, it follows that

$$M \otimes_A \hat{A} \cong [\hat{Q}_{\hat{B}}, M \otimes_A \hat{Q}_{\hat{B}}]_{\hat{B}} \cong [Q, M \otimes_A \hat{Q}]_B$$

which lies in mod-$(A,\kappa_A)$, so $\hat{M} \cong M \otimes_A \hat{A}$ for all $M \in$ mod-A! □

3.9. Corollary. If any couple of the following statements is true, then all of them hold :

1. $\kappa_A$ has property (T);
2. $\kappa_B$ has property (T);
3. $\hat{A}$ and $\hat{B}$ are Morita equivalent via the "quotient context". □

Consider the context derived from a finitely generated projective right A-module P, then $J_B = B$, hence $\hat{B} = B$, $\hat{Q}_{\hat{B}} = Q_B$, $\hat{P}_{\hat{A}} \cong [\hat{Q},\hat{B}]_{\hat{B}} = [Q,B]_B = Q^{\star}_B$ and $\kappa_B$ has property (T). Since ${}_AQ$ is finitely generated projective, $B \cong \mathrm{End}({}_AQ)$ and $\hat{A} \cong \mathrm{End}(Q_B)$, we obtain that the following statements are equivalent is this situation :

1. $\kappa_A$ has property (T);
2. $\hat{A}$ and B are Morita equivalent via the context determined by $Q^{\star}$ and Q;
3. Q is a finitely generated projective right B-module.

## B.4. Finiteness Conditions.

As before, let $\kappa$ be an idempotent kernel functor in A-mod. A left A-module M is said to be $\kappa$-finitely generated if there is an A-submodule $M' \subset M$, which is finitely generated and such that $M/M'$ is $\kappa$-torsion. Clearly, $Q_\kappa(M)$ is $\kappa$-finitely generated if M is, the converse being true also if A is left noetherian. Similarly, we say that M is $\kappa$-finitely presented if we may find a finitely presented A-module M' and an A-linear morphism $u:M' \to M$ such that Ker(u) and Coker(u) are $\kappa$-torsion. If then follows immediately that u induces an isomorphism $Q_\kappa(M') \cong Q_\kappa(M)$. Moreover, if A is left noetherian, then M is $\kappa$-finitely presented if and only if it is $\kappa$-finitely generated.

A left A-module is said to be $\kappa$-noetherian if $Q_\kappa(M)$ is a noetherian object in $(A,\kappa)$-mod. In particular, we call the ring A itself $\kappa$-noetherian, if ${}_AA$ is a $\kappa$-noetherian module. This coincides with the notion of being $\kappa$-noetherian, as defined in Section A.5. It is easy to see that if A is $\kappa$-noetherian,then a left A-module is $\kappa$-noetherian if and only if we may find a positive integer n and an exact sequence in $(A,\kappa)$-mod of the form

$$Q_\kappa(A)^n \longrightarrow Q_\kappa(M) \longrightarrow 0.$$

4.1. Lemma. A left A-module M is $\kappa$-noetherian if and only if every $N \subset M$ is $\kappa$-finitely generated.

Proof. Assume M is $\kappa$-noetherian and choose $N \subset M$. Consider an ascending chain $\{N_\alpha; \alpha \in I\}$ of finitely generated A-submodules of N with $\cup N_\alpha = N$, then $\{Q_\kappa(N_\alpha); \alpha \in I\}$ is an ascending chain of $\kappa$-closed A-submodules of M which stops after a finite number of steps, as M is $\kappa$-noetherian, i.e. there exists an index $\alpha_o$ such that $Q_\kappa(N_{\alpha_o}) = Q_\kappa(N)$. It follows that $N/N_{\alpha_o}$ is $\kappa$-torsion, so N is $\kappa$-finitely generated. Conversely, if $\{M_\alpha; \alpha \in I\}$ is an ascending chain of $\kappa$-closed A-submodules of $Q_\kappa(M)$, then we may choose an ascending chain of A-submodules $\{N_\alpha; \alpha \in I\}$ of M, such that $Q_\kappa(N_\alpha) = M_\alpha$. Let $N = \cup N_\alpha$ and let $N_1 \subset N$ be a finitely generated A-submodule such that $Q_\kappa(N_1) = Q_\kappa(N)$. Since $N_1 \subset N_{\alpha_o}$ for some $\alpha_o \in I$, it follows that
$M_{\alpha_o} = Q_\kappa(N_{\alpha_o}) \supset Q_\kappa(N_1) = Q_\kappa(N) \supset Q_\kappa(N_\alpha) = M_\alpha$ for every $\alpha$ and thus the chain $\{M_\alpha; \alpha \in I\}$ stops at $\alpha_o$. □

4.2. Lemma. Let A be $\kappa$-noetherian, then every $\kappa$-finitely generated left A-module is $\kappa$-finitely presented.

Proof. First assume that M is a finitely generated A-modules, i.e. there exists a positive integer n and a surjective map $\pi : A^n \to M$. Since $A^n$ is $\kappa$-noetherian, as A is, it follows from (4.1.) that $\mathrm{Ker}(\pi)$ is $\kappa$-finitely generated, hence M is $\kappa$-finitely presented. Secondly, if M is $\kappa$-finitely generated, consider a finitely generated $M_1$ in M with $M/M_1$ $\kappa$-torsion. There is a finitely presented A-module P and a morphism $u: P \to M_1$ with $\kappa$-torsion kernel and cokernel. Considering the composition $P \overset{u}{\to} M_1 \hookrightarrow M$ yields the assertion. □

4.3. Lemma. Let A be $\kappa$-noetherian, then any $\kappa$-finitely generated A-module M is $\kappa$-noetherian.

Proof. First assume that we have a surjection $\pi : A^n \to M$. If $\{M_\alpha;\alpha \in I\}$ is an ascending chain in $Q_\kappa(M)$ and $\{N_\alpha;\alpha \in I\}$ an ascending chain in M such that $Q_\kappa(N_\alpha) = M_\alpha$, for each $\alpha \in I$, then the chain $\{\pi^{-1}(N_\alpha);\alpha \in I\}$ has the property that all the elements in it are $\kappa$-torsion over some $\pi^{-1}(N_{\alpha_0})$, for some $\alpha_0 \in I$, since $A^n$ is $\kappa$-noetherian. Hence all $N_\alpha/N_{\alpha_0}$ are $\kappa$-torsion and therefore $\{M_\alpha\ \alpha \in I\}$ terminates. Now, in general, let $M_1 \subset M$ be finitely generated and such that $M/M_1$ is $\kappa$-torsion. If $N \subset M$, then $N \cap M_1$ is $\kappa$-finitely generated by (4.1.) applied to $M_1$. Since $N/N \cap M_1 \subset M/M_1$ it follows that N is $\kappa$-finitely generated, and hence again by (4.1), that M is $\kappa$-noetherian. □

4.4. Corollary If A is a $\kappa$-noetherian ring, then every $\kappa$-finitely generated A-algebra B is $\kappa$-noetherian as an A-module and $\overline{\kappa}$-noetherian as a ring. □

(Here $\overline{\kappa}$ is the idempotent kernel functor in B-mod canonically induced by $\kappa$, cf. Section A.5.)

4.5. Lemma Let A be $\kappa$-closed, then a $\kappa$-closed A-module M is $\kappa$-finitely presented, if and only if there exist free A-modules of finite rank $F_1$, $F_2$ and an exact sequence in $(A,\kappa)$-mod of the form $F_2 \to F_1 \to M \to 0$.

Proof. If M is $\kappa$-finitely presented, then $M = Q_\kappa(M_1)$ for some finitely presented $M_1$. If $A^n \to A^m \to M_1 \to 0$ is a presentation of $M_1$ in A-mod, then $A^n \to A^m \to M = Q_\kappa(M_1) \to 0$ is exact in $(A,\kappa)$-mod. Conversely, if $F_2 \overset{\varphi}{\to} F_1 \overset{\pi}{\to} M \to 0$ is exact in $(A,\kappa)$-mod, then $N = \mathrm{Coker}(\varphi)$ is finitely presented (in A-mod!) and $\pi$ induces a morphism $N \to M$ with $\kappa$-torsion kernel and cokernel. □

4.6. Lemma. If A is $\kappa$-noetherian, then $M \in$ A-mod is $\kappa$-finitely generated if and only if $Q_\kappa(M)$ is $\kappa$-finitely generated.

Proof. We know that M is $\kappa$-finitely generated if and only if M is $\kappa$-noetherian. Now, M or $Q_\kappa(M)$ is $\kappa$-noetherian if and only if $Q_\kappa(M)$ is a noetherian object in $(A,\kappa)$-mod. This proves the assertion. □

4.7. Remark. One actually may prove that if A is commutative it is sufficient to assume in the foregoing that $\kappa$ is a noetherian idempotent kernel functor. Indeed, if $Q_\kappa(M)$ is $\kappa$-finitely generated, let $M_1' \subset Q_\kappa(M)$ be finitely generated and such that $Q_\kappa(M)/M_1'$ is $\kappa$-torsion. As $\kappa$ is noetherian, there is a finitely generated left ideal $I \in \mathcal{L}(\kappa)$, such that $I M_1' \subset j_\kappa(M)$, where $j_\kappa : M \to Q_\kappa(M)$ is the localizing morphism. Lifting a finite set of A-generators of $IM_1'$ to M yields a finitely generated $M_1$ in M such that $M/M_1$ is $\kappa$-torsion. The other implication is trivial. The result is also valid in the noncommutative case (same proof!), if $\mathcal{L}(\kappa)$ possesses a basis of (left) finitely generated (twosided) ideals.

The property of being $\kappa$-finitely presented is useful if one wishes to apply local-global arguments. The reason for this may be found in consequent observations. Let S be a central multiplicative system in A, then we may associate to it an idempotent kernel functor $\kappa_S$ in A-mod, by defining for any $M \in$ A-mod the torsion $\kappa_S M$ to consist of the $m \in M$ such that $sm = 0$ for some $s \in S$. The localization functor at $\kappa_S$ is denoted by $S^{-1}(-)$. If the A-B-bimodule ${}_AM_B$ is $\kappa$-finitely presented over A and if N is $\kappa$-closed in A-mod, then for any S as above with $\kappa \leqslant \kappa_S$, we have that the canonical map

$$\varphi : S^{-1}({}_A[M,N]) \to {}_{S^{-1}A}[S^{-1}M, S^{-1}N]$$

is an isomorphism of left B-modules. Indeed, let M' be a finitely presented left A-module such that $Q_\kappa(M) = Q_\kappa(M')$, then

$$S^{-1}({}_A[M,N]) = S^{-1}({}_A[Q_\kappa(M),N]) = S^{-1}({}_A[Q_\kappa(M'),N]) = S^{-1}({}_A[M',N])$$

$$\overset{(1)}{=} {}_{S^{-1}A}[S^{-1}M',S^{-1}N] \overset{(2)}{=} {}_{S^{-1}A}[S^{-1}Q_\kappa(M'),S^{-1}N] = {}_{S^{-1}A}[S^{-1}Q_\kappa(M),S^{-1}N]$$

$$\overset{(2)}{=} {}_{S^{-1}A}[S^{-1}M,S^{-1}N],$$

where (1) holds, since M' is finitely presented, and (2), since $\kappa_S \geq \kappa$. The rest is just a trivial verification of compatibilities. Note also that if S is defined on a common central subring R of A and B and if M is defined over R, then $\varphi$ induces an isomorphism of left $S^{-1}B$-modules.

4.8. Lemma. If $\kappa$ is an idempotent kernel functor of finite type in R-mod and A is an R-algebra, then for any $p \in C(\kappa)$ and any pair of left A-modules M,N such that M is $\overline{\kappa}$-finitely presented and N is $\overline{\kappa}$-closed, for the induced idempotent kernel functor $\overline{\kappa}$ in A-mod, the canonical map

$$({}_A[M,N])_p \longrightarrow {}_{A_p}[M_p,N_p]$$

is bijective. □

Let us now turn to invertible bimodules and their finiteness properties.

4.9. Proposition. Let P be a $\kappa$-invertible A-bimodule and assume that $\kappa$ has finite type, then P is $\kappa$-finitely generated.

Proof. Let Q be an "inverse" for P and view the isomorphism $Q_\kappa(Q \otimes_A P) \cong Q_\kappa(A)$ as an identification. One may verify that this is allowed in the present proof by filling in the explicit isomorphisms in the appropriate places. Denote by $\overline{(-)}$ quotients modulo $\kappa$-torsion. Since $Q_\kappa(Q \otimes_A P) = Q_\kappa(A)$ and $1 \in Q_\kappa(A)$, it follows that for some finitely generated left A-ideal $I \in L(\kappa)$, we have $I \subset \overline{Q \otimes_A P}$. Assume then $I = \Sigma'_\alpha Ax_\alpha$ is of the form $\Sigma'_i \overline{q_{i,\alpha} \otimes p_{i,\alpha}}$ for some $q_{i,\alpha} \in Q$ and $p_{i,\alpha} \in P$.

Put $P_1 = \sum_{i,\alpha} Ap_{i,\alpha}$, then $P_1$ is a finitely generated left A-submodule of P. Denote by $\varphi : Q \underset{A}{\otimes} P_1 \to \overline{Q \underset{A}{\otimes} P}$ the map

$$\begin{array}{ccc} Q \underset{A}{\otimes} P_1 & \longrightarrow & Q \underset{A}{\otimes} P \\ \downarrow & \overset{\varphi}{\searrow} & \downarrow \\ \overline{Q \underset{A}{\otimes} P_1} & \longrightarrow & \overline{Q \underset{A}{\otimes} P} \end{array}$$

Clearly $I \subset \mathrm{Im}(\varphi) \subset \overline{Q \underset{A}{\otimes} P}$ and so $\mathrm{Im}(\varphi)/I$ is $\kappa$-torsion, since $\overline{Q \underset{A}{\otimes} P}/I$ is. It follows that $\mathrm{Coker}\,(Q \underset{A}{\otimes} P_1 \to Q \underset{A}{\otimes} P)$ is $\kappa$-torsion. On the other hand, as Q is $\kappa$-invertible, it is $\kappa$-flat, hence $\mathrm{Ker}(Q \underset{A}{\otimes} P_1 \to Q \underset{A}{\otimes} P)$ is $\kappa$-torsion too, so $Q_\kappa(Q \underset{A}{\otimes} P_1) = Q_\kappa(Q \underset{A}{\otimes} P) = Q_\kappa(A)$.

We thus finally obtain that $Q_\kappa(P_1) = Q_\kappa(A \underset{A}{\otimes} P_1) = Q_\kappa(Q_\kappa(A) \underset{A}{\otimes} P_1) =$
$= Q_\kappa(Q_\kappa(P \underset{A}{\otimes} Q) \underset{A}{\otimes} P_1) = Q_\kappa(P \underset{A}{\otimes} Q \underset{A}{\otimes} P_1) = Q_\kappa(P \underset{A}{\otimes} Q_\kappa(Q \underset{A}{\otimes} P_1)) =$
$= Q_\kappa(P \underset{A}{\otimes} Q_\kappa(A)) = Q_\kappa(P)$, proving the assertion. □

4.10. Corollary. (of the proof) Let $(A,\lambda)$ and $(B,\mu)$ be couples consisting of a ring and an idempotent kernel functor in the category of left modules over that ring. Let P be a $(\lambda,\mu)$-invertible A-B-bimodule and assume that $\mu$ has finite type, then ${}_AP$ is a $\lambda$-finitely generated left A-module. □

4.11. Corollary. If A is $\kappa$-noetherian, then any $\kappa$-invertible A-bimodule is $\kappa$-finitely presented.

Proof. Immediately by (4.2.) and (4.9.). □

Assume now that A is an algebra over a commutative ring R and let $\kappa$ be an idempotent kernel functor on R-mod. We may then induce the idempotent kernel functor $\overline{\kappa}$ in A-mod. It is easily verified that $\mathcal{L}(\overline{\kappa})$ consists of all left ideals L of A with the property that $L \cap R \in \mathcal{L}(\kappa)$. Moreover, we may compare for any $M \in$ A-mod the R-module $Q_\kappa(M)$ and the left A-module

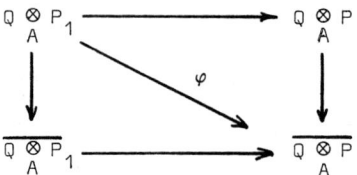

$Q_{\overline{\kappa}}(M)$. Indeed, first note that $Q_{\kappa}(M)$ is an object in A-mod; this may be proved in the usual way, noting that left multiplication by $a \in A$ is an R-linear endomorphism of M, which thus extends to $Q_{\kappa}(M)$, endowing $Q_{\kappa}(M)$ with a left A-module structure, extending that of M uniquely. A similar argument shows that if M is an A-bimodule, then so is $Q_{\kappa}(M)$.

Let us now prove that $Q_{\kappa}(M)$ is $\overline{\kappa}$-closed in A-mod, when endowed with this A-module structure. Consider a diagram of the form

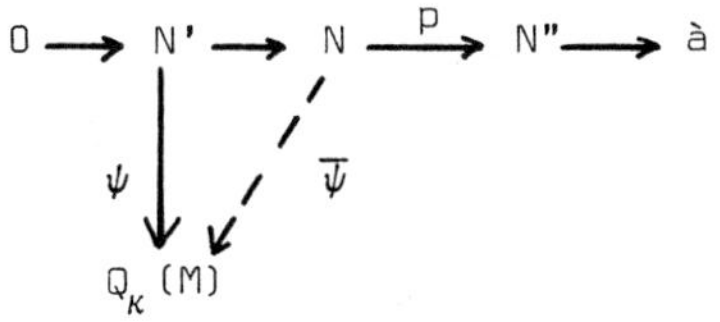

where all modules and morphisms are in A-mod and where N" is $\overline{\kappa}$-torsion Since the maps are certainly R-linear by restriction of scalars and since $N" = \overline{\kappa}N" = \kappa N"$, it follows that $\psi$ extends to an R-linear map $\overline{\psi} : N \to Q_{\kappa}(M)$. Let us check that $\overline{\psi}$ is A-linear as follows. Define for any $a \in A$ the map $\varphi : N \to Q_{\kappa}(M)$ by $\varphi(n) = \overline{\psi}(an) - a\overline{\psi}(n)$, then, since $\overline{\psi}$ is R-linear, and since R maps into the center of A, the morphism $\varphi$ is R-linear. Moreover, for $n \in N'$ we have $\varphi(n) = \overline{\psi}(an) - a\overline{\psi}(n) = \psi(an) - a\psi(n) = 0$ since $\psi$ is A-linear, so $\varphi \mid N'$ is the zero map. But, $Q_{\kappa}(M)$ being a $\kappa$-closed R-module, $\varphi$ extends to N in a unique way, so $\varphi$ has to be the zero-map on N, showing that $\overline{\psi}$ is A-linear indeed. This proves that $Q_{\kappa}(M)$ is $\overline{\kappa}$-closed. Since the A-module structure on $Q_{\kappa}(M)$ is canonically induced from that of M, it is clear that the localizing morphism $j_{\kappa} : M \to Q_{\kappa}(M)$ is left A-linear (and right A-linear, if M is an A-bimodule). Moreover, $Ker(j_{\kappa}) = \kappa M = \overline{\kappa}M$ and $Coker(j_{\kappa})$ is $\kappa$-torsion, hence $\overline{\kappa}$-torsion, so it follows that, upto canonical isomorphism, we have $Q_{\overline{\kappa}}(M) = Q_{\kappa}(M)$. We may rephrase this as follows :

4.12. Proposition. Let R be a commutative ring and $\kappa$ an idempotent kernel functor in R-mod. Let A be an R-algebra and $\overline{\kappa}$ the induced idempotent kernel functor in A-mod, then

1. for any left A-module (resp. A-bimodule) M the R-module $Q_\kappa(M)$ is canonically endowed with a unique left A-module (resp. A-bimodule) structure extending that of M and for this structure the A-modules $Q_\kappa(M)$ and $Q_{\overline{\kappa}}(M)$ are canonically isomorphic;

2. in particular, $Q_\kappa(A)$ possesses a unique canonical ringstructure making it isomorphic to $Q_{\overline{\kappa}}(A)$. □

4.13. Corollary. If A is an R-algebra and $\kappa$ an idempotent kernel functor of finite type in R-mod, then any $\overline{\kappa}$-invertible A-bimodule is $\overline{\kappa}$-finitely generated (on the left).

Proof. This follows immediately from the fact that $\mathcal{L}(\overline{\kappa})$ has a basis of finitely generated left A-ideals, as one easily verifies. □

The foregoing shows that if A is an R-algebra and $\kappa$ an idempotent kernel functor in R-mod, then we cause no ambiguity by denoting $\overline{\kappa}$ by $\kappa$. If it becomes necessary to specify, we will explicitly write $\mathcal{L}(A,\kappa)$ for $\mathcal{L}(\overline{\kappa})$, $\mathcal{L}(R,\kappa)$ for $\mathcal{L}(\kappa)$.
In the sequel of this section $\kappa$ will denote an idempotent kernel functor of finite type in R-mod and A, B will denote R-algebras..
Call a left (resp. right) A-module M left (resp. right) $\kappa$-quasiprojective (over R) if $M_p$ is a left (resp. right) projective $A_p$-module for all $p \in \mathcal{C}(R,\kappa)$.

4.14. Lemma. Any A-B-bimodule ${}_AM_B$ defined over R, with the property that $M_B$ is right $\kappa$-quasiprojective is $\kappa$-flat.

Proof. First note that if N" is a $\kappa$-torsion left B-module, then $M \underset{B}{\otimes} N"$ is a $\kappa$-torsion left A-module, since M is defined over R. On the other hand let $i : N' \to N$ be a monomorphism of left B-modules, then $i_p : N'_p \to N_p$ is a monomorphism for each $p \in C(\kappa)$, hence so is $M_p \otimes i_p : M_p \otimes N'_p \to M_p \otimes N_p$, since $M_p$ is right projective by assumption. It follows that $Q_\kappa(M \otimes i)$ is a monomorphism of left A-modules, hence Ker$(M \otimes i)$ is $\kappa$-torsion. □

4.15. Proposition. Let P be an A-B-bimodule which is defined over R and which is $\kappa$-finitely presented on the left and on the right (as an A-resp. B-module), then the following assertions are equivalent :

1. ${}_AP_B$ is $\kappa$-invertible;

2. $P_p$ is an invertible $A_p$-$B_p$-bimodule for all $p \in C(R,\kappa)$.

Proof. (1) ⇒ (2) holds without any finiteness condition on ${}_AP_B$; indeed, let Q be an inverse for P, i.e. $Q_\kappa(P \underset{B}{\otimes} Q) \xrightarrow{\sim} Q_\kappa(A)$ and $Q_\kappa(Q \underset{A}{\otimes} P) \xrightarrow{\sim} Q_\kappa(B)$, then for all $p \in C(\kappa)$ we obtain bimodule isomorphisms $P_p \underset{B_p}{\otimes} Q_p \cong (P \underset{B}{\otimes} Q)_p \cong$ $\cong Q_\kappa(P \underset{B}{\otimes} Q)_p \cong Q_\kappa(A)_p \cong A_p$ and similarly $Q_p \underset{A_p}{\otimes} P_p \cong B_p$. This proves that $P_p$ is an invertible $A_p$-$A_p$-bimodule, indeed. Conversely, if we assume (2.), then the foregoing lemma shows that P is certainly $\kappa$-flat. Now, since P is $\kappa$-finitely presented, it follows that $({}_A[P,Q_\kappa(A)])_p \cong {}_{A_p}[P_p,A_p]$, hence ${}_A[P,Q_\kappa(A)]$ is $\kappa$-flat too, by the same argument. Moreover, the canoncial map $\varphi : P \underset{B}{\otimes} {}_A[P,Q_\kappa(A)] \to Q_\kappa(A)$ yields for each $p \in C(\kappa)$ an isomorphism $\varphi_p : P_p \underset{B_p}{\otimes} ({}_A[P,Q_\kappa(A)])_p \cong P_p \underset{B_p}{\otimes} {}_{A_p}[P_p,A_p] \xrightarrow{\sim} A_p$, hence an isomorphism $Q_\kappa(\varphi) : P \perp {}_A[P,Q_\kappa(A)] \xrightarrow{\sim} Q_\kappa(A)$. From this it follows easily that ${}_A[P,Q_\kappa(A)] \cong [P,Q_\kappa(B)]_B$ as B-A-bimodules and that ${}_AP_B$ is $\kappa$-invertible. □

4.16. Corollary. If P is an A-bimodule which is defined over R and which is left and right $\kappa$-finitely presented over A, then P is $\kappa$-invertible if and only if $P_p$ is an invertible $A_p$-bimodule for all $p \in C(R,\kappa)$. □

4.17. Lemma. Let P be a $\kappa$-finitely generated $\kappa$-quasiprojective left A-bimodule, then P is $\kappa$-finitely presented if for all $f \neq 0$ in R, there is only a finite number of $p \in C(R,\kappa)$ containing f.

Proof. We may assume P to be a finitely generated left A-module, i.e. fitting into an exact sequence of left A-modules

$$0 \to K \to A^n \to P \to 0,$$

for some positive integer n. If $p \in C(\kappa)$, then this yields an exact sequence of $A_p$-modules

$$0 \to K_p \to A_p^n \to P_p \to 0,$$

which splits, as $P_p$ is a projective left $A_p$-module by assumption. Hence we may find a left $A_p$-module $L^{(p)}$ such that $K_p \oplus L^{(p)} = A_p^n$. Let $\{e_i\}$ be the canonical basis for $A_p^n$, then for each $1 \leqslant i \leqslant n$, we may find $k_i \in K$, $s \in R-p$, $\lambda_i \in L^{(p)}$ such that $e_i = k_i/s+\lambda_i$. Moreover, $\lambda_i$ is of the form $l_i/s$ for some $l_i \in A^n$ (it is clear that a *single* $s \in R-p$ does the trick !). Let $L = \sum_i Al_i$, then we claim that $A_s^n = K_s \oplus L_s$. Indeed, first note that $K_s \cap L_s = 0$, since if $\alpha = \sum_i a_i \frac{l_i}{s^n} \in K_s$, then $\alpha \in K_p$ and $\alpha \in L_p$ as well, so $\alpha = 0$. Moreover, obviously $K_s + L_s \subset A_s^n$ and since $e_i = k_i/s + l_i/s \in K_s + L_s$, we get equality as well. This shows our claim to be true. But then it follows that $A_s^n$ maps onto $K_s$, showing that $K_s$ is finitely generated, say $K_s = \Sigma R_s x_\mu$, where $x_\mu \in K$. Let $\{p_1,\dots,p_m\}$ be the primes in $C(\kappa)$ containing s, then for each of these primes we have $K_{p_i} \oplus L^{(p_i)} = A^n_{p_i}$, so K is a finitely generated left $A_{p_i}$-module, say $K_{p_i} = \Sigma A_{p_i} y_{\nu,i}$, where $y_{\nu,i} \in K$. Let $K' = \sum_\mu Ax_\mu + \sum_{\nu,i} Ay_{\nu,i} \subset K$, then we claim that $K'_p = K_p$ for all $p \in C(\kappa)$. Indeed,

a. let $p \in C(\kappa)$ with $s \in p$, then $p = p_1$ say, and so $K'_{p_1} \supset$

$(\sum_\nu Ay_{\nu,1})_{p_1} = \sum_\nu A_{p_1} y_{\nu,1} = K_{p_1}$ ;

b. let $p \in C(\kappa)$ with $s \notin p$, then $K'_p \supset (\sum_\mu Ax_\mu)_s = \sum_\mu A_s \kappa_\mu = K_s$, so $K_p = (K_s)_p \subset K'_p$.

We thus obtain equality in both cases. This shows that K/K' is $\kappa$-torsion, i.e. K is $\kappa$-finitely generated. It now easily follows that P is $\kappa$-finitely presented. □

4.18. Remark. We may often avoid having to recur to finiteness properties of P. Indeed, the reason why we usually need, or at least *seem* to need P to be $\kappa$-finitely presented, is when localizing "Hom-sets", i.e. we want to know whether $S^{-1}({}_A[P,M]) \cong {}_{S^{-1}A}[S^{-1}P,S^{-1}M]$. If P is $\kappa$-finitely presented, this works. We do not need this assumption in the following case. Let P be a $\kappa$-invertible A-B-bimodule defined over R, then for any $p \in C(R,\kappa)$ and any $\kappa$-closed left A-module M we claim that $({}_A[P,M])_p = {}_{A_p}[P_p,M_p]$ as left $B_p$-modules. Indeed, ${}_A[P,-]$ is an "inverse" for $\underline{\underline{a}}_\kappa(P \otimes_B -) : (B,\kappa)\text{-mod} \to (A,\kappa)\text{-mod}$, so for any $M \in (A,\kappa)\text{-mod}$, there is a canonical isomorphism $Q_\kappa(P \otimes_B {}_A[P,M]) = M$, i.e. for all $p \in C(R,\kappa)$ we have

$$(\star)\ P_p \otimes_{B_p} ({}_A[P,M])_p \cong M_p$$

as left $A_p$-modules. On the other hand, $P_p$ is an invertible $A_p$-$B_p$-bimodule (with inverse $Q_p$, if ${}_BQ_A$ is an "inverse" for P), so if $N \in A_p\text{-mod}$, we have $P_p \otimes_{B_p} {}_{A_p}[P_p,N] \cong N$ and in particular

$$(\star\star)\ P_p \otimes_{B_p} {}_{A_p}[P_p,M_p] \cong M_p$$

The isomorphisms $(\star)$ and $(\star\star)$ are canonical, hence tensoring them by $Q_p \otimes_{A_p} -$ on the left we find an isomorphism of left $B_p$-modules

$$({}_A[P,M])_p \cong B_p \otimes_{B_p} ({}_A[P,M])_p \cong Q_p \otimes_{A_p} P_p \otimes_{B_p} ({}_A[P,M])_p \cong Q_p \otimes_{A_p} M_p \cong$$

$$\cong Q_p \otimes_{A_p} P_p \otimes_{B_p} {}_{A_p}[P_p,M_p] \cong B_p \otimes_{B_p} {}_{A_p}[P_p,M_p] \cong {}_{A_p}[P_p,M_p] .$$

B.5. Relative Azumaya Algebras.

We have already explained in [162] that if R is a commutative ring and $\kappa$ an idempotent kernel functor, then a $\kappa$-Azumaya algebra over R is closely related to its center $Q_\kappa(R)$, as far as information upto $\kappa$-torsion is concerned. In this section we show that this is also true on the level of relative Picard groups.

For any commutative ring R and any idempotent kernel functor $\kappa$ in R-mod, introduce the categories $\mathcal{A}(R)$ and $\mathcal{M}(R,\kappa)$ as follows. The objects in both categories are the R-algebras. Morphisms in $\mathcal{A}(R)$ are isomorphisms of R-algebras, morphisms $B \to A$ in $\mathcal{M}(R,\kappa)$ are relative Morita equivalences $B \overset{P}{\underset{\kappa}{\sim}} A$ over R, where ${}_A P_B$ is a $\kappa$-closed $\kappa$-invertible A-B-bimodule defined over R (we use notations of the foregoing section, as far as induced idempotent kernel functors are concerned). Note that these may be composed in the obvious way, as we have pointed out before. Consequently, $\mathrm{Aut}_{\mathcal{A}(R)}(A) = \mathrm{Aut}_R(A)$ resp. $\mathrm{Aut}_{\mathcal{M}(R,\kappa)}(A) = \mathrm{Pic}_R(A,\kappa)$. Clearly, $\mathcal{A}(R)$ is a category with product : the product $A \perp B$ of two objects A and B in $\mathcal{A}(R)$ is just $Q_\kappa(A \underset{R}{\otimes} B) = A \perp B$ (which is also defined over R!) and the product of morphisms $\alpha_1 : B_1 \to A_1$ resp. $\alpha_2 : B_2 \to A_2$ in A(R) is $Q_\kappa(\alpha_1 \otimes \alpha_2) =$
$= \alpha_1 \perp \alpha_2 : B_1 \perp B_2 \to A_1 \perp A_2$, which is well-defined.

Similarly, $\mathcal{M}(R,\kappa)$ is a category with product, on objects given by $A \perp B$ and on morphisms $\alpha_i : B_i \overset{P_i}{\underset{\kappa}{\frown}} A_i$ by $\alpha_1 \perp \alpha_2 : B_1 \perp B_2 \overset{P_1 \perp P_2}{\underset{\kappa}{\frown}} A_1 \perp A_2$.
We thus obtain morphisms $\mathrm{Pic}_R(A_1,\kappa) \times \mathrm{Pic}_R(A_2,\kappa) \to \mathrm{Pic}_R(A_1 \perp A_2,\kappa)$ and similarly for $\mathrm{Aut}_R(-)$. It is also clear that we have homomorphisms $\mathrm{Pic}_R(A,\kappa) \to \mathrm{Pic}_R(A \underset{R}{\otimes} B,\kappa) : [P] \to [P \perp B]$ for any $A,B \in \mathcal{M}(R,\kappa)$.
Recall that a $\kappa$-closed R-module which is faithful over $Q_\kappa(R)$ is said to be a $\kappa$-progenerator if it is $\kappa$-finitely presented and $\kappa$-quasiprojective.

For more details, see F. Van Oystaeyen, A. Verschoren [162] .

An R-algebra A is a $\kappa$-Azumaya algebra if it is a $\kappa$-progenerator such that the canonical map $A^e = A \underset{R}{\otimes} A^{opp} \to End_R(A)$ induces an isomorphism $A^f = Q_\kappa(A^e) = A \perp A^{opp} \to End_R(A)$. Note that it follows from (1.13.) that $End_R(A)$ is $\kappa$-closed. If A is a $\kappa$-closed $\kappa$-finitely presented R-algebra and if $\kappa$ has finite type, then A is a $\kappa$-Azumaya algebra if and only if $A_p$ is an Azumaya algebra over $R_p$ for all $p \in \mathcal{C}(\kappa)$. Examples of $\kappa$-Azumaya algebras arise in geometric contexts, cf. [162] . Moreover, if E is a $\kappa$-progenerator then $End_R(E)$ is a $\kappa$-Azumaya algebra. If $A_1$ and $A_2$ are $\kappa$-Azumaya algebras, then so is $A_1 \perp A_2$.

For simplicity's sake we will assume from now on R to be $\kappa$-noetherian; note that most results remain valid under the weaker assumption : $\kappa$ has finite type.

Let A be a $\kappa$-Azumaya algebra over R, then we claim that the $A^e$-R-bimodule A is $\kappa$-invertible. We will prove this in a series of lemmas.

5.1. Lemma. Let M be a bimodule over some R-algebra A. If M is $\kappa$-closed and defined over R, then so is $M^A$.

Proof. As $M^A \subset M$, clearly $M^A$ is $\kappa$-torsion free. Let us show that it is $\kappa$-injective. Consider an exact diagram

$$\begin{array}{ccccccccc} 0 & \longrightarrow & I & \longrightarrow & R & \longrightarrow & R/I & \longrightarrow & 0 \\ & & {\scriptstyle f}\downarrow & & \downarrow{\scriptstyle \overline{f}} & & & & \\ & & M^A & \hookrightarrow & M & & & & \end{array}$$

with $I \in \mathcal{L}(\kappa)$. Since M is $\kappa$-closed by assumption, the map $f: I \to M^A$ extends to an R-linear $\overline{f} : R \to M$. To show that $\overline{f}$ factorizes through $M^A$, it suffices to show that $\overline{f}(1) \in M^A$. Pick $i \in I$, then for all $a \in A$ we have

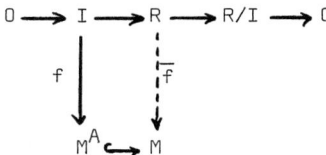

$i\,\bar{f}(1)\,a = \bar{f}(i)\,a = f(ia) = af(i) = a\,i\,\bar{f}(1)$, so $I(\bar{f}(1)a - a\,\bar{f}(1)) = 0$ hence $\bar{f}(1)a - a\bar{f}(1) \in \kappa M = 0$, and as a is arbitrary, this finishes the proof. □

5.2. Lemma. If A is a $\kappa$-finitely generated R-algebra and M is a $\kappa$-torsion free A-bimodule over R, then for all $p \in \mathcal{C}(\kappa)$, we have $(M^A)_p = M_p^{A_p}$.

Proof. Clearly $(M^A)_p \subset M_p^{A_p}$, so it suffices to show that $M_p^{A_p} \subset (M^A)_p$. Let B be a finitely generated R-submodule of A such that A/B is $\kappa$-torsion. Clearly, $M^A \subset M^B$. On the other hand, if $m \in M^B$, then for all $b \in B$ we have $mb = bm$. Take $a \in A$, then for some $I \in \mathcal{L}(\kappa)$ we have $Ia \subset B$. Let $i \in I$, then $ia \in B$ so $(ia)m = m(ia)$, hence $i(am-ma) = 0$. As i is arbitrary, we find $I(am-ma) = 0$, so $am-ma \in \kappa M = 0$, i.e. we have proved that $M^A = M^B$. It thus suffices to verify that $M_p^{A_p} \subset (M^B)_p$. Let $m/t \subset M_p^{A_p}$, then $m/1 \in M_p^{A_p}$, hence for all $a \in A$ we obtain that $s(am-ma) = 0$ for $s \in R-p$. Since B is finitely generated over R, we may find a single $s \in R-p$ with $s(bm-mb) = 0$ for all $b \in B$, i.e. $b(sm) = (sm)b$ for all $b \in B$, so $sm \in M^B$ and $m/t \in (M^B)_p$, indeed. □

5.3. Proposition. Let A be a $\kappa$-Azumaya algebra over R and M a $\kappa$-closed A-bimodule defined over R, then the canonical map $A \perp M^A \to M$ is an isomorphism.

Proof. The above morphism is induced by $A \otimes_R M^A \to M : a \otimes m \to am = ma$.

Now, for all $p \in \mathcal{C}(\kappa)$, we have $(M^A)_p = M_p^{A_p}$, so it follows that $(A \otimes_R M^A)_p = A_p \otimes_{R_p} M_p^{A_p}$, which is isomorphic to $M_p$ under the canonical morphism $A_p \otimes_{R_p} M_p^{A_p} \to M_p$, since $A_p$ is an Azumaya algebra over $R_p$. The assertion follows from an easy local-global argument. □

5.4. Proposition. Under the same assumption, if N is a $\kappa$-closed R-module then the canonical map $N \to (A \perp N)^A$ is an isomorphism.

Proof. Do it locally, as in (5.3.). □

5.5. Corollary. If A is a $\kappa$-Azumaya algebra over R, then the $A^e$-R-bimodule A is $\kappa$-invertible.

Proof. It suffices to exhibit an inverse equivalence for

$$A \perp - = Q_\kappa(A \otimes_R -) : (R,\kappa)\text{-mod} \to (A^e,\kappa)\text{-mod}.$$

Now, one easily checks that $(-)^A : (A^e,\kappa)\text{-mod} \to (R,\kappa)\text{-mod}$ does the trick, in view of (5.3.) and (5.4.). As ${}_{A^e}A_R$ arises from an equivalence, this proves the assertion. □

5.6. Corollary. Let B be an R-algebra, then for any $\kappa$-Azumaya algebra A over R, there is a canonical isomorphism

$$\mathrm{Pic}_R(B,\kappa) \xrightarrow{\sim} \mathrm{Pic}_R(B \perp A,\kappa).$$

Proof. Since $\kappa$ is defined on a commutative central subring, we clearly have : $\mathrm{Pic}_R(B \otimes_R A,\kappa) = \mathrm{Pic}_R(B \perp A,\kappa)$. The morphism we aim to show to be an isomorphism has been defined above. Since ${}_{A^e}A_R$ is $\kappa$-invertible, the statement now follows immediately from the theory of categories with product and the fact that for any $B \in \mathcal{M}(R,\kappa)$ we have $\mathrm{Aut}_{\mathcal{M}(R,\kappa)}(B) = \mathrm{Pic}_R(B,\kappa)$. More precisely, we use the well-known (and easy) fact that if C and C' are isomorphic in $(\mathcal{C},\perp)$, then for any $D \in \mathcal{C}$ the groups $\mathrm{Aut}_{\mathcal{C}}(D \perp C)$ and $\mathrm{Aut}_{\mathcal{C}}(D \perp C')$ are isomorphic. □

Consider a $\kappa$-closed R-algebra A. Let $f \in \mathrm{Aut}_R(A)$ and put $\varphi(f) = [A_f] = [{}_1A_f]$ as before, where [ - ] denotes the class in $\mathrm{Pic}_R(A,\kappa)$. Of course $A_f$ *is* $\kappa$-invertible, since it is invertible with "inverse" ${}_{f^{-1}}A = {}_{f^{-1}}A_1$ . Moreover the map $\varphi : \mathrm{Aut}_R(A) \to \mathrm{Pic}_R(A,\kappa)$ thus defined is easily verified to be a group homomorphism. Let Inn(A) denote the group of inner automorphisms of A. We then have the following result :

5.7. Proposition. Let A be a $\kappa$-closed R-algebra, then the following sequence of groups is exact :

$$1 \longrightarrow \mathrm{Inn}(A) \xrightarrow{i} \mathrm{Aut}_R(A) \xrightarrow{\varphi} \mathrm{Pic}_R(A,\kappa)$$

Proof. It is clear that the canonical map i is a monomorphism and that $\varphi\, i = 1$. Assume $\varphi(f) = 1$, then A and $A_f$ are isomorphic A-bimodules. Now, if $g : A \cong A_f$ is an isomorphism of left A-modules, then for some $b \in A^{\star}$, we have $g(a) = (ab)_f$ for all $a \in A$, where $(\star)_f$ denotes the element $\star$, viewed as an element in $A_f$. If g is an isomorphism of A-bimodules, it has to be right linear as well so for all $a, c \in A$ we should have $g(ac) = g(a)c$; i.e. $(acb)_f = (ab)_f c$, so $acb = ab\, f(c)$ in A. Let $a = 1$, then we find that for all $c \in A$, we have $f(c) = b^{-1} c\, b$, i.e. $f \in \mathrm{Inn}(A)$. □

5.8. Corollary. The following sequence is exact for any R-algebra A :

$$1 \longrightarrow \mathrm{Inn}(Q_\kappa(A)) \longrightarrow \mathrm{Aut}_R(Q_\kappa(A)) \xrightarrow{\varphi} \mathrm{Pic}_R(A,\kappa). \qquad \square$$

5.9. Corollary. Let A be a $\kappa$-Azumaya algebra over R, then there is an exact sequence

$$1 \longrightarrow \mathrm{Inn}(A) \longrightarrow \mathrm{Aut}_R(A) \longrightarrow \mathrm{Pic}(R,\kappa).$$

Proof. This result may be viewed as a relative version of the Rosenberg-Zelinski sequence. The result follows immediately from (5.5.) and (5.7.) □

5.10. Example. If R is a (not necessarily noetherian) Krull domain, then we define the idempotent kernel functor $\sigma$ by $\sigma_1 = \inf\{\kappa_{R-p}; p \in X^{(1)}(R)\}$ Since R is $\sigma_1$-noetherian, it follows that a $\sigma_1$-Azumaya algebra A over R in this case just an "iv-algebra" or "reflexive Azumaya algebra" in the sense of S. Yuan [178] and M. Orzech [112] , i.e. a divisorial R-lattice such that $A \otimes_R A^{opp}$ is canonically isomorphic to $End_R(A)$. Here the <u>modified tensorproduct</u> $M \tilde{\otimes}_R N$ of two (divisorial) R-lattices is defined by $M \tilde{\otimes}_R N = (M \otimes_R N)^{**} = \bigcap_{p \in X^{(1)}(R)} (M \otimes_R N)_p$, the last intersection being taken within $(M \otimes_R N) \otimes_R K = (M \otimes_R K) \otimes_K (N \otimes_R K)$, where K is the field of fractions of R. The foregoing corollary yields that for any reflexive Azumaya algebra A over R we then have an exact sequence

$$1 \longrightarrow Inn(A) \longrightarrow Aut_R(A) \longrightarrow Cl(R)$$

Here $Cl(R) = Pic(R,\sigma_1) = Pic_R(A,\sigma_1)$, the <u>(divisor)class group</u> of R.

5.11. <u>Example</u>. Let A be a prime PI ring of PI degree n, which is finitely generated and flat over its noetherian center R. Let $Spec_n(A)$ be the *open* subset of Spec(A) consisting of all prime ideals P of A with maximal PI-degree, i.e. such that A/P has PI degree n. The map $P \to P \cap R$ defines a homeomorphism between $Spec_n(A) \subset Spec(A)$ and an open subset $X(I) = \{p \in Spec(R);\ I \not\subset p\} \subset Spec(R)$ for some ideal I of R. One may check locally that A is a $\kappa_I$-Azumaya algebra over R, hence it follows that $Pic_R(A,\kappa_I) = Pic(R,\kappa_I)$. Now it has been proved in A. Verschoren [166] and F. Van Oystaeyen, A. Verschoren [162] that $Pic(R,\kappa_I)$ is just the Picard group of the quasi-affine open subscheme X(I) of Spec(R) and it may be calculated by different methods, i.e. it is essentially well-known. The above corollary then shows that the sequence

$$1 \longrightarrow Inn(A) \longrightarrow Aut_R(A) \longrightarrow Pic(X(I))$$

is exact.

5.12. <u>Proposition</u>. For any $\kappa$-torsionfree R-algebra A, there exists a canonical group homomorphism

$$\alpha : Pic(A,\kappa) \to Aut(Z(Q_\kappa(A))).$$

Proof. For notational convenience we will assume A to be $\kappa$-closed. If $c \in Z(A)$ and P is a $\kappa$-invertible A-bimodule with class $[P] \in \text{Pic}(A,\kappa)$ then define $f : P \to P$ by sending p to c p. There is a unique $\alpha_P(c) \in R$ such that $c\,p = p\,\alpha_P(c)$. On one hand, we have $cpa = (cp)a = p\alpha_P(c)a$, on the other hand $(cpa = c(pa) = pa\alpha_P(c)$, therefore $(a\alpha_P(c)-\alpha_P(c)a)$ annihilates P and so $a\alpha_P(c) = \alpha_P(c)a$ for all $a \in A$, yielding $\alpha_P(c) \in Z(A)$. That $\alpha_P$ is injective is clear. Moreover, multiplication by $c \in Z(A)$ on the left extends to left multiplication by c in $Q_\kappa(P \otimes_A Q)$ for any Q and $\alpha_{P \perp Q} = \alpha_Q\,\alpha_P$, thus, if $[Q] = [P]^{-1}$, then $\alpha_Q\,\alpha_P = \alpha_{P\perp Q} = \alpha_{Q \perp P} = \alpha_P\,\alpha_Q = 1_{Z(A)}$. This shows $\alpha_P$ is surjective and that that $\alpha$ is a group-homomorphism. Note that $\alpha_P$ only depends upon the class of P in $\text{Pic}(A,\kappa)$. Indeed, if $[P'] = [P'']$ in $\text{Pic}(A,\kappa)$, then there exists an A-bimodule isomorphism $f : P' \tilde{\to} P''$. By definition, $cp' = p'\alpha_{P'}(c)$ resp. $cp'' = p''\alpha_{P''}(c)$ for all $p' \in P'$ and $p'' \in P''$. But, if $p'' \in P''$, then we have by assumption that there is $p' \in P'$ with $f(p') = p''$, so

$$cp'' = cf(p') = f(cp') = f(p'\alpha_{P'}(c)) = f(p')\alpha_{P'}(c) = p''\alpha_{P'}(c),$$

for all $p'' \in P''$ and then unicity implies that $\alpha_{P''}(c) = \alpha_{P'}(c)$. □

Define Picent $(A,\kappa)$ to be the group $\text{Picent}_{Z(Q_\kappa(A))}(A,\kappa)$. This implies that $\kappa$ lives on $Z(Q_\kappa(A))$. We will assume that $\kappa$ is defined on $R \subset Z(A)$ and is induced on $Z(Q_\kappa(A))$ through the canonical map $Z(A) \to Z(Q_\kappa(A))$. We call Picent $(A,\kappa)$ the relative central Picard group of A with respect to $\kappa$. One defines similarly Autcent(A) to be $\text{Aut}_{Z(A)}(A)$, the central automorphism group of A. From the foregoing, we retain :

5.13. Corollary. There is an exact sequence of groups :

$$1 \longrightarrow \text{Inn}(A) \longrightarrow \text{Autcent}(A) \longrightarrow \text{Picent}(A,\kappa)$$

for any $\kappa$-torsion free R-algebra A.

Proof. The lemma below shows that our assumption implies that $Z(Q_\kappa(A)) = Q_\kappa(Z(A))$. Now, we claim that $\mathrm{Pic}_{Z(A)}(A,\kappa) = \mathrm{Pic}_{Q_\kappa(Z(A))}(A,\kappa)$, where $\kappa$ is defined on $R \subset Z(A)$. Indeed, assume that P is a $\kappa$-closed A-bimodule such that $zp = pz$ for all $z \in Z(A)$ and $p \in P$, then, if $s \in Q_\kappa(Z(A))$ we may find $I \in \mathcal{L}(\kappa)$ such that $I\,s \subset Z(A)$, hence $I(sp-ps) = 0$, i.e. $sp = ps$ as P is $\kappa$-closed, hence in particular $\kappa$-torsionfree. As $\mathrm{Pic}_{Q_\kappa(Z(A))}(A,\kappa) = \mathrm{Pic}_{Z(Q_\kappa(A))}(A,\kappa)$, this proves the result. □

5.14. Lemma. Let A be an R-algebra which is $\kappa$-torsionfree for some idempotent kernel functor $\kappa$ in R-mod, then $Z(Q_\kappa(A)) = Q_\kappa(Z(A))$.

Proof. Clearly $Z(A) \subset Z(Q_\kappa(A))$, hence $Q_\kappa(Z(A)) \subset Z(Q_\kappa(A))$, as $Z(Q_\kappa(A)) = Q_\kappa(A)^A$ is $\kappa$-closed by (5.1.). Conversely, if $z \in Z(Q_\kappa(A))$, then $Iz \subset A$ for some $I \in \mathcal{L}(\kappa)$. Moreover, $Iz \subset Z(A)$, as I and z commute with $Q_\kappa(A)$, hence a fortiori with A. It follows that $z \in Q_\kappa(Z(A))$. □

Define $P(A,\kappa)$ to consist of all left isomorphism classes of $\kappa$-closed left $\kappa$-finitely generated $\kappa$-quasiprojective A-modules.
The class of such a module E in $P(A,\kappa)$ will be denoted by $\{E\}$. As $\kappa$ has finite type, every $\kappa$-invertible A-bimodule P over R yields an element $\{P\}$ in $P(A,\kappa)$ and this defines a map

$$\gamma : \mathrm{Pic}_R(A,\kappa) \to P(A,\kappa).$$

Moreover, $\mathrm{Pic}_R(A,\kappa)$ acts on $P(A,\kappa)$ by $\mathrm{Pic}_R(A,\kappa) \times P(A,\kappa) \to P(A,\kappa)$ : $([P],\{E\}) \to \{P \underset{A}{\perp} E\}$. It is now easy to give a more precise description of the above sequences. Indeed, we have

5.15. Lemma. The left cosets in $\mathrm{Pic}_R(A,\kappa)/\mathrm{Im}(\varphi)$ are the fibres of $\gamma$.

Proof. Fix $[P] \in \mathrm{Pic}_R(A,\kappa)$ and denote by $\widetilde{[P']}$ the left class of $[P']$ in $\mathrm{Pic}_R(A,\kappa)/\mathrm{Im}(\varphi)$, then $\widetilde{[P]} = \widetilde{[P']}$ if and only if $Q_\kappa(P \otimes Q_\kappa(A)_f) \cong P'$

for some $f \in \mathrm{Aut}_R(Q_\kappa(A))$. But $Q_\kappa(P \otimes Q_\kappa(A)_f) = Q_\kappa(P_f) = P_f$, so $P' \cong P_f$, hence in particular P' is *left* isomorphic to P as an A-module. Conversely, if we are given an isomorphism of left A-modules $u : P \to P'$, then we obtain an automorphism $f : Q_\kappa(A) \overset{\sim}{\to} \mathrm{End}({}_AP') \overset{\hat{u}}{\overset{\sim}{\to}} \mathrm{End}({}_AP) \overset{\sim}{\to} Q_\kappa(A)$ of $Q_\kappa(A)$ and u now has the property that $u(pf(a)) = u(p)a$, so $P' \cong P_f$ as A-bimodules defined over R. □

5.16. Lemma. Let P and P' be $\kappa$-invertible A-bimodules over R and let E be a $\kappa$-progenerator in R-mod, then the following assertions are equivalent

1. $P \perp \mathrm{End}_R(E) \cong P' \perp \mathrm{End}_R(E)$ in $A \perp \mathrm{End}_R(E)$-mod;

2. $P \perp {}_R[E,R] \cong P' \perp {}_R[E,R]$ in A-mod.

Proof. This reduces essentially to the commutative theory developed in F. Van Oystaeyen, A. Verschoren [162] and the theory of categories with product. Indeed, view E as a left $B = \mathrm{End}_R(E)$-module, then $E^\star = {}_R[E,R]$ is an R-B-bimodule and $E^\star$ is $\kappa$-invertible as such. Moreover, since it follows that $A \perp E^\star$ is then a $\kappa$-invertible A-A ⊥ B-bimodule, it is clear that there is a bijective correspondence between the isomorphism classes of objects $M \in (A \perp B,\kappa)$-mod and the isomorphism classes of objects $(A \perp E^\star) \underset{A \perp B}{\perp} M \in (A,\kappa)$-mod. Since the class of $N \perp B$ corresponds to $N \perp E^\star$ for any $N \in (A,\kappa)$-mod, this proves the assertion. □

Recall from above that for $F = \mathrm{Pic}_R(-,\kappa)$ or $\mathrm{Aut}_R(-)$ or Inn(-), we get for any pair of $\kappa$-Azumaya algebras A', A" over R a homomorphism $F^{A'}_{A' \perp A''} : F(A \perp A') \to F(A \perp A' \perp A'')$. These maps do satisfy many nice properties, in particular, they form a direct system and we may form $\varinjlim F(A \perp A') = F^\infty_R(A)$, where A' runs over all (isomorphism classes of) $\kappa$-Azumaya algebras over R. The groups thus obtained for particular choices of F will be denoted by $\mathrm{Inn}^\infty_R(A,\kappa)$ resp. $\mathrm{Aut}^\infty_R(A,\kappa)$. Note that the analogous group $\mathrm{Pic}^{\infty}_R(A,\kappa)$ is just $\mathrm{Pic}_R(A,\kappa)$ due to the fact that all $F^{A'}_{A'\perp A''}$ are bijective for $F = \mathrm{Pic}_R(-,\kappa)$! We thus obtain the following :

5.17. <u>Proposition.</u> There is a canonical exact sequence of groups

$$1 \longrightarrow \mathrm{Inn}^{\infty}_R(Q_\kappa(A),\kappa) \longrightarrow \mathrm{Aut}^{\infty}_R(Q_\kappa(A),\kappa) \xrightarrow{\tilde{\varphi}} \mathrm{Pic}_R(A,\kappa).$$

<u>Proof.</u> We may assume A to be $\kappa$-closed. Just note that everything behaves well, i.e. product preserving, so we get a commutative exact diagram(★) for all A', A" :

$$(\star)\qquad \begin{array}{ccccc} \mathrm{Inn}(A\perp A') & \longrightarrow & \mathrm{Aut}_R(A\perp A') & \longrightarrow & \mathrm{Pic}_R(A,\kappa) \\ \downarrow & & \downarrow & & \| \\ \mathrm{Inn}(A\perp A'\perp A'') & \longrightarrow & \mathrm{Aut}_R(A\perp A'\perp A'') & \longrightarrow & \mathrm{Pic}_R(A,\kappa) \end{array}$$

The assertion follows by taking the limit over all $\kappa$-Azumaya algebras over R. □

5.18. <u>Proposition.</u> If $[P'], [P''] \in \mathrm{Pic}_R(A,\kappa)$ then equivalently :

1. $[P']$ and $[P'']$ are in the same left coset of $\mathrm{Pic}_R(A,\kappa)/\mathrm{Im}\,\tilde{\varphi}$ ;
2. $P'\perp A' \cong P''\perp A'$ in $A\perp A'$-mod, for some $\kappa$-Azumaya algebra A' over R;
3. $P'\perp E \cong P''\perp E$ in A-mod, for some $\kappa$-progenerator E in R-mod.

<u>Proof</u>. The equivalence of 1. and 2. has essentially been pointed out before. The equivalence of 2. and 3. may be checked as follows : clearly 2. is equivalent to the same statement, but with A' of the particular form $A' = \mathrm{End}_R(E)$, where E is a $\kappa$-progenerator in R-mod. To see this, just "tensor" ($\perp$) with $E^{opp}$ and recall that for any $\kappa$-Azumaya algebra B over R we have $B\perp B^{opp} \cong \mathrm{End}_R(B)$ and that A' is a $\kappa$-progenerator in R-mod. The equivalence of 2. and 3. now follows immediately from (5.16.). □

Let $Z(R)$ denote the category of commutative R-algebras and their isomorphisms, then taking centers defines a covariant functor

$Z : \mathcal{A}(R) \to \mathcal{Z}(R)$. it has been pointed out by A. Fröhlich [52] that this functor possesses the following remarkable (although trivial) property : let A be an R-algebra and f an automorphism of the R-algebra Z(A), then we may find an R-algebra B with center Z(B) = Z(A) and an isomorphism $\overline{f} : A \to B$ of R-algebras, such that $f = Z(\overline{f})$. Moreover, this determines the Z(A)-algebra B uniquely upto isomorphism, So, we may write $B = A^f$. We then obtain, with $\mathcal{A}(R,\kappa) = \mathcal{A}(R) \cap (R,\kappa)$-mod the following commutative diagram :

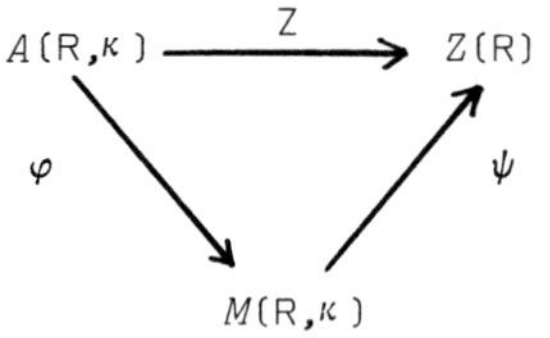


(★)

where $\varphi$ is defined as before, i.e. the identity on R-algebras and associating to an isomorphism $f : B \tilde{\to} A$ the $\kappa$-invertible A-B-bimodule $A_f$. The map $\psi$ sends an algebra in $\mathcal{M}(R,\kappa)$ to its center and a $\kappa$-invertible A-B-bimodule P over R to the isomorphism $\alpha_P : Z(B) \to Z(C)$, defined by $\alpha_P(b)p = pb$ for all $p \in P$ and $b \in Z(B)$.

Let us now adopt the following notational conventions :

a. $\mathrm{Aut}^a_R(Z(A)) \subset \mathrm{Aut}_R(Z(A))$ is the subgroup of all R-autcmorphisms f of Z(A) such that $A^f \cong A$ as Z(A)-algebras (i.e. in $\mathcal{A}(Z(A))$!).

b. $\mathrm{Aut}^m_R(Z(A),\kappa) \subset \mathrm{Aut}_R(Z(A))$ is the subgroup of all R-automorphisms f of Z(A) such $A^f \cong A$ in $\mathcal{M}_R(Z(A),\kappa)$ (i.e. A and $A^f$ are $\kappa$-Morita equivalent over R!).

We then have :

5.19. <u>Proposition</u>. Assume that A is $\kappa$-closed (for simplicity's sake) then the following diagram of groups is commutative and exact :

$$\begin{array}{ccccccccc}
1 & \longrightarrow & \mathrm{Autcent}(A) & \longrightarrow & \mathrm{Aut}_R(A) & \longrightarrow & \mathrm{Autcent}^a_R((Z(A)) & \longrightarrow & 1 \\
 & & \downarrow & & \downarrow & & \downarrow & & \\
1 & \longrightarrow & \mathrm{Picent}(A,\kappa) & \longrightarrow & \mathrm{Pic}_R(A,\kappa) & \longrightarrow & \mathrm{Aut}^m_R\,(Z(A),\kappa) & \longrightarrow & 1
\end{array}$$

Proof. One easily sees that there are canonical maps

$$\mathrm{Aut}_R(A) \longrightarrow \mathrm{Aut}_R(A)/\mathrm{Autcent}(A) \longrightarrow \mathrm{Autcent}_R(Z(A))$$

resp.

$$\alpha : \mathrm{Pic}_R(A,\kappa) \rightarrow \mathrm{Pic}_R(A,\kappa)/\mathrm{Picent}(A,\kappa) \longrightarrow \mathrm{Aut}_R(Z(A))$$

It also follows easily from the preceding remarks that the diagram commutes, so let us just check that the images of both maps just defined are as claimed. For the former map, there is no problem : the result is actually already stated in A. Fröhlich's [ 52 ]. Indeed, an R-automorphism $f$ of $Z(A)$ may be lifted to an R-automorphism of A if and only if A is isomorphic to $A^f$ as a $Z(A)$-algebra. The image of $\alpha$ may be obtained as follows : let $P = \varphi(\overline{f})$ be the $\kappa$-invertible $A^f$-A-bimodule over R which corresponds to a lifting $\overline{f} : A \to A^f$ of $f \in \mathrm{Aut}_R(Z(A))$, then $\psi(P) = f$ by the commutativity of the diagram (★). If $f \in \mathrm{Aut}^m_R(Z(A),\kappa)$ and Q a $\kappa$-invertible $A$-$A^f$-bimodule over $Z(A)$, then $U = Q \perp_{A^f} P$ is a $\kappa$-invertible A-bimodule A-bimodule over R with $\psi(U) = f$. On the other hand, if U is a $\kappa$-invertible A-bimodule over R and $\psi(U) = f$, then, taking P' to be an "inverse" for P, we have that $Q = U \perp_A P'$ is a $\kappa$-invertible $A$-$A^f$-bimodule over $Z(A)$, and this finishes the proof. □

5.20. Corollary. If A is commutative, then there is a split exact sequence of groups

$$1 \longrightarrow \mathrm{Picent}(A,\kappa) \longrightarrow \mathrm{Pic}_R(A,\kappa) \longrightarrow \mathrm{Aut}_R(A) \longrightarrow 1. \qquad \square$$

5.21. Corollary. If A and B are $\kappa$-Morita equivalent over R, then Picent $(A,\kappa) \cong$ Picent$(B,\kappa)$. □

5.22. Corollary. Picent$(A,\kappa)$ is a normal subgroup of $Pic_R(A,\kappa)$. □

## B.6. Changes.

In order to calculate explicitly some relative Picard groups, we want to compare them to other related groups. The main purpose of this section is to study exact sequences involving relative Picard groups and other groups which are generally better understood. In particular, we are interested in the behaviour of $Pic_R(A,\kappa)$, when some of the data R, $\kappa$ or A are changed in certain specific ways.

Let us first briefly sketch what happens when $\lambda$ in A-mod is transformed to another $\mu$ in A-mod such that either $\lambda \leqslant \mu$ or $\mu \leqslant \lambda$ . It is clear that an arbitrary $\mu$ may then be considered in two steps : $\mu' = \mu \wedge \lambda$ and $\mu' \leqslant \mu$. Now, the case $\mu \leqslant \lambda$ has actually been treated in the section on Morita equivalence. So, let us consider $\lambda \leqslant \mu$ and relate $\lambda$-invertibility to $\mu$-invertibility.

Obviously, we have $Q_\lambda(P \otimes_A Q) \cong Q_\lambda(A)$, etc. hence also $Q_\mu(P \otimes_A Q) \cong Q_\mu(A)$, etc, but neither P nor Q has to be $\mu$-flat if P and Q are $\lambda$-flat. If we consider the canonical inclusion $(A,\mu)$-mod $\to (A,\lambda)$-mod, then a $\lambda$-invertible A-bimodule P induces an equivalence $Q_\lambda(P \otimes_A -) : (A,\lambda)\text{-mod} \to (A,\lambda)\text{-mod}$, hence an equivalence (by restriction), $Q_\lambda(P \otimes_A -) : (A,\mu)\text{-mod} \to (A,\mu')\text{-mod}$ between $(A,\mu)$-mod and a quotient category $(A,\mu')$-mod in $(A,\lambda)$-mod. If Q is inverse for P, then $\mathcal{L}(\mu')$ consists of all left ideals L of A such that Q/QL is $\mu$-torsion. So, if $\mu' = \mu$, we are certainly done. On the other hand, this yields that $Q_\lambda(P \otimes_A -)$ is already an equivalence, whereas it seems to be sufficient that $Q_\mu(P \otimes_A -)$ be an autoequivalence of $(A,\mu)$-mod. However the latter condition is not really weaker, as we will show in the following.

6.1. Lemma. If for some A-bimodule P we have that P/PL is $\mu$-torsion for any $L \in \mathcal{L}(\mu)$, then $P \otimes_A M''$ is $\mu$-torsion for any $\mu$-torsion left A-module M.

Proof. This is a relative version of the fact that flatness may be checked on ideals. Indeed, pick $m'' \in M''$ and consider an exact sequence

$$0 \longrightarrow L \longrightarrow A \xrightarrow{\varphi} M'',$$

where $\varphi$ is right multiplication by m" and $L = \mathrm{Ker}(\varphi)$. Since A/L injects into M", which is $\mu$-torsion by assumption, it is $\mu$-torsion, hence $L \in \mathcal{L}(\mu)$. Moreover, $\varphi$ extends to an isomorphism $A/L \xrightarrow{\sim} Am''$. As P/PL is $\mu$-torsion, so is $P \otimes_A Am''$, hence so is $P \otimes_A (\oplus_{m''} Am'') = \oplus_{m''} (P \otimes_A Am'')$. Finally, since $M'' = \Sigma_{m''} Am''$ is a quotient of $\oplus_{m''} Am''$, it follows that $P \otimes_A M''$ is $\mu$-torsion too, proving the assertion. □

It now easily follows that the above statement (just preceding the Lemma) is valid. Indeed, if P and Q are $\mu$-invertible, "inverse" to each other then for all $L \in \mathcal{L}(\mu)$ we have that P/PL and Q/QL are $\mu$-torsion. But the converse is true also : if for any left ideal L of A, we have that P/PL is $\mu$-torsion, then $L \in \mathcal{L}(\mu)$. Indeed, since P/PL is $\mu$-torsion, then so is $Q \otimes_A (P/PL) = Q \otimes_A P \otimes_A A/L$, i.e. $0 = Q_\mu(Q \otimes_A P \otimes_A (A/L)) = Q_\mu(Q_\mu(Q \otimes P) \otimes_A (A/L)) = Q_\mu(Q_\mu(A) \otimes_A (A/L))$, i.e. $L \in \mathcal{L}(\mu)$.
On the other hand, if $P \perp_\mu Q \cong Q_\mu(A) \cong Q \perp_\mu P$ and if P/PL resp. Q/QL is $\mu$-torsion for all $L \in \mathcal{L}(\mu)$, then from (6.1.) and the usual trick, it follows that P resp. Q is $\mu$-flat.

Finally, let us note that if P is $\mu$-invertible and if Q is an A-bimodule such that $P \perp_\mu Q \cong Q_\mu(A) \cong Q \perp_\mu P$, then Q (or $Q_\mu(Q)$- is an "inverse" for P. Indeed, let Q' be an inverse for P, then we find that $Q_\mu(Q) = Q_\mu(Q_\mu(A) \otimes_A Q) = Q_\mu(Q_\mu(Q' \otimes_A P) \otimes_A Q) = Q_\mu(Q' \otimes_A P \otimes_A Q) = Q_\mu(Q' \otimes_A Q_\mu(P \otimes_A Q)) =$

$= Q_\mu(Q' \otimes_A Q_\mu(A)) = Q_\mu(Q')$!

We thus have proved :

6.2. Proposition. Let P be a $\lambda$-closed $\lambda$-invertible A-bimodule with inverse $Q = {}_A[P, Q_\lambda(A)]$, then the following are equivalent for $\lambda \leqslant \mu$ :

1. P is $\mu$-invertible;
2. P/PL and Q/QL are $\mu$-torsion for all $L \in \mathcal{L}(\mu)$. □

Note. If $\lambda$, $\mu$ and P are defined over the commutative subring R of A, then P/PL = P/LP is automatically $\mu$-torsion for $L \in \mathcal{L}(\mu)$, whence :

6.3. Corollary. Let A be an R-algebra and $\lambda \leqslant \mu$ idempotent kernel functors in R-mod, then there is a canonical grouphomomorphism $\mathrm{Pic}_R(A,\lambda) \to \mathrm{Pic}_R(A,\mu) : [P] \to [Q_\mu(P)]$ . □

Let us now consider the behaviour of $\mathrm{Pic}_R(A,\kappa)$ under a change of the base-ring R. If $f : T \to R$ is a morphism of commutative rings and if A is an R-algebra provided with an idempotent kernel functor $\kappa$ in A-mod, then there is a canonical monomorphism of groups $\theta : \mathrm{Pic}_R(A,\kappa) \to \mathrm{Pic}_T(A,\kappa)$. We claim that $\theta$ is an isomorphism if f is epimorphic. Indeed, let $i : R \to Z(Q_\kappa(A))$ be the canonical inclusion, and for $[P] \in \mathrm{Pic}_T(A,\kappa)$, let $\varphi = \varphi_P : Z(Q_\kappa(A)) \to Z(Q_\kappa(A))$ denote the endomorphism of $Z(Q_\kappa(A))$ defined by $\varphi(z)m = mz$ for all $z \in Z(Q_\kappa(A))$ and $m \in P$. Now, $\varphi(c)if = if$ for each $c \in Z(Q_\kappa(A))$ by the choice of P, hence $\varphi(c)i = i$, as f is an epimorphism. But this states exactly that $[P] \in \mathrm{Pic}_R(A,\kappa)$.

Let us recall some results and constructions, which are essentially due to A. Fröhlich [52], but which we will need here in a relative context. Fix notations as follows : A is an arbitrary T-algebra, where T is a commutative R-algebra and $\kappa$ is an idempotent kernel functor in R-mod. Furthermore, there will be given a ringhomomorphism $\psi : R \to S$ of R into another commutative ring S. All unadorned tensorproducts are over R. We write A(S),... for $Q_\kappa(A \otimes S) = A \perp S, \ldots$ . Clearly $\kappa$ induces an idempotent kernel functor in $A \otimes S$-mod, which we will denote by the same symbol $\kappa$. Note also that it frequently happens that $Q_\kappa(A) \otimes S = A(S)$. Let us point out two examples:

a. A is a noetherian ring and $\Sigma$ a multiplicative subset of R, consisting of regular elements, and we take $\psi : R \to S = \Sigma^{-1}R$;

b. $S = R[T_1,\ldots,T_n]$ and $\psi : R \to R[T_1,\ldots,T_n]$ is the canonical inclusion.

6.4. Lemma. With these notations, there are grouphomomorphisms

$$\mathrm{Pic}_T(A,\kappa) \to \mathrm{Pic}_{T \otimes S}(A \otimes S,\kappa),$$

$$\mathrm{Aut}_T(A) \to \mathrm{Aut}_{T \otimes S}(A \otimes S),$$

$$\mathrm{Aut}_T(Q_\kappa(A)) \to \mathrm{Aut}_{T \otimes S}(Q_\kappa(A \otimes S) = A(S)).$$

Proof. All of this is rather obvious. For example, the first map is given by sending $[P]$ to $[P \perp S]$, the last map is given by sending the automorphism f of $Q_\kappa(A)$ to the automorphism $f \otimes S$ of $Q_\kappa(A) \otimes S$ and this last automorphism to $Q_\kappa(f \otimes S)$, which is an automorphism of $Q_\kappa(Q_\kappa(A) \otimes S) = Q_\kappa(A \otimes S) = A(S)$. Details are easy to check and left to the reader. □

Consider pairs (P,f) of a $\kappa$-closed $\kappa$-invertible A-bimodule P over R and an isomorphism $f : P(S) \xrightarrow{\sim} A(S)$ of $A \otimes S$-bimodules (or equivalently of A(S)-bimodules, as one easily verifies). An isomorphism $(P,f) \xrightarrow{\sim} (Q,g)$ of such couples is an isomorphism of A-bimodules $h : P \to Q$ such that the induced map $h(S) = Q_\kappa(h \otimes S)$ makes the following diagram commutative :

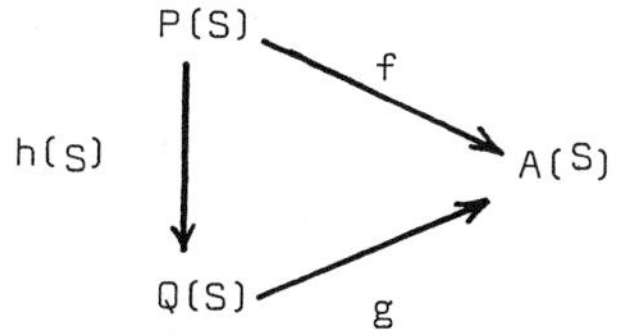

We may view this as a category with product, by defining

$(P_1, f_1) \perp (P_2, f_2) = (P_1 \underset{A}{\perp} P_2, f_1 \star f_2)$, where $P_1 \underset{A}{\perp} P_2 = Q_\kappa(P_1 \otimes P_2)$

and where $f_1 \star f_2$ is the composition

$$(P_1 \underset{A}{\perp} P_2)(S) = Q_\kappa(P_1(S) \underset{A \otimes S}{\otimes} (P_2 \otimes S)) \xrightarrow{f_1'} Q_\kappa(A(S) \underset{A \otimes S}{\otimes} (P_2 \otimes S)) =$$

$$= Q_\kappa(P_2 \otimes S) = P_2(S) \xrightarrow{f_2} A(S),$$

where $f_1' = Q_\kappa(f_1 \underset{A \otimes S}{\otimes} (P_2 \otimes S))$. The isomorphism classes $[P, f]$ form a group $\mathrm{Pic}_\psi(A, \kappa)$ with multiplication induced by $\perp$. Moreover, the isomorphism classes with P defined over T form a subgroup $\mathrm{Pic}_{\psi,T}(A, \kappa)$. We have a canonical map $\mathrm{Pic}_\psi(A, \kappa) \to \mathrm{Pic}_R(A, \kappa) : [P, f] \to [P]$. Finally, if $u \in U(Z(A(S))) = Z(A(S))^\star$, then we associate to it the pair $(Q_\kappa(A), g_u)$, where $g_u : A(S) \to A(S)$ is just multiplication by u, on the left say. We thus obtain a homomorphism $U(Z(A(S))) \to \mathrm{Pic}_\psi(A, \kappa) : u \to [Q_\kappa(A), g_u]$.

6.5. Proposition. With maps defined as above, there is a commutative exact diagram :

$$\begin{array}{ccccccccc}
U(Z(Q_\kappa(A))) & \to & U(Z(A(S))) & \to & \mathrm{Pic}_{\psi,T}(A,\kappa) & \to & \mathrm{Pic}_T(A,\kappa) & \to & \mathrm{Pic}_{T \otimes S}(A \otimes S, \kappa) \\
\| & & \| & & \downarrow & & \downarrow & & \downarrow \\
U(Z(Q_\kappa(A))) & \to & U(Z(A(S))) & \to & \mathrm{Pic}_\psi(A,\kappa) & \to & \mathrm{Pic}_R(A,\kappa) & \to & \mathrm{Pic}_S(A \otimes S, \kappa)
\end{array}$$

Proof : Straightforward. □

Let us denote by $\mathcal{D}(A, \kappa)$ this diagram depending on A and $\kappa$.

6.6. Corollary. Let A and B be T-algebras and $\lambda$ resp. $\mu$ idempotent kernel functor in A-mod resp. B-mod, then every $(\lambda, \mu)$-invertible A-B-bimodule P over T induces an isomorphism of diagrams $\mathcal{D}(A, \lambda) \xrightarrow{\sim} \mathcal{D}(B, \mu)$.

Proof. It actually suffices to exhibit a canonical isomorphism $\mathrm{Pic}_\psi(A, \lambda) \xrightarrow{\sim} \mathrm{Pic}_\psi(B, \mu)$, compatible with the maps considered and induced by P. Let ${}_BQ_A$ be an "inverse" for ${}_AP_B$, then to every pair $[M,f]$ for A and $\lambda$ we associate the pair $[N,g]$ for B and $\mu$, where $N = Q \underset{A}{\perp} M \underset{A}{\perp} P$ and g is given by $(Q \underset{A}{\perp} M \underset{A}{\perp} P)(S) = Q_\mu(Q \underset{A}{\otimes}(M \otimes S)\underset{A}{\otimes} P) \cong Q_\mu(Q \underset{A}{\otimes} M(S) \underset{A}{\otimes} P) \xrightarrow{\sim}$ $Q_\mu(Q \underset{A}{\otimes} A(S)\underset{A}{\otimes} P) \cong Q_\mu(Q \underset{A}{\otimes} (A \otimes S)\underset{A}{\otimes} P) \cong Q_\mu((Q \underset{A}{\otimes} P)\otimes S) \cong B(S)$. Note that there is a slight ambiguity in notation here, as $B(S), (Q \underset{A}{\perp} M \underset{A}{\perp} P)(S), \ldots$ stand for $Q_\mu(- \otimes S)$, whereas $A(S)$, $M(S), \ldots$ stand for $Q_\lambda(- \otimes S)$. The fact that P and Q are invertible, hence relatively flat with respect to the correct idempotent kernel functors, permits us to fiddle around as we did. □

The above constructions may be adapted to the case of automorphisms. Assume $A = Q_\kappa(A)$ for convenience's sake. We let $Aut_\psi(A,\kappa)$ be the fibre product of the diagram

$$\begin{array}{ccc} & & U(A(S)) \\ & & \downarrow{\scriptstyle \beta} \\ Aut_R(A) & \xrightarrow[\alpha]{} & Aut_S(A(S)) \end{array}$$

This group consists of all $(f,u) \in Aut_R(A) \times U(A(S))$ for which $\alpha(f) = \beta(u)$. Here $\alpha$ is the map which to $f \in Aut_R(A)$ associates the automorphism $\tilde{f} = Q_\kappa(f \otimes S)$ of $A(S)$-which is defined over T- and $\beta$ is the map which to $u \in U(A(S))$ associates the inner automorphism $a \mapsto u a u^{-1}$ of $A(S)$. We have to mention the idempotent kernel functor $\kappa$ in the notations because there also exists a group $Aut_\psi(A)$, which is the pull-back

$$\begin{array}{ccccc} & & U(A \otimes S) & & \\ & \nearrow & & \searrow & \\ Aut_\psi(A) & & & & Aut_S(A \otimes S) \\ & \searrow & & \nearrow & \\ & & Aut_R(A) & & \end{array}$$

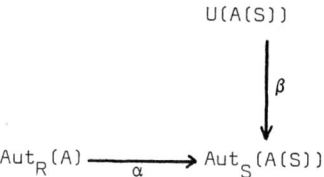

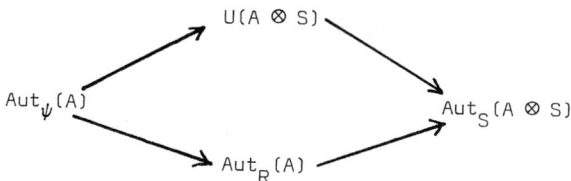

There is a map $\delta$ : $Aut_\psi(A) \to Aut_\psi(A,\kappa)$, which to $(f,u) \in Aut_\psi(A)$ associates $(f,u') \in Aut_\psi(A,\kappa)$. Moreover, $(f,u) \in Ker(\delta)$ if and only if $u' \in Z(A(S))$.
Define canonical maps $I_{\psi,A}$ : $U(A) \to Aut_\psi(A)$ and $I^\kappa_{\psi,A}$: $U(A) \to Aut_\psi(A,\kappa)$
One may define canonical maps $I_{\psi,A}$ : $U(A) \to Aut_\psi(A)$ and $I^\kappa_{\psi,A}$ : $U(A) \to Aut_\psi(A,$
in the obvious way. Of course, we may restrict to $Aut_T(A)$, automorphisms over T and we then obtain subgroups $Aut_{\psi,T}(A)$ and $Aut_{\psi,T}(A,\kappa)$ of the above groups. Note that in the examples a. and b. given above, we obtain $Aut_\psi(A) = Aut_\psi(A,\kappa)$ resp. $Aut_{\psi,T}(A) = Aut_{\psi,T}(A,\kappa)$. Finally, define $Out_\psi(A)$ resp. $Out_\psi(A,\kappa)$ as $Aut_\psi(A)/Im(I_{\psi,A})$ resp. $Aut_\psi(A,\kappa)/Im(I^\kappa_{\psi,A})$.
We may again define a diagram as above. The morphisms are the expected ones: to $u \in U(Z(A(S)))$ we associate $(1,u) \in Aut_\psi(A,\kappa)$ and the map $(f,u) \mapsto f$ yields a homomorphism $Aut_\psi(A,\kappa) \to Aut_R(A)$. It is clear that these morphisms factorize through $U(Z(A(S))) \to Out_\psi(A,\kappa)$ and $Out_\psi(A,\kappa) \to Out_R(A)$, where $Out_R(A)$ is defined as $Out_R(A) = Aut_R(A)/Inn(A)$. Finally, we define a homomorphism $Aut_\psi(A,\kappa) \to Pic_\psi(A,\kappa)$ as follows. Let $(f,u) \in Aut_\psi(A,\kappa)$, then we first construct the $\kappa$-invertible A-bimodule $A_f$ and then, fpr all $x \in A_f \perp S$, we define $g(x) = x\,u \in A \perp S$, thus obtaining an isomorphism of $A \otimes S$-bimodules. To check that this is well-defined, consider the canonical map $A_f \otimes S \to A_f \perp S$. Restricting g to $A_f \otimes S$ should give an $A \otimes S$-bimodule morphism $A_f \otimes S \to A \perp S$. It is clearly left $A \otimes S$-linear, so it suffices to verify right $A \otimes S$-linearity. Let $a \otimes s \in A \otimes S$ and $b_f \otimes t \in A_f \otimes S$, then a routine calculation shows that $g((b_f \otimes t)(a \otimes s))$ $= g(b_f \otimes t)(a \otimes s)$. So, the restriction $g|A_f \otimes S$ is well-defined and $A \otimes S$-linear on both sides, hence so is g.
It is easy to check that the homomorphism $Aut_\psi(A,\kappa) \to Pic_\psi(A,\kappa)$ defined above, factorizes through $Out_\psi(A,\kappa) \to Pic_\psi(A,\kappa)$. Indeed, if $(f,u) \in Im(I^\kappa_{\psi,A})$, then $(A_f,g) \cong (A,1)$! With notations as before, we have

6.7. Proposition. There is an exact commutative diagram :

$$\begin{array}{ccccccccc}
U(Z(A)) & \longrightarrow & U(Z(A(S))) & \longrightarrow & Out_{\psi,T}(A,\kappa) & \longrightarrow & Out_T(A) & \longrightarrow & Out_{T\otimes S}(A(S)) \\
\| & & \| & & \downarrow & & \downarrow & & \downarrow \\
U(Z(A)) & \longrightarrow & U(Z(A(S))) & \longrightarrow & Out_{\psi}(A,\kappa) & \longrightarrow & Out_R(A) & \longrightarrow & Out_S(A(S))
\end{array}$$

Let us denote by $E(A,\kappa)$ this diagram, depending on A and $\kappa$.

6.8. Corollary. The maps $Out_{\psi}(A,\kappa) \to Pic_{\psi}(A,\kappa)$ and $Out_{\psi,T}(A,\kappa) \to Pic_{\psi,T}(A,\kappa)$ are injective and yield a morphism $E(A,\kappa) \to D(A,\kappa)$ of commutative diagrams.

If $T = Z(A)$, then we write Outcent$(A,\kappa)$ for $Out_T(A,\kappa)$, $Picent_{\psi}(A,\kappa)$ for $Pic_{\psi,T}(A,\kappa)$, $Autcent_{\psi}(A,\kappa)$ for $Aut_{\psi,T}(A,\kappa)$ and $Outcent_{\psi}(A,\kappa)$ for $Out_{\psi,T}(A,\kappa)$. We then have :

6.9. Corollary. If $Z(A) \otimes S$ maps onto $Z(A \otimes S)$, then we have exact sequences :

$$U(Z(A)) \to U(Z(A(S))) \to Picent_{\psi}(A,\kappa) \to Picent(A,\kappa) \to Pic_{Z(A\otimes S)}(A \otimes S,\kappa)$$

resp.

$$U(Z(A)) \to U(Z(A(S))) \to Outcent_{\psi}(A,\kappa) \to Outcent(A,\kappa) \to Out_{Z(A\otimes S)}(A(S),\kappa)$$

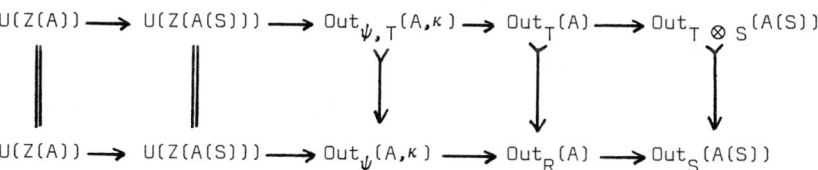

We will see below that the surjectivity condition in this corollary is rather mild and usually satisfied in the applications, we have in mind. The relative groups $Pic_{\psi}(A,\kappa)$ have an ideal-theoretic interpretation. Assume that $A \to A(S)$ is injective and view A as a subalgebra of A(S). We define multiplication of A-subbimodules V and W of A(S) by putting $V.W = Q_{\kappa}(VW)$, where VW is the usual product of V and W, consisting of all sums $\sum_i v_i w_i$, with $v_i \in V, w_i \in W$. In general this product is *not* associative. By abuse of language, we call V as above a $\kappa$-invertible ideal (of A in A(S)) if V is $\kappa$-flat and we may find another $\kappa$-flat A-subbimodule W of A(S), together with A-bimodule isomorphisms $V.W \cong A \cong W.V$. This implies that we may identify VW and WV with ideals in $L(\kappa)$. Note that the term "$\kappa$-invertible ideal" is meant as a single expression. It is made clear by the following, that there cannot result any confusion.

6.10. Lemma. Let V be a $\kappa$-invertible ideal of A in A(S), then V is a $\kappa$-invertible A-bimodule.

Proof. Let $V_1$ and $V_2$ be $\kappa$-flat subbimodules of A(S) and assume that $V_1$ is a $\kappa$-invertible ideal with inverse $W_1$, i.e. view $V_1W_1$ as an A-ideal in $L(\kappa)$. Consider the multiplication map $m : V_1 \underset{A}{\otimes} V_2 \to V_1V_2$, which is an A-bimodule morphism, which is surjective as such. If $\sum_i v_i' \otimes v_i'' \in \mathrm{Ker}(m)$, for some $v_i' \in V_1$ and $v_i'' \in V_2$, then $\sum_i v_i' v_i'' = 0$ in A(S).

Now, $(V_1W_1)(\Sigma v_i' \otimes v_i'') \subset \Sigma V_1 \otimes W_1 v_i' v_i'' = 0$, so $\sum_i v_i' \otimes v_i' \in \kappa(V_1 \underset{A}{\otimes} V_2)$, i.e. Ker(m) is $\kappa$-torsion and $Q_\kappa(m) : V_1 \underset{A}{\perp} V_2 \to V_1.V_2$ is an isomorphism of A-bimodules. Choosing $V_2 = W_1$ proves the assertion. □

The set of all $\kappa$-invertible ideals, which are $\kappa$-closed forms a group under the product; this group is denoted by $I_\psi(A,\kappa)$ or $I(A,\kappa)$ when no ambiguity arises. Moreover, one easily sees that each element in $I_\psi(A,\kappa)$ is defined over R and we may define a nice grouphomomorphism $I_\psi(A,\kappa) \to \mathrm{Pic}_R(A,\kappa)$ by associating $[V] \in \mathrm{Pic}_R(A,\kappa)$ to $V \in I_\psi(A,\kappa)$.
Actually, more is true.

6.11. Proposition. With the above notations and assumptions, associating to any $V \in I_\psi(A,\kappa)$ the pair $[V,f_V]$ where $f_V: V(S) \to A(S)$ is induced by $V \subset A(S)$ defines an isomorphism $I_\psi(A,\kappa) \xrightarrow{\sim} \mathrm{Pic}_\psi(A,\kappa)$, and the following diagram is commutative :

$$\begin{array}{ccccc} U(Z(A(S))) & \longrightarrow & \mathcal{I}_{\psi}(A,\kappa) & \longrightarrow & \mathrm{Pic}_R(A,\kappa) \\ & \searrow & & \nearrow & \\ & & \mathrm{Pic}_{\psi}(A,\kappa) & & \end{array}$$

<u>Proof</u>. The map $f_V : V(S) \to A(S)$ is induced by $f'_V : V \underset{A}{\otimes} S \cong V \otimes (A \otimes S) \to A(S)$ : $v \otimes b \mapsto vb$, or equivalently, by $\alpha_V : V \underset{A}{\otimes} A(S) \to A(S)$: $v \otimes b \mapsto vb$, since of course $Q_\kappa(f'_V) = Q_\kappa(\alpha_V)$, by the $\kappa$-flatness of V. Choose an "inverse" W for V in A(S), and, view VW = I as an ideal in $L(\kappa)$.

Consider $b \in A(S)$ and pick $i \in I$. Then i may be written as a sum $\sum_i v_i w_i$ for certain $v_i \in V$, $w_i \in W$, so $ib = \alpha_V(\Sigma\, v_i \otimes w_i b)$, i.e. $ib \in \mathrm{Im}(\alpha_V)$ and $Ib \subset \mathrm{Im}(\alpha_V)$, since i is arbitrary in I, thus $\mathrm{Coker}(\alpha_V)$ is $\kappa$-torsion. On the other hand, assume that $\sum_i v_i \otimes b_i \in \mathrm{Ker}(\alpha_V)$ for certain $v_i \in V$ and $b_i \in A(S)$, then $\sum_i v_i b_i = 0$ in A(S) so $VW \sum_i v_i \otimes b_i \subset \Sigma\, V \otimes W v_i b_i = 0$, i.e. $\sum_i v_i \otimes b_i \in \kappa\ \mathrm{Ker}(\alpha_V)$ and $\mathrm{Ker}(\alpha_V)$ is $\kappa$-torsion too. It follows that $f_V$ is an isomorphism indeed !

So we have defined a map $\omega : \mathcal{I}_{\psi}(A,\kappa) \to \mathrm{Pic}_{\psi}(A,\kappa) : V \to [V, f_V]$, which we will presently show to be an isomorphism. If suffices to exhibit an inverse for $\omega$, so pick $[P,f] \in \mathrm{Pic}_{\psi}(A,\kappa)$ and assume that $[Q,g]$ is an inverse for $[P,f] \in \mathrm{Pic}_{\psi}(A,\kappa)$. Let "f(P)" be the image of the map $P \to P \otimes S \to P(S) \xrightarrow{f} A(S)$ and put $\omega'([P,f]) = Q_\kappa(f(P))$. This is a well-defined map $\mathrm{Pic}_{\psi}(A,\kappa) \to \mathcal{I}_{\psi}(A,\kappa)$, since f(P) and hence $\omega'([P,f])$ *only* depend upon the class of (P,f) and since by $\kappa$-flatness one has $\omega'([P,f]).\omega'([Q,g]) \cong A \cong \omega'([Q,g]).\omega'([P,f])$ .By construction we have $\omega'\omega = 1$. Let us check that $\omega\,\omega' = 1$ as well.

Pick $[P] \in \mathrm{Pic}_{\psi}(A,\kappa)$. In order to have $P \cong Q_\kappa(f(P))$, it is sufficient to prove that $P \to P(S)$ is injective. But, this is obvious, since the injectivity of $A \to A(S)$ entails that the kernel of $A \to A \otimes S$ is $\kappa$-torsion.

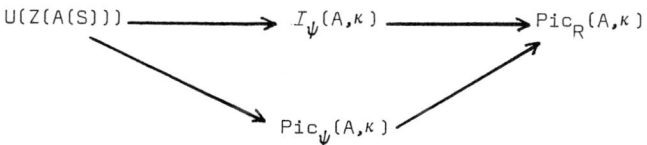

As P is $\kappa$-flat, it follows that the kernel of $P = P \otimes_A A \to P \otimes_A (A \otimes S) = P \otimes S$ is $\kappa$-torsion too, proving the assertion. □

6.12. Corollary. (of the proof). Let V be an A-subbimodule of A(S), then the following are equivalent :

1. V is a $\kappa$-invertible ideal;
2. V is a $\kappa$-invertible A-bimodule. □

A similar result may be derived for automorphisms, etc. as well. Indeed, define $N_\psi(A,\kappa)$ or $N(A,\kappa)$ to be the subgroup of U(A(S)) consisting of all u such that $uAu^{-1} = A$. We may then prove exactly as above that $N_\psi(A,\kappa)$ and $\mathrm{Aut}_\psi(A,\kappa)$ are isomorphic, by sending $u \in N_\psi(A,\kappa)$ to $(u,f) \in \mathrm{Aut}_\psi(A,\kappa)$, where f is the automorphism $a \mapsto u\,a\,u^{-1}$ of A. This yields an isomorphism $N_\psi(A,\kappa)/U(A) \cong \mathrm{Out}_\psi(A,\kappa)$. It is also clear that associating Au = uA to $u \in N_\psi(A,\kappa)$ defines a homomorphism $N_\psi(A,\kappa) \to I_\psi(A,\kappa)$, with kernel U(A), making the following diagram commutative :

$$\begin{array}{ccc} N_\psi(A,\kappa) & \longrightarrow & \mathrm{Aut}_\psi(A,\kappa) \\ \downarrow & & \downarrow \\ I_\psi(A,\kappa) & \longrightarrow & \mathrm{Pic}_\psi(A,\kappa) \end{array}$$

Remark. It is perhaps not completely satisfactory to impose the $\kappa$-flatness condition in the definition of $\kappa$-invertible ideals V of A in A(S). Indeed, it would have been nicer, if $\kappa$-flatness followed from the condition $V.W \cong A \cong W.V$. Anyway, the above proofs show that the latter condition implies that $V \perp W \cong A \cong W \perp V$, so if $\kappa$ is induced from R-mod, then $\kappa$-flatness follows automatically, as we have pointed out before.

Let us now reconsider the exact sequences defined in (6.5.) and (6.7.), but do not assume A to be $\kappa$-closed, here.

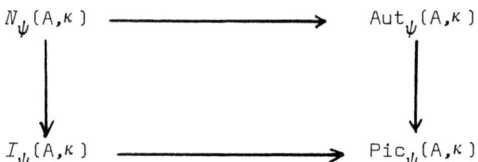

Recall from (5.12.) that there is a canonical group homomorphism $\alpha : \mathrm{Pic}(A,\kappa) \to \mathrm{Aut}(Z(Q_\kappa(A)))$. It is defined as follows : if $z \in Z(Q_\kappa(A))$ and if $[P] \in \mathrm{Pic}(A,\kappa)$, then we put $f : P \to P : p \to zp$ and one easily verifies that there is a unique $\alpha_P \in \mathrm{Aut}(Z(Q_\kappa(A)))$ such that $f(p) = zp = p\alpha_P(z)$ for all $p \in P$. We put $\alpha([P]) = \alpha_P$ and thus obtain a well-defined exact sequence of groups

$$1 \to \mathrm{Picent}(A,\kappa) \to \mathrm{Pic}_R(A,\kappa) \to \mathrm{Aut}_R(Z(Q_\kappa(A))).$$

Now, assume that $Z(Q_\kappa(A))$ maps *into* $Z(A(S))$ and that $V = \mathrm{Aut}_{Z(Q_\kappa(A)) \otimes S}(Z(Q_\kappa(A \otimes S))) = 1$, e.g. $Z(Q_\kappa(A \otimes S)) = Q_\kappa(Z(Q_\kappa(A)) \otimes S)$, then we claim that $\mathrm{Picent}_\psi(A,\kappa) = \mathrm{Pic}_\psi(A,\kappa)$.

Indeed, $V = 1$ implies that $\mathrm{Picent}(A \otimes S,\kappa) = \mathrm{Pic}_{Z(Q_\kappa(A)) \otimes S}(A \otimes S,\kappa)$, so we obtain a commutative exact diagram

$$\begin{array}{ccccccc} 1 \to & \mathrm{Picent}(A,\kappa) & \longrightarrow & \mathrm{Pic}_R(A,\kappa) & \longrightarrow & \mathrm{Aut}_R(Z(Q_\kappa(A))) \\ & \downarrow & & \downarrow & & \downarrow \\ 1 \to & \mathrm{Picent}(A \otimes S,\kappa) & \to & \mathrm{Pic}_S(A \otimes S,\kappa) & \longrightarrow & \mathrm{Aut}_S(Z(A(S))) \end{array}$$

where the last vertical map is injective by our assumption. It follows that the first two vertical maps have isomorphic kernels, hence that $\mathrm{Picent}_\psi(A,\kappa) = \mathrm{Pic}_\psi(A,\kappa)$.

6.13. Corollary. Assume that the following conditions are satisfied :

1. $Z(Q_\kappa(A))$ maps into $Z(A(S))$;
2. $\mathrm{Aut}_{Z(Q_\kappa(A)) \otimes S}(Z(Q_\kappa(A \otimes S))) = 1$;
3. $\mathrm{Picent}(A \otimes S,\kappa) = 1$,

then the following sequences are exact :

a. $1 \longrightarrow \mathrm{Picent}(A,\kappa) \longrightarrow \mathrm{Pic}_R(A,\kappa) \longrightarrow \mathrm{Pic}_S(A \otimes S,\kappa)$;

b. $U(Z(A(S))) \longrightarrow \mathrm{Pic}_\psi(A,\kappa) \longrightarrow \mathrm{Picent}(A,\kappa) \longrightarrow 1$;

c. $1 \longrightarrow U(Z(A(S))) \longrightarrow \mathrm{Aut}_{\psi}(A,\kappa) \longrightarrow \mathrm{Autcent}(A,\kappa) = \mathrm{Autcent}(A)$;

d. $1 \longrightarrow U(Z(A(S)))/U(Z(A)) \longrightarrow \mathrm{Aut}_{\psi}(A,\kappa)/U(A) \longrightarrow \mathrm{Outcent}(A,\kappa) = \mathrm{Outcent}(A)$

Proof. It will be sufficient to give a proof of the first two statements, i.e. exactness of a. and b.

Since $\mathrm{Picent}(A \otimes S,\kappa) = 1$ and $\mathrm{Aut}_{Z(Q_{\kappa}(A)) \otimes S}(Z(Q_{\kappa}(A \otimes S))) = 1$, it follows that $\mathrm{Pic}_{Z(Q_{\kappa}(A)) \otimes S}(A(S),\kappa) = 1$, so from the results above, we may deduce that $\mathrm{Picent}_{\psi}(A,\kappa) = \mathrm{Pic}_{\psi}(A,\kappa) \to \mathrm{Picent}(A,\kappa)$ is surjective. For each $R \to T$ we then have a commutative diagram

$$\begin{array}{ccc} \mathrm{Pic}_{\psi,T}(A,\kappa) & \longrightarrow & \mathrm{Pic}_{T}(A,\kappa) \\ \downarrow & & \downarrow \\ \mathrm{Pic}_{\psi}(A,\kappa) & \longrightarrow & \mathrm{Pic}_{R}(A,\kappa) \end{array}$$

so for $T = Z(Q_{\kappa}(A))$, it follows from the fact that $\mathrm{Pic}_{\psi,T}(A,\kappa) = \mathrm{Pic}_{\psi}(A,\kappa)$ that the map $\mathrm{Pic}_{\psi}(A,\kappa) \to \mathrm{Pic}_{R}(A,\kappa)$ factorizes through the injective map $\mathrm{Picent}(A,\kappa) \to \mathrm{Pic}_{R}(A,\kappa)$. Now, we have a long exact sequence

$$\begin{array}{ccccccccc} U(Z(Q_{\kappa}(A))) & \to & U(Z(A(S))) & \to & \mathrm{Pic}_{\psi}(A,\kappa) & & \to & \mathrm{Pic}_{R}(A,\kappa) & \to \mathrm{Pic}_{S}(A \otimes S,\kappa) \\ & & & & & \searrow & \nearrow & & \\ & & & & & \mathrm{Picent}(A,\kappa) & & & \\ & & & & \nearrow & & \searrow & & \\ & & & & 0 & & 0 & & \end{array}$$

This yields the assertion. □

Note that if $A \to A(S)$ is injective, then the foregoing exact sequences may be interpreted in terms of $I_{\psi}(A,\kappa)$ and $N_{\psi}(A,\kappa)$, e.g. the exact sequences b. and d. become :

$$U(Z(A(S))) \longrightarrow I_{\psi}(A,\kappa) \longrightarrow \mathrm{Picent}(A,\kappa) \longrightarrow 1$$

resp.

$$1 \longrightarrow U(Z(A(S)))/U(Z(A)) \longrightarrow N_{\psi}(A,\kappa)/U(A) \longrightarrow \mathrm{Outcent}(A).$$

Before considering specific applications, let us study the condition $\mathrm{Aut}_{Z(Q_\kappa(A))\otimes S}(Z(Q_\kappa(A\otimes S))) = \mathrm{Aut}_{Q_\kappa(Z(Q_\kappa(A))\otimes S)}(Z(Q_\kappa(A\otimes S))) = 1$. It is certainly satisfied if $Z(Q_\kappa(A)) \perp S = Z(A \perp S) = Z(Q_\kappa(A) \perp S)$. Let us start with the following, straightforward generalization of (5.2.) :

6.14. <u>Lemma.</u> Let A be an R-algebra which is $\kappa$-finitely generated as an R-module and let M be a $\kappa$-torsionfree A-bimodule defined over R. For any idempotent kernel functor $\lambda \geqslant \kappa$ in R-mod, we have $Q_\lambda(M^A) = Q_\lambda(M)^{Q_\lambda(A)}$.

<u>Proof</u>. Clearly, $Q_\lambda(M^A) \subset Q_\lambda(M)^{Q_\lambda(A)}$. Conversely, let $m \in Q_\lambda(M)^{Q_\lambda(A)}$, then for some $I \in \mathcal{L}(\lambda)$, we have $Im \subset (M/\lambda M)^A$. Let $B \subset A$ be a finitely generated R-module, such that A/B is $\kappa$-torsion. For all $i \in I$ we may find $n_i \in M$ such that $im = \overline{n}_i$, the image of $n_i$ in $M/\lambda M$, hence for all $b \in B$ we have $n_i b - bn_i \in \lambda M$, so we may find a single $J \in \mathcal{L}(\lambda)$ such that for <u>all</u> $b \in B$ we have $(jn_i)b = b(jn_i)$, for any $j \in J$. It follows that $Jn_i \subset M^B$, hence $\overline{n}_i \in Q_\lambda(M^B)$, so $Im \subset Q_\lambda(M^B)$ and finally $m \in Q_\lambda(M^B)$. As M is $\kappa$-torsionfree and A/B is $\kappa$-torsion, one verifies exactly as in (5.2.) that $M^B = M^A$. □

6.15. <u>Corollary</u>. Let A be an R-algebra which is $\kappa$-finitely generated as an R-module and let M be a $\kappa$-torsion free A-bimodule defined over R. If $\kappa$ has finite type and $R \to S$ is a $\kappa$-flat extension of commutative rings, then $M^A \perp S = (M \perp S)^{A \perp S}$.

<u>Proof</u>. Since $R \to S$ is $\kappa$-flat, we have an inclusion $M^A \perp S \to M \perp S$ which factorizes through $M^A \perp S \to (M \perp S)^{A \perp S}$. It suffices to show that this inclusion is bijective at each $p \in \mathcal{C}(\kappa)$ (since $(M \perp S)^{A \perp S}$ is $\kappa$-closed!). Now, by (5.2.) or the foregoing result, we have that $(M^A \perp S)_p = M_p^{A_p} \otimes S_p$ for each $p \in \mathcal{C}(\kappa)$. On the other hand, the same argument yields that $((M \perp S)^{A \perp S})_p = (M \perp S)_p^{(A \perp S)_p} = (M_p \otimes S_p)^{A_p \otimes S_p}$. Now

$A_p$ is finitely generated over $R_p$ and $R_p \to S_p$ is flat, hence $M_p^{A_p} \otimes S_p = (M_p \otimes S_p)^{A_p \otimes S_p}$ by the lemma below. □

6.16. Lemma. (A. Fröhlich [52]). Let A be an algebra over a commutative ring S, finitely generated as S-module, and let T be a commutative flat S-algebra, then for any A-bimodule M we have $M^A \underset{S}{\otimes} T \cong (M \underset{S}{\otimes} T)^{A \underset{S}{\otimes} T}$.

Proof. Consider the exact sequence $0 \to \mathrm{Ker}(f) \to A^e = A \underset{S}{\otimes} A^{opp} \xrightarrow{f} A \to 0$ where $f(a \otimes b) = ab$. If $\{a_i\}$ is a set of generators of A over S, then $\{a_i \otimes 1 - 1 \otimes a_i\}$ is a set of generators of Ker(f). Hence A is a finitely presented $A^e$-module, so $\mathrm{Hom}_{A^e}(A,M) \underset{S}{\otimes} T \cong \mathrm{Hom}_{A^e \underset{S}{\otimes} T}(A \underset{S}{\otimes} T, M \underset{S}{\otimes} T)$. □

6.17. Corollary. Let A be a $\kappa$-torsion free algebra, which is $\kappa$-finitely generated as an R-module, then $Z(A) \perp S = Z(A \perp S)$ for all $\kappa$-flat extensions $R \to S$. In particular, for any $\kappa$-finitely generated R-algebra A, we have $Z(Q_\kappa(A)) \perp S = Z(Q_\kappa(A \otimes S))$. □

Surjectivity of the map $Z(A) \perp S \to Z(A \perp S)$ is a phenomenon which occurs more frequently than surjectivity of the map $Z(A) \otimes S \to Z(A \otimes S)$. Yet, in order to derive the characteristic exact sequences above, we do need the latter condition in general, due to the fact that for arbitrary $\kappa$ we do not have $\mathrm{Pic}_{Z(A)}(A,\kappa) = \mathrm{Pic}_{Z(Q_\kappa(A))}(A,\kappa)$. Note that Picent $(A,\kappa)$ is defined intrinsically in terms of $(A,\kappa)$-mod. However, if $\kappa$ is central, then $\mathrm{Pic}_{Z(A)}(A,\kappa) = \mathrm{Picent}(A,\kappa)$, even if A is not $\kappa$-closed. We thus have :

6.18. Corollary. Let A be a $\kappa$-finitely generated R-algebra and let S be a $\kappa$-flat commutative overring of R, then the following sequence of groups is exact :

$U(Z(Q_\kappa(A))) \to U(Z(A(S))) \to \mathrm{Picent}_\psi(A,\kappa) \to \mathrm{Picent}(A,\kappa) \to \mathrm{Picent}(A \otimes S,\kappa)$. □

6.19. Corollary. Let $\lambda \geqslant \kappa$ in R-mod and assume that $\lambda$ induces a perfect localization. Denote by $\psi : R \to Q_\lambda(R)$ the canonical morphism and write $\mathrm{Pic}_\lambda(A,\kappa)$, $\mathrm{Picent}_\lambda(A,\kappa), \dots$ for $\mathrm{Pic}_\psi(A,\kappa)$, $\mathrm{Picent}_\lambda(A,\kappa), \dots$ . We then have :

1. for any algebra A over the commutative R-algebra T, there is an exact sequence :

$$U(Z(Q_\kappa(A))) \to U(Z(Q_\lambda(A))) \to \mathrm{Picent}_{\lambda,T}(A,\kappa) \to \mathrm{Pic}_T(A,\kappa) \to \mathrm{Pic}_{Q_\lambda(T)}(Q_\lambda(A)),$$

the first map being injective, if A is $\lambda$-torsion free ;

2. if A is $\lambda$-torsion free, then there is an exact sequence :

$$1 \to U(Z(Q_\kappa(A))) \to U(Z(Q_\lambda(A))) \to \mathrm{Picent}_\lambda(A,\kappa) \to \mathrm{Picent}(A,\kappa) \to \mathrm{Picent}(Q_\lambda(A))$$

and $\mathrm{Picent}_\lambda(A,\kappa) = \mathrm{Pic}_\lambda(A,\kappa) = \mathrm{Ker}(\mathrm{Pic}_R(A,\kappa) \to \mathrm{Pic}_R(A,\lambda))$.

Proof. First note that if $M \in Q_\lambda(A)$-mod, then $\lambda M = 0$ as $\lambda$ induces a perfect localization, so $\kappa M = 0$ as well and $\kappa$ is trivial in $Q_\lambda(A)$-mod. This immediately yields 1. : in particular we then have $\mathrm{Pic}(Q_\lambda(A),\kappa) = \mathrm{Pic}(Q_\lambda(A))$. If $\lambda A = 0$, then $Q_\kappa(A) \subset Q_\lambda(A)$, so certainly $Z(Q_\kappa(A)) \subset Z(Q_\lambda(A)) = Z(Q_\kappa(Q_\lambda(A)))$ and $Q_\lambda(Z(Q_\kappa(A))) = Z(Q_\kappa(A))$. Since $Z(Q_\kappa(A)) \to Q_\lambda(Z(Q_\kappa(A)))$ is epimorphic, this yields the exactness of the second sequence. Finally if $[P,f] \in \mathrm{Pic}_\lambda(A,\kappa)$, then by definition $[P] \in \mathrm{Pic}_R(A,\kappa)$ and f is an isomorphism of A-bimodules between $P(S) = Q_\kappa(P \otimes Q_\lambda(A)) = Q_\lambda(P)$ and $A(S) = Q_\lambda(A)$, i.e. $[P]$ lies in the kernel of the canonical map $\mathrm{Pic}_R(A,\kappa) \to \mathrm{Pic}_R(A,\lambda)$ defined in (6.3.). The converse is obvious. □

6.20. Example. Choose $\lambda$ such that A is $\lambda$-torsion free and $\mathrm{Picent}(Q_\lambda(A)) = 1$, e.g. $Q_\lambda(A)$ is an Azumaya algebra over a local domain, then from the foregoing we may derive exact sequences

$$1 \to \mathrm{Picent}(A,\kappa) \to \mathrm{Pic}_R(A,\kappa) \to \mathrm{Pic}_{Q_\lambda(R)}(Q_\lambda(A))$$

and

$$U(Z(Q_\lambda(A))) \to \mathcal{I}_\lambda(A,\kappa) \to \mathrm{Picent}(A,\kappa) \to 1$$

where $\mathcal{I}_\lambda(A,\kappa)$ is the group of all $\kappa$-flat A-bimodules V in $Q_\lambda(A)$ with the property that for some similar W we have $Q_\kappa(VW) \cong Q_\kappa(A) \cong Q_\kappa(A) \cong Q_\kappa(WV)$ For example, if R is an integral domain with field of fractions K, then we may choose $\lambda$ associated to the multiplicative system R-{0}. If A is finitely generated and torsion free as an R-module, and if $A \otimes_R K$ is semisimple, then we recover a result of A. Fröhlich's by choosing $\kappa = \hat{o}$ : there are exact sequences

$$U(Z(A \otimes K)) \to I(A) \to \mathrm{Picent}(A) \to 1$$

and

$$1 \to \mathrm{Picent}(A) \to \mathrm{Pic}_R(A) \to \mathrm{Pic}_K(A \otimes_R K),$$

where I(A) is the group of invertible ideals of A in $A \otimes K$.

<u>Assume from now on that the center C is a $\kappa$-noetherian R-algebra.</u>

6.21. <u>Lemma</u>. (A. Fröhlich [ 52 ] ). If [ M ] ∈ Picent(Z(B)), then $M \cong (M \otimes_{Z(A)} B)^B$ for any ring B.

<u>Proof</u>. The canonical map $M \to (M \otimes_{Z(A)} A)^A : x \mapsto x \otimes 1$ is an isomorphism for M = Z(A), hence for all finitely generated projective Z(A)-modules M. □

6.22. <u>Lemma</u>. Let A be $\kappa$-finitely generated and $\kappa$-torsion free over its center C and let [ P ] ∈ Picent(C,$\kappa$), then the C-modules P and $(P \perp_C A)^A$ are isomorphic.

<u>Proof</u>. Let $p \in \mathcal{C}(C,\kappa)$, then $(P \perp_C A)^A \otimes_C C_p = (P \perp_C A)_p^{A_p}$ by (6.15). But, $(P \perp_C A)_p^{A_p} = (P_p \otimes_{C_p} A_p)^{A_p}$, and $C_p = Z(A_p)$ by (6.14.) so it follows that $(P_p \otimes_{C_p} A_p)^{A_p} = P_p$ by the foregoing lemma. □

Let $T_\kappa$ : Picent(C,$\kappa$) → Picent(A,$\kappa$) be defined by sending [ P ] to $[P \perp_C A]$. This map is well defined, compatible with $\kappa$-Morita equivalence over C, and if E is a $\kappa$-Azumaya algebra over C, then we obtain a commutative diagram

$$\begin{array}{ccc} & & \mathrm{Picent}(A,\kappa) \\ & \nearrow & \downarrow \\ \mathrm{Picent}(C,\kappa) & & \\ & \searrow & \mathrm{Picent}(A \perp_C E,\kappa) \end{array}$$

We claim that if A is as in the foregoing lemma, then for $[Q] \in \mathrm{Pic}(A,\kappa)$, we have that $[Q] \in \mathrm{Im}(T_\kappa)$ if and only if the following conditions are satisfied :

1. $[Q^A] \in \mathrm{Picent}(C,\kappa)$;
2. $\omega_Q : Q^A \perp_A A \to Q$ is an isomorphism.

Indeed, if $[Q] \in \mathrm{Im}(T_\kappa)$, then $[Q] = [P \perp_C A]$ for some $[P] \in \mathrm{Picent}(C,\kappa)$, so $Q^A \cong (P \perp_C A)^A \cong P$ by the lemma and $\omega_Q$ is of course an isomorphism. The converse is obvious !

6.23. <u>Lemma.</u> If A is $\kappa$-finitely generated and $\kappa$-torsion free over C, then $T_\kappa$ is injective and $\mathrm{Im}(T_\kappa) \subset Z(\mathrm{Picent}(A,\kappa))$.

<u>Proof</u>. If $[P_1]$, $[P_2] \in \mathrm{Picent}(C,\kappa)$ and $T_\kappa([P_1]) = T_\kappa([P_2])$, then $P_1 \perp_C A \cong P_2 \perp_C A$, so $P_1 \cong (P_1 \perp_C A)^A \cong (P_2 \perp_C A)^A \cong P_2$ proving injectivity. Moreover, if $M \in C\text{-mod}$ and N is an A-bimodule over C, then $M \perp_C N \cong N \perp_C M$, so, since $(M \perp_C A) \perp_A N \cong M \perp_C N$ and similarly on the other side, $\mathrm{Im}(T_\kappa)$ is central indeed. □

We are now able to provide a local-global description of $\mathrm{Im}(T_\kappa)$. First, note that for every $p \in \mathcal{C}(\kappa)$ we have a morphism

$$\mathrm{Pic}_R(A,\kappa) \longrightarrow \mathrm{Pic}_{R_p}(A_p,\kappa) = \mathrm{Pic}_{R_p}(A_p) = \mathrm{Pic}_R(A_p),$$

compatible with Morita equivalence. Let us denote by $\xi$ or by $\xi_R(A,\kappa)$, if we want to be precise, the induced map

$$\xi : \mathrm{Pic}_R(A,\kappa) \longrightarrow \prod_{p \in \mathcal{C}(\kappa)} \mathrm{Pic}_R(A_p).$$

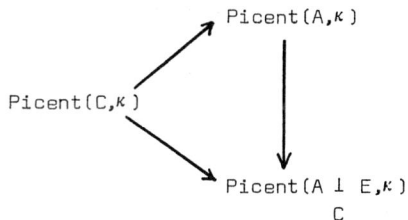

It is clear that $\mathrm{Ker}(\xi) \subset \mathrm{Picent}(A,\kappa)$. Indeed, let $[P] \in \mathrm{Ker}(\xi)$, pick $c \in C$ and consider the morphism $f_c : P \to P : p \to pc - cp$. Clearly $f_c$ is A-linear, hence for $p \in \mathcal{C}(\kappa)$, we have a local morphism $f_{c,p} : P_p \to P_p$. Now, $Z(A)_p = C_p \subset Z(A_p)$ and $P_p \cong A_p$-bimodules, so $f_{c,p} = 0$. Since this holds for all $p \in \mathcal{C}(\kappa)$ and since P is $\kappa$-closed, we find that $f_c = 0$, i.e. $[P] \in \mathrm{Picent}(A,\kappa)$.

Let us now compare the subgroups $\mathrm{Im}(T_\kappa)$ and $\mathrm{Ker}(\xi)$ within $\mathrm{Picent}(A,\kappa)$.

6.24. Lemma. If for all $p \in \mathcal{C}(\kappa)$ the ring $C_p = C \otimes_R R_p$ is semi local, then $\mathrm{Im}(T_\kappa) \subset \mathrm{Ker}(\xi)$.

Proof. For all $p \in \mathcal{C}(\kappa)$, there is a commutative diagram

$$\begin{array}{ccc} \mathrm{Picent}(C,\kappa) & \xrightarrow{\;T_\kappa\;} & \mathrm{Picent}(A,\kappa) \\ \downarrow & & \downarrow \\ \mathrm{Picent}(C_p) & \longrightarrow & \mathrm{Pic}_R(A_p) \end{array}$$

But, $\mathrm{Picent}(C_p) = 0$ by assumption, so $\mathrm{Im}(T_\kappa) \subset \mathrm{Ker}(\xi)$. □

6.25. Lemma. Assume that A is $\kappa$-finitely generated and $\kappa$-torsion free over C, then $\mathrm{Ker}(\xi) \subset \mathrm{Im}(T_\kappa)$.

Proof. Assume that $[P] \in \mathrm{Picent}(A,\kappa)$ is in the kernel of $\xi$ then for all $p \in \mathcal{C}(\kappa)$ there is an isomorphism $P_p \cong A_p$ as $A_p$-bimodules, hence, using (6.15.) we obtain that $(P^A)_p = P_p^{A_p} = C_p$. Moreover, as A is $\kappa$-finitely generated over C an C is $\kappa$-noetherian, P and hence $P^A$ is $\kappa$-finitely presented over C by (4.4.) and (4.11.). From (4.15.) it thus follows that $P^A$ is a $\kappa$-invertible C-bimodule, i.e. $[P^A] \in \mathrm{Picent}(C,\kappa)$. Now, in order to show that $[P] \in \mathrm{Im}(T_\kappa)$, it suffices to check that $\omega_P : P^A \perp_A A \to P$ is an isomorphism. But, as it is an isomorphism at each $p \in \mathcal{C}(\kappa)$, since $(P^A)_p = P_p^{A_p}$, the assertion follows. □

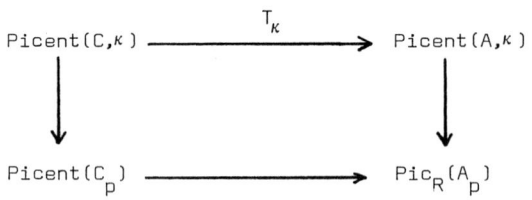

6.26. Corollary. If $C_p$ is semilocal for all $p \in C(\kappa)$ and A is $\kappa$-finitely generated and $\kappa$-torsion free over C, then there are exact sequences

$$1 \to \mathrm{Picent}(C,\kappa) \to \mathrm{Pic}_R(A,\kappa) \to \prod_{p \in C(\kappa)} \mathrm{Pic}_R(A_p)$$

resp.

$$1 \to \mathrm{Picent}(C,\kappa) \to \mathrm{Picent}(A,\kappa) \to \prod_{p \in C(\kappa)} \mathrm{Picent}(A_p)$$

Proof. From the foregoing lemmas, it follows that the sequence

$$1 \to \mathrm{Picent}(C,\kappa) \xrightarrow{T_\kappa} \mathrm{Picent}(A,\kappa) \xrightarrow{\xi} \prod_{p \in C(\kappa)} \mathrm{Pic}_R(A_p)$$

is exact. The exactness of the first sequence above follows from $\mathrm{Picent}(A,\kappa) \subset \mathrm{Ker}(\xi : \mathrm{Pic}_R(A,\kappa) \to \prod \mathrm{Pic}_R(A_p))$. Moreover the second sequence is also exact, since $C_p = Z(A_p)$ for all $p \in C(\kappa)$. □

Let us assume that A is $\kappa$-finitely generated and torsion free over a central noetherian domain R. We say that R has $\kappa$-dimension. n if $\sup\{ht(p);\ p \in C(\kappa)\} = n$. Assume that $A \perp K = A \otimes K$ is semisimple and let $I(A,\kappa)$ be the group of all A-bimodules V of $A \perp K$ which are $\kappa$-closed and have the property that for some similar W we have $V.W = Q_\kappa(A) = W.V$, where $V.V' = Q_\kappa(VV') \subset A \perp K$ by definition. The group law is given by . ; as we have pointed out before, the objects in $I(A,\kappa)$ are automatically $\kappa$-flat. For $p \in C(\kappa)$ we also define $I(A_p)$ as $I(A_p,\kappa)$ and it is clear that $I(A_p)$ consists of all $A_p$-subbimodules V of $A \perp K$ with $VW = A_p = WV$ for some similar W. There is a map $I(A,\kappa) \to \prod_{p \in C(\kappa)} I(A_p)$.

6.27. Proposition. With the above assumptions, if the $\kappa$-dimension of R is 1, then there is an isomorphism of groups $I(A,\kappa) \cong \coprod_{p \in C(\kappa)} I(A_p)$.

Proof. Our assumptions imply that A may be identified with a subring of $A \perp K = \Gamma$ and that V,... is $\kappa$-finitely presented. Let us show that $V_p \cong A_p$ for almost all p in $C(\kappa)$. Indeed, pick $V' \subset V$ finitely generated over A such that $V/V'$ is $\kappa$-torsion. In particular, we thus have $V' \otimes K = V \otimes K = \Gamma$, i.e. $1 \in \Gamma$ may be written as a sum $\Sigma v_i f^{-1}$ for some $f \in R-\{0\}$. It follows that $1 \in V' \otimes R_f$, so $V' \otimes R_f \cong A \otimes R_f$. In particular, for all $p \in C(\kappa)$ not containing $f$, which is the complement of a finite set in $C(\kappa)$, we have $V_p \cong A_p$. This shows that $I(A,\kappa)$ injects into $\bigsqcup_{p \in C(\kappa)} I(A_p)$. Now, everything being torsion free, we find that $V = \bigcap_{p \in C(\kappa)} V_p \subset V \otimes K = \Gamma$.

So, let there be given $V(p) \in I(A_p)$ for all $p \in C(\kappa)$ and assume that $V(p) = A_p$ for almost all $p \in C(\kappa)$. Since A is $\kappa$-finitely generated, there is $A' \subset A$ finitely generated over R such that $A/A'$ is $\kappa$-torsion. In particular, $A' \otimes_R K = \Gamma$, i.e. $A'$ is an R-lattice in $\Gamma$. For all $p \in C(\kappa)$, we have $V(p) \in I(A_p)$, with $V(p) = A_p$ for almost all $p \in C(\kappa)$. In particular, $V(p)$ is an $R_p$-lattice for each $p \in C(\kappa)$. Upon replacing A by $x^{-1}A$ for some $x \neq 0$ in R, we may assume that $V(p) \subset A_p$ for all $p \in C(\kappa)$ and still have $V(p) = A_p$ for almost all $p \in C(\kappa)$. Let $\{p_1,\dots,p_k\}$ denote the exceptional primes in $C(\kappa)$, and let $Q = A \cap V(p_1) \cap \dots \cap V(p_k) \subset A$, then Q is $\kappa$-finitely generated since each factor is an R-submodule of A, hence Q is an R-submodule of A and A is $\kappa$-finitely generated over R.

It is also clear that $Q \otimes K = \Gamma$. Now, recall that if p and p' are prime ideals of R, with (0) being the only prime of R contained in $p \cap p'$, then for all R-modules $E \subset K$ we have $(E_p)_{p'} = KE$. It now easily follows, that $Q_p = V(p)$ for all $C(\kappa)$. Indeed, if $p \neq p_i$ $(i = 1,\dots,k)$ for $p \in C(\kappa)$, then we have $V(p_i)_p = \Gamma$, so $Q_p = A_p \cap \Gamma \cap \dots \cap \Gamma = A_p = V(p)$. On the other hand if $p = p_i$ for some $1 \leq i \leq k$, then $Q_{p_i} = A_{p_i} \cap V(p_1)_{p_i} \cap \dots \cap V(p_i)_{p_i} \cap \dots \cap V(p_k)_{p_i} = A_{p_i} \cap \Gamma \cap \dots \cap V(p_i) \cap \dots \cap \Gamma = V(p_i)$. So, if we put $V = Q_\kappa(Q)$, then V is $\kappa$-closed, $\kappa$-finitely generated (hence $\kappa$-finitely presented) and $V_p = V(p)$ for all $p \in C(\kappa)$. Finally, $V \in I(A,\kappa)$! Indeed, for each $V(p)$, pick $W(p) \in I(A_p)$ such that $V(p)\,W(p) = A_p = W(p)\,V(p) \subset \Gamma$. Do the same reasoning with $W(p)$, i.e. find a $\kappa$-finitely generated $W \subset \Gamma$ with $W_p = W(p)$

for all $p \in C(\kappa)$, then within $\Gamma$ we have $Q_\kappa(WV) = \bigcap_{p \in C(\kappa)} (WV)_p =$
$\bigcap_{p \in C(\kappa)} W_p V_p = \bigcap_{p \in C(\kappa)} W(p)V(p) = \bigcap_{p \in C(\kappa)} A_p = Q_\kappa(A)$ and so $W.V = Q_\kappa(A)$.
One proves similarly that $V.W = Q_\kappa(A)$, which finishes the proof. □

Exercise. Prove that (6.27.) and the following results remain valid if one replaces the assumption "R is noetherian" by "R is $\kappa$-noetherian".

6.28. Proposition. Under the same assumptions, $\text{Picent}(A_p) = 1$ for almost all $p \in C(\kappa)$.

Proof. As before, let us write $A \otimes_R K$ and if B is a ring, let us write $B^e$ for $B \otimes_{Z(B)} B^{opp}$. If B is an R-algebra, then we write $B^f$ for $Q_\kappa(B^e)$.
Our assumptions imply A to be a $\kappa$-finitely presented $A^e$-module, i.e. there is a map $\psi : A' \to A$ in $A^e$-mod, where A' is finitely presented and $\text{Ker}(\psi)$ and $\text{Coker}(\psi)$ are $\kappa$-torsion. Consider the canonical map $\theta : {}_{A^e}[A', A^f] \to {}_{A^e}[A', A]$. This commutes with flat ring extensions, so in particular for any flat $C \to D$ we obtain an induced map
$\theta \otimes_C D : {}_{A^e \otimes_C D}[A' \otimes_C D, A^f \otimes_C D] \to {}_{A^e \otimes_C D}[A' \otimes_C D, A \otimes_C D]$. Note also that
$A^e \otimes_C D = (A \otimes_C D) \otimes_D (A \otimes_C D)^{opp}$. Now, recall that if $A$ is an algebra over a ring $R$, then $A$ is separable over $R$ if and only if for $A^\star = A \otimes_R A^{opp}$, the map $(A^\star)^A \to A^A$ is surjective, this map being just ${}_{A^\star}[A, A^\star] \to {}_{A^\star}[A, A]$.
So, taking $D = C \otimes_R K$, which is certainly flat over C, we get $\theta \otimes_C D = \theta \otimes_R K$, which *is* surjective. Indeed, $A^e \otimes_R K = \Gamma \otimes_D \Gamma^{opp}$ and $D = C \otimes_R K = Z(\Gamma)$ as K is flat over R, so $A^e \otimes_R K = \Gamma^e$. Similarly, $A' \otimes_C D = \Gamma$ and $A^f \otimes_R K = A^e \otimes_R K = \Gamma^e$. With $D = C \otimes_R K$ we see that $\theta \otimes_R K$ is just the canonical map $(\Gamma^e)^\Gamma \to \Gamma^\Gamma = Z(\Gamma)$, which is surjective as $\Gamma$ is central separable.

Let us now conclude by showing that the surjectivity of $\theta \underset{R}{\otimes} K$ implies that $\theta \underset{R}{\otimes} R_p$ is surjective for almost all $p \in \mathcal{C}(\kappa)$. Changing notations, we have that the C-linear map $\theta : M \to C$ induces a surjective $C \underset{R}{\otimes} K$-linear map $\theta \underset{R}{\otimes} K : M \underset{R}{\otimes} K \to C \underset{R}{\otimes} K$, i.e. $(\theta \underset{R}{\otimes} K)(\Sigma\, m_i \otimes \frac{r_i}{f}) = 1$ for some $m_i \in M$, $r_i, f \in R$, $f \neq 0$. But then, for all $p \not\ni f$ we have that $(\theta \underset{R}{\otimes} R_p)(\Sigma\, m_i \otimes \frac{r_i}{f}) = 1 \in C \underset{R}{\otimes} R_p$, i.e. for $D = C \underset{R}{\otimes} R_p = C_p$, the map $\theta \underset{C}{\otimes} D$ is surjective as well. Now, $R_p$ is flat over R, so for $p \in \mathcal{C}(\kappa)$, not containing f we find that $A_p$ is separable over $Z(A_p)$. We will finish the proof by showing that $Z(A_p)$ is semilocal, for then $\mathrm{Picent}(A_p) = \mathrm{Picent}(Z(A_p)) = 1$, indeed. Now, first note that with A' as above and with $D = A' \cap C$ (which is an R-module, but not a ring in general), we have that $C_p = D_p$, hence $C_p$ is finitely generated as a module over the local noetherian ring $R_p$, hence $C_p$ is semilocal ! □

6.29. <u>Corollary</u>. In the situation of (6.28 .), there is an exact sequence:

$$1 \longrightarrow \mathrm{Picent}(C,\kappa) \longrightarrow \mathrm{Picent}(A,\kappa) \longrightarrow \coprod_{p \in \mathcal{C}(\kappa)} \mathrm{Picent}(A_p) \longrightarrow 1$$

<u>Proof</u>. This follows immediately from the fact that there is an exact commutative diagram

$$\begin{array}{ccccc}
\mathcal{I}(A,\kappa) & \longrightarrow & \mathrm{Picent}(A,\kappa) & & \\
\downarrow & & \downarrow & & \\
\coprod_{p \in \mathcal{C}(\kappa)} (A_p) & \longrightarrow & \coprod_{p \in \mathcal{C}(\kappa)} \mathrm{Picent}(A_p) & \longrightarrow & 1 \\
\downarrow & & & & \\
1 & & & &
\end{array}$$

So, the right vertical map is surjective! □

Let us now give some elementary examples. We denote by $(X, \mathcal{O}_X)$ or X a locally noetherian scheme and by $\mathcal{A}$ a coherent sheaf of $\mathcal{O}_X$-algebras.

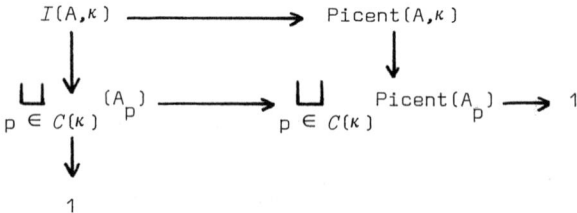

A typical example arises as follows : choose an arbitrary commutative noetherian ring R and let X be the affine scheme Spec(R). For any finite R-algebra A we then put $\mathcal{A} = \tilde{A}$. A coherent $\mathcal{A}$-Bimodule over $\mathcal{O}_X$ is then just a sheaf of $\mathcal{A}$-bimodules $\mathcal{M}$ which is coherent as an $\mathcal{O}_X$-module and which has the property that the left and right actions of $\mathcal{O}_X$ on $\mathcal{M}$ coincide. Such a sheaf $\mathcal{M}$ is said to be an invertible $\mathcal{A}$-bimodule over $\mathcal{O}_X$ if there exists a similar sheaf $\mathcal{N}$ and an $\mathcal{A}$-Bimodule isomorphism $\mathcal{M} \otimes_{\mathcal{A}} \mathcal{N} \cong \mathcal{A} \cong \mathcal{N} \otimes_{\mathcal{A}} \mathcal{M}$. The set of isomorphism classes $[\mathcal{M}]$ of invertible $\mathcal{A}$-Bimodules over $\mathcal{O}_X$ forms a group with multiplication induced by the tensorproduct over $\mathcal{O}_X$ and is denoted by $\mathrm{Pic}_{\mathcal{O}_X}(X,\mathcal{A})$ of $\mathrm{Pic}_{\mathcal{O}_X}(\mathcal{A})$. If A is a finite R-algebra then $\mathrm{Pic}_{\tilde{R}}(\tilde{A}) = \mathrm{Pic}_R(A)$. More generally, if we denote by $X(I) \subset \mathrm{Spec}(R)$ the open subset consisting of all $p \in \mathrm{Spec}(R)$ with $I \not\subset p$, then we have :

6.30. Proposition. Let R be a noetherian ring and I an ideal of R; if $\mathcal{A}$ is a coherent sheaf of $\tilde{R}|X(I)$-algebras, then $\mathrm{Pic}_{\tilde{R}|X(I)}(\mathcal{A}) = \mathrm{Pic}_R(\Gamma(X(I),\mathcal{A}),\kappa_I)$.

Proof. Recall from [162] that under our assumptions, for any coherent sheaf of modules $\mathcal{M}$ on X(I), we may find a coherent sheaf of $\mathcal{O}_X$-modules $\mathcal{M}'$ on Spec(R) with $\mathcal{M}'|X(I) = \mathcal{M}$. Moreover, if M' and M" in R-mod satisfy $\tilde{M}'|X(I) = \tilde{M}''|X(I)$, then $Q_I(M') \cong Q_I(M'')$. It follows easily that $A = \Gamma(X(I),\mathcal{A})$ is $\kappa_I$-finitely generated, hence a straightforward verification yields that there is a one-to-one correspondence between invertible $\mathcal{A}$-Bimodules over $\tilde{R}|X(I)$ and $\kappa_I$-invertible A-bimodules over R. The rest of the proof is left to the reader. □

Let us consider a particular case in somewhat more detail. Assume R is a noetherian domain of Krull dimension 1 with field of fractions K and $\mathcal{A}$ be a coherent sheaf of $\tilde{R}|X(I)$-algebras on some open $X(I) \subset \mathrm{Spec}(R)$ such that $A = \Gamma(X(I),\mathcal{A})$ is torsion free with center $Q_I(R)$ and $A \otimes_R K$ is semisimple, then we have the following exact sequence :

$$1 \longrightarrow \mathrm{Pic}(X(I)) \longrightarrow \mathrm{Pic}_{\underset{\sim}{R}|X(I)}(A) \longrightarrow \coprod_{p \in X(I)} \mathrm{Picent}(A_p) \longrightarrow 1$$

Now, let R be a noetherian Krull domain with field of fractions K and let A be a maximal R-order in a central simple K-algebra $\Sigma$. As in J. Riley [124] call two left A-lattices L and M quasi-equal and write $L \sim M$ if $L_p = M_p$ for all $p \in X^{(1)}(R)$. If we write $L^{-1}$ for the set of all $s \in \Sigma$ with $LsL \subset L$, then clearly $L \sim M$ if and only if $L^{-1} = M^{-1}$. Let $\sigma_1 = \inf\{\kappa_{R-p}; p \in X^{(1)}(R)\}$, then it follows that $L \sim M$ if and only if $Q_{\sigma_1}(L) = Q_{\sigma_1}(M)$, as $(L^{-1})^{-1} = \bigcap_{p \in X^{(1)}(R)} L_p = Q_{\sigma_1}(L)$ and $((L^{-1})^{-1})^{-1} = L^{-1}$ for any left A-lattice L. It is now easy to see that the collection G(A) in loc. cit of equivalence classes modulo ~ of *two sided* A-lattices forms an (abelian) group in the obvious way. The proof of the next result is an easy exercise :

6.31. Proposition. With the above assumption, sending the equivalences class module ~ of a two sided A-lattice L to $Q_{\sigma_1}(L)$ yields an isomorphism $G(A) \tilde{\to} \mathcal{I}(A,\sigma_1)$. □

Of course, this generalizes to maximal orders over arbitrary Krull domains. Note also that $G(A)/K^{\star}$ (= Cl(A)) $\cong \mathrm{Picent}(A,\sigma_1)$, by (6.13.) We will come back to this in more detail in chapter E.

## B.7. Mayer-Vietoris Sequences for Relative Picard Groups.

We will start by describing a construction of relative objects in the vein of Milnor's constructions, cf. [ 101 ] . Consider the following commutative diagram of rings :

$$\begin{array}{ccc} A & \xrightarrow{h_1} & A_1 \\ {\scriptstyle h_2}\downarrow & & \downarrow{\scriptstyle f_1} \\ A_2 & \xrightarrow[f_2]{} & A' \end{array} \qquad (\star)$$

Assume that $(\star)$ is a pull-back diagram, then we may identify A and the subring of $A_1 \times A_2$ consisting of all $(a_1,a_2)$ with $f_1(a_1) = f_2(a_2)$. This cartesian diagram induces a square of functors :

$$\begin{array}{ccc} A\text{-mod} & \xrightarrow{\eta_1} & A_1\text{-mod} \\ {\scriptstyle \eta_2}\downarrow & & \downarrow{\scriptstyle \varphi_1} \\ A_2\text{-mod} & \xrightarrow[\varphi_2]{} & A'\text{-mod} \end{array} \qquad (\star\star)$$

together with a functorial isomorphism $\tau : \varphi_1 \eta_1 \xrightarrow{\sim} \varphi_2 \eta_2$ . Here, we have written $\eta_i = A_i \underset{A}{\otimes} -$ resp. $\varphi_i = A' \underset{A_i}{\otimes} -$ for the canonical functors and $\tau$ for the canonical isomorphism induced by the associative and cancellative properties of the tensorproduct. Let us write $\mu(A)$ for the category $A_1\text{-mod} \underset{A'\text{-mod}}{\times} A_2\text{-mod}$. Recall that in general a product category $\mathcal{C}_1 \underset{\mathcal{C}'}{\times} \mathcal{C}_2$ with fibre maps $F_i : \mathcal{C}_i \to \mathcal{C}'$ is defined as follows : its objects are triples $(C_1,\alpha,C_2)$ , where $C_i \in \mathcal{C}_i$ and where $\alpha : F_1 C_1 \xrightarrow{\sim} F_2 C_2$ is an isomorphism in $\mathcal{C}'$; its morphisms $(C_1,\alpha,C_2) \to (D_1,\beta,D_2)$ are couples $(\mu_1,\mu_2)$, where $\mu_i : C_i \to D_i$ is a morphism in $\mathcal{C}_i$ and such that the following diagram commutes :

$$\begin{array}{ccc} F_1C_1 & \xrightarrow{F_1 \mu_1} & F_1D_1 \\ {\scriptstyle \alpha}\downarrow & & \downarrow{\scriptstyle \beta} \\ F_2C_2 & \xrightarrow[F_2 \mu_2]{} & F_2 D_2 \end{array}$$

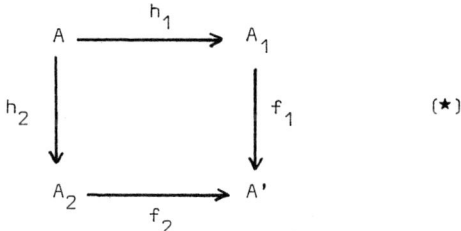

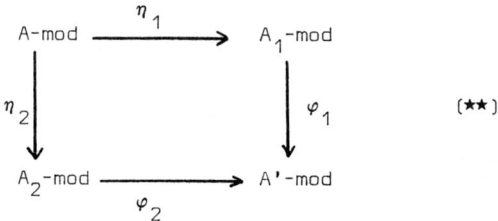

Let $U : A\text{-mod} \to \mu(A) : M \to (\eta_1 M, \tau_M, \eta_2 M)$ denote the canonical functor. From H. Bass [18] or J. Milnor [101], we know that U possesses an adjoint S, which is defined on each $(M_1, \alpha, M_2) \in \mu(\Lambda)$ by the cartesian diagram

$$\begin{array}{ccccc} S(M_1,\alpha,M_2) & \longrightarrow & & & M_2 \\ \downarrow & & & & \downarrow \\ M_1 & \longrightarrow & \varphi_1 M_1 & \xrightarrow[\alpha]{\sim} & \varphi_2 M_2 \end{array}$$

A cartesian square (★) is called a <u>Milnor square</u> if $f_1$ or $f_2$ at least is surjective. In this case, it has been proved in loc. cit. that one then obtains an exact sequence, which is a formal analogue of the Mayer Vietoris sequence in algebraic topology :

$$K_1A \to K_1A_1 \oplus K_1A_2 \to K_1A' \to K_oA \to K_oA_1 \oplus K_oA_2 \to K_oA'.$$

In particular, in the commutative case, one may derive by means of the determinant map, an exact sequence :

$$1 \to U(A) \to U(A_1) \times U(A_2) \to U(A') \to \mathrm{Pic}(A) \to \mathrm{Pic}(A_1) \times \mathrm{Pic}(A_2) \to \mathrm{Pic}(A')$$

We will see below how to generalize this to the relative situation. Let us already note here, that without adequate restrictions one should not expect to obtain an exact sequence involving Picard groups. Indeed, if the base rings A and A' are noncommutative, an arbitrary morphism $A \to A'$ does not necessarily induce a morphism $\mathrm{Pic}(A) \to \mathrm{Pic}(A')$.

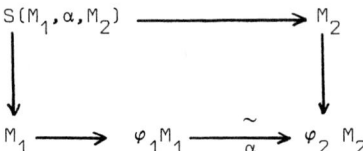

We will need the following general construction which is the relative counterpart of the one due to J. Milnor [101]. Let us start from a cartesian square (★) and assume that all occuring rings in this section are algebras over a fixed commutative $\kappa$-noetherian ring R whereas ring morphisms, modules, etc. are defined over R. Let $\kappa$ be an idempotent kernel functor in R-mod; define $\psi_i : (A_i,\kappa)\text{-mod} \to (A',\kappa)\text{-mod}$, by mapping $M_i \in (A_i,\kappa)$-mod to $Q_\kappa(A' \underset{A_i}{\otimes} M_i)$ and $\theta_i$ : A-mod $\to (A_i,\kappa)$-mod by mapping $M \in$ A-mod to $Q_\kappa(A_i \underset{A}{\otimes} M)$. We then have a functorial isomorphism $\gamma : \psi_1\theta_1 \tilde{\to} \psi_2\theta_2$ given by $\psi_1\theta_1(M) = \psi_1(Q_\kappa(A_1 \underset{A}{\otimes} M)) = Q_\kappa(A' \underset{A_1}{\otimes} Q_\kappa(A_1 \underset{A}{\otimes} M)) =$ $Q_\kappa(A' \underset{A_1}{\otimes} A_1 \underset{A}{\otimes} M) \cong Q_\kappa(A' \underset{A_2}{\otimes} A_2 \underset{A}{\otimes} M) = Q_\kappa(A' \underset{A_2}{\otimes} Q_\kappa(A_2 \underset{A}{\otimes} M)) = \psi_2\theta_2(M)$.

Let us form the fibre product square

$$\begin{array}{ccc} \Delta(A,\kappa) & \xrightarrow{\ \xi_1\ } & (A_1,\kappa)\text{-mod} \\ {\scriptstyle \xi_2}\downarrow & & \downarrow{\scriptstyle \psi_1} \\ (A_2,\kappa)\text{-mod} & \xrightarrow[\ \psi_2\ ]{} & (A',\kappa)\text{-mod}. \end{array} \qquad \delta : \psi_1\xi_1 \tilde{\to} \psi_2\xi_2$$

There is a canonical functor $T : \Lambda\text{-mod} \to \Delta(\Lambda,\kappa) : M \to (\theta_1 M, \gamma_M, \theta_2 M)$ such that $\gamma = \delta T$ and $\theta_i = \xi_i T$. We claim that T possesses a right adjoint $S : \Delta(A,\kappa) \to$ A-mod. For any $\underline{M} = (M_1,\mu,M_2) \in \Delta(A,\kappa)$ define $S(\underline{M})$ as follows by the cartesian square :

$$\begin{array}{ccccc} S(\underline{M}) & & \longrightarrow & & M_2 \\ \downarrow & & & & \downarrow \\ M_1 & \longrightarrow & \psi_1 M_1 & \xrightarrow[\ \mu\ ]{\sim} & \psi_2 M_2 \end{array}$$

It is clear that $S(\underline{M})$ is an A-module and that associating $S(\underline{M})$ to $\underline{M}$ yields an additive functor $S : \Delta(A,\kappa) \to$ A-mod. Let us check the adjointness property, i.e. pick $N \in$ A-mod and $\underline{M} \in \Delta(A,\kappa)$, then by definition we have $\mathrm{Hom}_A(N,S(\underline{M})) = \mathrm{Hom}_A(N,M_1) \underset{\mathrm{Hom}_A(N,\psi_2 M_2)}{\times} \mathrm{Hom}_A(N,M_2)$. But, it is clear that $\mathrm{Hom}_A(N,M_i) = \mathrm{Hom}_{A_i}(A_i \underset{A}{\otimes} N, M_i) = \mathrm{Hom}_{A_i}(Q_\kappa(A_i \underset{A}{\otimes} N), M_i) =$ $= \mathrm{Hom}_{A_i}(\theta_i N, M_i)$ since $M_i$ is $\kappa$-closed in $A_i$-mod, so we find that $\mathrm{Hom}_A(N,S(M))$ consists exactly of the $(u_1,u_2) \in \mathrm{Hom}_{A_1}(\theta_1 N, M) \times \mathrm{Hom}_{A_2}(\theta_2 N, M)$, such that the following diagram commutes :

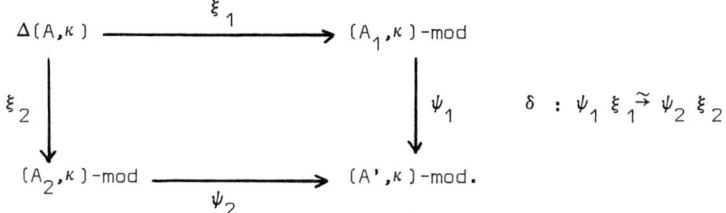

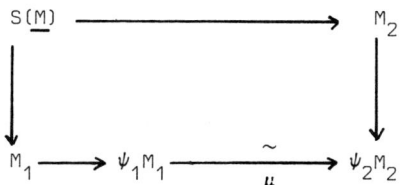

$$\begin{array}{ccc} \psi_1\,\theta_1\,N & \xrightarrow{\ \psi_1\,u_1\ } & \psi_1\,M_1 \\ \downarrow{\scriptstyle \gamma_N} & & \downarrow{\scriptstyle \mu} \\ \psi_2\,\theta_2 N & \xrightarrow{\ \psi_2\,u_2\ } & \psi_2\,M_2 \end{array}$$

But, since $T\,N = (\theta_1 N,\ \gamma_N,\ \theta_2 N)$ and $\underline{M} = (M_1, \mu, M_2)$ this is exactly the condition stating that $(u_1, u_2)$ belongs to $\mathrm{Hom}_{\Delta(A,\kappa)}(TN, \underline{M})$. Note that S actually maps $\Delta(A,\kappa)$ to $(A,\kappa)$-mod. This is due to the following remark :

7.1. Lemma. If A is an arbitrary ring and $\kappa$ an idempotent kernel functor in A-mod, then for any pair of morphisms $M_i \to N (i = 1,2)$ in $(A,\kappa)$-mod, the fibre product $M_1 \underset{N}{\times} M_2$ in A-mod lies in $(A,\kappa)$-mod.

Proof. Just note that for any $I \in \mathcal{L}(\kappa)$, we have $\mathrm{Hom}_A(I, M_1 \underset{N}{\times} M_2) =$

$\mathrm{Hom}_A(I,M_1) \underset{\mathrm{Hom}_A(I,N)}{\times} \mathrm{Hom}_A(I,M_2) = \mathrm{Hom}_A(A,M_1) \underset{\mathrm{Hom}_A(A,N)}{\times} \mathrm{Hom}_A(A,M_2) =$

$\mathrm{Hom}_A(A, M_1 \underset{N}{\times} M_2)$. □

Pick $p \in C(\kappa)$ : then to the cartesian square $(\star)$ we may associate a cartesian square $(\star_p)$ and a commutative diagram $(\star\star_p)$ due to the fact that localizing at p is exact :

$(\star_p)$

$$\begin{array}{ccc} A_p & \xrightarrow{\ h_{1,p}\ } & A_{1,p} \\ \downarrow{\scriptstyle h_{2,p}} & & \downarrow{\scriptstyle f_{1,p}} \\ A_{1,p} & \xrightarrow[\ f_{2,p}\ ]{} & A'_p \end{array}$$

$(\star\star_p)$

$$\begin{array}{ccc} A_p\text{-mod} & \xrightarrow{\ \theta_{1,p}\ } & A_{1,p}\text{-mod} \\ \downarrow{\scriptstyle \theta_{2,p}} & & \downarrow{\scriptstyle \varphi_{1,p}} \\ A_{2,p}\text{-mod} & \xrightarrow[\ \varphi_{2,p}\ ]{} & A'_p\text{-mod} \end{array}$$

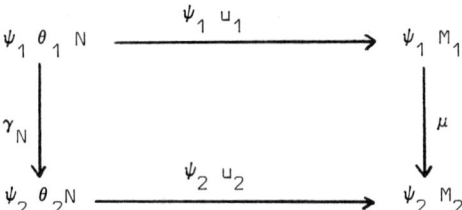

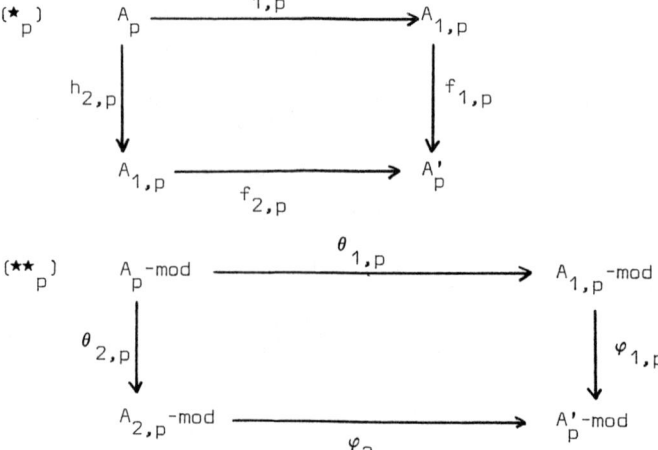

together with a functorial isomorphism $\gamma_p : \varphi_{1,p}\,\theta_{1,p} \overset{\sim}{\to} \varphi_{2,p}\,\theta_{2,p}$.

Consider also

$$\Delta_p = A_{1,p}\text{-mod} \underset{\Lambda'_p\text{-mod}}{\times} A_{2,p}\text{-mod}$$

and the canonical map $T_p : A_p\text{-mod} \to \Delta_p : M \mapsto (\theta_{1,p}M,\ \gamma_M, \theta_{2,p}M)$.

Finally, let $S_p$ be Milnor's adjoint of $T_p$.

We may extend the localization functor at p to $\Delta(A,\kappa)$ by mapping $(M_1,\mu,M_2) \in \Delta(A,\kappa)$ to $(M_{1,p},\mu_p,M_{2,p}) \in \Delta_p$. Denote this functor by $Q_p : \Delta(A,\kappa) \to \Delta_p$ - there is no possible confusion with the actual localization $Q_p : (A,\kappa)\text{-mod} \to A_p\text{-mod}$. Since $\mu$ is an isomorphism between $\psi_1 M_1 = Q_\kappa(A' \otimes_{A_1} M_1)$ and $\psi_2 M_2 = Q_\kappa(A' \otimes_{A_2} M_2)$, we obtain an isomorphism between $\varphi_{1,p} M_{1,p} = A'_p \otimes_{A_{1,p}} M_{1,p}$ and $\varphi_{2,p} M_{2,p} = A'_p \otimes_{A_{2,p}} M_{2,p}$ by localizing at p.

7.2. <u>Proposition</u>. The following diagrams commute

1.
$$\begin{array}{ccc} \Delta(A,\kappa) & \xrightarrow{\ S\ } & (A,\kappa)\text{-mod} \\ \downarrow{\scriptstyle Q_p} & & \downarrow{\scriptstyle Q_p} \\ \Delta_p & \xrightarrow[\ S_p\ ]{} & A_p\text{-mod} \end{array}$$

2.
$$\begin{array}{ccc} (A,\kappa)\text{-mod} & \xrightarrow{\ T\ } & \Delta(A,\kappa) \\ \downarrow{\scriptstyle Q_p} & & \downarrow{\scriptstyle Q_p} \\ A_p\text{-mod} & \xrightarrow[\ T_p\ ]{} & \Delta_p \end{array}$$

<u>Proof</u>. Let us verify the commutativity of 2., say. Pick $M \in (A,\kappa)\text{-mod}$, then $Q_pTM = Q_p((\theta_1M,\gamma_M,\theta_2M)) = (\theta_{1,p}M_p,\gamma_{M,p},\theta_{2,p}M_p)$ and $T_pQ_pM = T_pM_p = (\theta_{1,p}M_p,\gamma_{p,M_p},\theta_{2,p}M_p)$; that $\gamma_{M,p} = \gamma_{p,M_p}$, follows from the definitions and the rest is obvious. □

Let $\underline{M} = (M_1,\mu,M_2)$ and $\underline{N} = (N_1,\nu,N_2)$ be objects in $\Delta(A,\kappa)$. We claim that if $\varphi : \underline{M} \to \underline{N}$ is a morphism in $\Delta(A,\kappa)$ with the property that $Q_p(\varphi):Q_p\underline{(M)} \to Q_p\underline{(N)}$ is an isomorphism for all $p \in \mathcal{C}(\kappa)$, then $\varphi$ is an isomorphism too. Indeed, suppose $\varphi = (u_1,u_2)$, where $u_i : M_i \to N_i$. $(i = 1,2)$ makes the following diagram commutative :

$$\begin{array}{ccc} Q_\kappa(A' \otimes_{A_1} M_1) & \xrightarrow{\;Q_\kappa(A' \otimes_{A_1} u_1)\;} & Q_\kappa(A' \otimes_{A_1} N_1) \\ \big\downarrow \mu & & \big\downarrow \nu \\ Q_\kappa(A' \otimes_{A_2} M_2) & \xrightarrow[\;Q_\kappa(A' \otimes_{A_2} u_2)\;]{} & Q_\kappa(A' \otimes_{A_2} N_2) \end{array}$$

Now, since $Q_p(u_1,u_2) = (u_{1,p},u_{2,p})$ is an isomorphism for all $p \in \mathcal{C}(\kappa)$, and since $M_i$ and $N_i$ are $\kappa$-closed for $i = 1,2$, it follows in particular that $u_1$ and $u_2$ are isomorphisms, say with inverses $v_i$, $i = 1,2$. That $(v_1,v_2)$ is a morphism from $\underline{N}$ to $\underline{M}$ in $\Delta(A,\kappa)$, which is inverse to $(u_1,u_2)$, follows immediately from the commutativity of the above diagram. □

7.3. <u>Corollary</u>. For any $M \in (A,\kappa)$-mod resp. $\underline{M} \in \Delta(A,\kappa)$,

1. the canonical morphism $\varphi_M : M \to STM$ is an isomorphism if and only if $\varphi_{p,M_p} : M_p \to S_pT_pM_p$ is an isomorphism for all $p \in \mathcal{C}(\kappa)$ ;
2. the canonical morphism $\psi_{\underline{M}} : TS\underline{M} \to \underline{M}$ is an isomorphism if and only if $\psi_{p,Q_p\underline{M}} : T_pS_pQ_p\underline{M} \to Q_p\underline{M}$ is an isomorphism for all $p \in \mathcal{C}(\kappa)$.

<u>Proof</u>. Let us show that $M \to STM$ is isomorphic if and only if $M_p \to S_pT_pM_p$ is an isomorphism for all $p \in \mathcal{C}(\kappa)$. By the foregoing remarks, we know that $\varphi_M: M \to STM$ is an isomorphism if for all $p \in \mathcal{C}(\kappa)$, the map $Q_p\varphi_M : M_p \to (STM)_p$ is an isomorphism. But $(STM)_p = Q_pSTM$ and $Q_pST = S_pQ_pT = S_pT_pQ_p$, so $Q_p\varphi_M = \varphi_{p,M_p}$, proving the assertion. The proof of the second part is similar to this. □

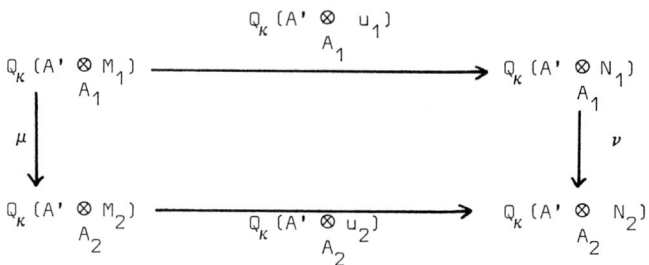

A cartesian diagram (★) is called a Milnor square if one of the maps $f_1$ or $f_2$ is surjective. it is said to be locally a Milnor square (with respect to $\kappa$) if it is cartesian and if for all $p \in \mathcal{C}(\kappa)$ the square $(\star_p)$ is a Milnor square. Since localizing at a central prime ideal is an exact functor, it is clear that any Milnor square is locally a Milnor square.

Another example occurs as follows. Let R be a commutative ring and let $S_1$ resp. $S_2$ be multiplicative subsets of R. We call $S_1$ and $S_2$ comaximal if for all $(s_1,s_2) \in S_1 \times S_2$ we have $Rs_1 + Rs_2 = R$. Write $S = S_1S_2$ for the multiplicative system in R consisting of all $s_1s_2$ with $s_i \in S_i$, then we have the following :

7.4. Lemma (and example). Let R be a central subring of A and $S_1, S_2$ two comaximal multiplicative systems in R. Let M be a left A-module, then the following square $\delta(M)$ of localizations is cartesian

$$\begin{array}{ccc} M & \longrightarrow & M_{S_1} \\ \downarrow & & \downarrow \\ M_{S_2} & \longrightarrow & M_S \end{array}$$

Moreover, in particular $\delta(A)$ is locally a Milnor square for any idempotent kernel functor $\kappa$ in R-mod.

Proof. The first statement follows from the fact that $\Gamma = A_{S_1} \times A_{S_2}$ is faithfully flat over A. Indeed, in order to show that $\delta(M)$ is cartesian, we have to show that the sequence

$$0 \longrightarrow M \longrightarrow M_{S_1} \oplus M_{S_2} \longrightarrow M_S$$

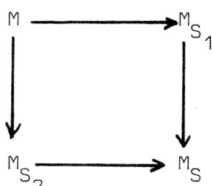

is exact, so by faithful flatness, it suffices to check that the sequence obtained by tensoring with $\Gamma$ is exact. This in turn is equivalent to the exactness of the two sequences obtained by tensoring with $A_{S_1}$ resp. $A_{S_2}$, i.e. we may assume $S_i \subset U(R) = R^{\star}$ and $S = S_{3-i}(i = 1,2)$. In this case, the statement becomes obvious. Let $p \in Spec(R)$ be an arbitrary prime ideal of R, then $p \subset m$ for some maximal ideal m of R and for this m we have $m \cap S_1 = \emptyset$ or $m \cap S_2 = \emptyset$, by the comaximality of $S_1$ and $S_2$. Assume the former, then $(A_{S_1})_m = A_m$ and $(A_{S_2})_m = (A_S)_m$. But then certainly $(A_{S_2})_p = ((A_{S_2})_m)_p = ((A_S)_m)_p = (A_S)_p$, i.e. $f_{2,p}$ is an isomorphism and $(\star_p)$ is a Milnor square. □

7.5. Note. The foregoing lemma may easily be generalized as follows : let $\mu,\nu$ be idempotent kernel functors in R-mod and assume A to be prime for simplicity's sake, then there is a cartesian diagram

$$\begin{array}{ccc} Q_{\mu\wedge\nu}(A) & \longrightarrow & Q_{\mu}(A) \\ \downarrow & & \downarrow \\ Q_{\nu}(A) & \longrightarrow & Q_{\mu\vee\nu}(A) \end{array}$$

In particular, if $\mu\wedge\nu = \hat{0}$, then this is locally a Milnor square for any idempotent kernel functor in R-mod. More generally, it is locally a Milnor square for any $\kappa \geqslant \mu\wedge\nu$ in R-mod. Indeed, let $X(\kappa)$ denote the set of all $p \in Spec(R)$ with $p \notin \mathcal{L}(\kappa)$; equivalently : such that $\kappa(R/p) = 0$, then for any $p \in X(\kappa)$ we have $p \in X(\nu)$ and we may easily verify that this yields the conclusion.

Recall that a left A-module M is said to be $\kappa$-quasiprojective resp. $\kappa$-quasifree (over R) if for all $p \in \mathcal{C}(R,\kappa)$ the module $M_p$ is projective resp. free over $A_p$. Let us denote by $\mathbb{P}(A,\kappa)$ resp. $\mathbb{F}(A,\kappa)$ the full subcategory of $(A,\kappa)$-mod consisting of $\kappa$-finitely generated $\kappa$-quasi projective resp. $\kappa$-quasifree left A-modules.

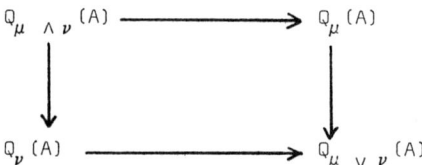

7.6. Lemma. Assume that we have a pair of module morphisms $M_1 \xrightarrow{u_1} P \xleftarrow{u_2} M_2$ compatible with $(f_1, f_2)$. If $A_i$ is $\kappa$-finitely generated over R for i = 1,2 then $M_1 \underset{P}{\times} M_2$ is $\kappa$-finitely generated as an A-module if $M_i$ is a $\kappa$-finitely generated $A_i$-module.

Proof. Let $N_i \subset M_i$ be a finitely generated $A_i$-submodule such that $M_i/N_i$ is $\kappa$-torsion. Since $A_i$ is $\kappa$-finitely generated over R, it follows that $N_i$ is a $\kappa$-finitely generated R-module, hence so is $N_1 \times N_2$ and since R is $\kappa$-noetherian, so is the R-submodule $N_1 \underset{P}{\times} N_2 \subset N_1 \times N_2$. Therefore $N_1 \underset{P}{\times} N_2$ is a $\kappa$-finitely generated $A = A_1 \underset{A'}{\times} A_2$-module. In order to finish the proof, it suffices to verify that $N_1 \underset{P}{\times} N_2$ is $\kappa$-cotorsion in $M_1 \underset{P}{\times} M_2$ as a left A-module, but the proof of this is just a straightforward, componentwise argument. □

7.7. Corollary. Assume that $A_1$ and $A_2$ are $\kappa$-finitely generated over R and that (★) is locally a Milnor square, then the following diagram is cartesian

$$\begin{array}{ccc} \mathbb{P}(A,\kappa) & \xrightarrow{\eta_1} & \mathbb{P}(A_1,\kappa) \\ \downarrow{\scriptstyle \eta_2} & & \downarrow{\scriptstyle \varphi_1} \\ \mathbb{P}(A_2,\kappa) & \xrightarrow[\varphi_2]{} & \mathbb{P}(A',\kappa) \end{array}$$

Proof. The functors $\eta_i$ and $\varphi_i$ are well-defined, because any R-algebra morphism $u : A \to B$ induces a functor $\mathbb{P}(A,\kappa) \to \mathbb{P}(B,\kappa)$, by mapping $P \in \mathbb{P}(A,\kappa)$ to $Q_\kappa(B \underset{A}{\otimes} P) = B \underset{A}{\perp} P$. In order to verify this, note that for any $p \in \mathcal{C}(\kappa)$ we have that $Q_\kappa(B \underset{A}{\otimes} P)_p = B_p \underset{A_p}{\otimes} P_p$ is a left projective $B_p$-module, since $P_p$ is a left projective $A_p$-module. Thus $B \underset{A}{\perp} P$ is $\kappa$-quasiprojective. Moreover, let $Q \subset P$ be a finitely generated A-submodule of P such that P/Q is $\kappa$-torsion, then the image of $B \underset{A}{\otimes} Q$ in $B \underset{A}{\otimes} P$

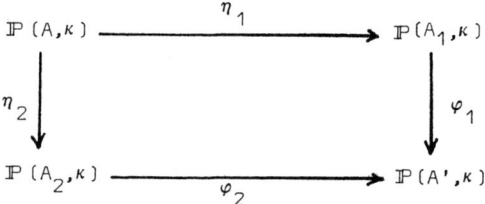

is finitely generated over B and the quotient, which is isomorphic to $B \underset{A}{\otimes} (P/Q)$ is certainly $\kappa$-torsion. Hence $B \underset{A}{\otimes} P$ is $\kappa$-finitely generated over B and so is $B \underset{A}{\perp} P$. It remains to verify that the restrictions of S and T to $\mathbb{P}$ are mutually inverse to each other. But, Milnor's theorem [100] yields that for each $p \in C(\kappa)$ the functors $S_p$ and $T_p$ are inverse equivalences (since $(\star_p)$ is a Milnor square) when restricted to projective, finitely generated left modules, so the foregoing immediately yields the result. □

7.8. Corollary. With assumptions as before, the following diagram is cartesian :

$$\begin{array}{ccc} \mathbb{F}(A,\kappa) & \xrightarrow{\eta_1} & \mathbb{F}(A_1,\kappa) \\ \downarrow{\scriptstyle \eta_2} & & \downarrow{\scriptstyle \varphi_1} \\ \mathbb{F}(A_2,\kappa) & \xrightarrow[\varphi_2]{} & \mathbb{F}(A',\kappa) \end{array}$$

Proof. Upto some easy verifications, we only have to verify that the "S-construction" applied to $\mathbb{F}(A_1,\kappa) \underset{\mathbb{F}(A',\kappa)}{\times} \mathbb{F}(A_2,\kappa)$ yields an object in $\mathbb{F}(A,\kappa) \subset \mathbb{P}(A,\kappa)$.

This may be seen as follows. Choose $p \in C(\kappa)$, localize at p everywhere and simplify notation by writing $(-)$ for $(-)_p$ again. If in $(M_1,\mu,M_2) = \underline{M}$ both $M_1$ and $M_2$ are free, then $S(\underline{M})$ is free as well. These data fit into a cartesian diagram :

$$\begin{array}{ccccc} S(\underline{M}) & & \longrightarrow & & M_2 \\ \downarrow & & & & \downarrow \\ M_1 & \longrightarrow & \psi_1 M_1 & \xrightarrow{\sim} & \psi_2 M_2 \end{array}$$

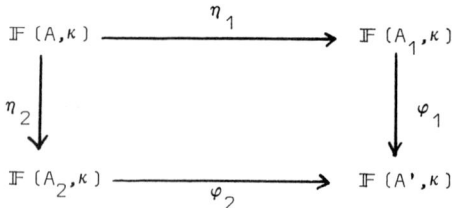

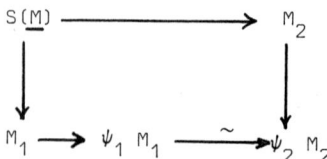

After localizing, $\psi_i$ is just tensoring with $A' \underset{A_i}{\otimes} -$. Since $M_i$ is free, we may find a basis $\{u_{i,j};\ j = 1,\dots,n_i\}$ of $M_i$ as a left $A_i$-module, and then, (denoting by $\overline{\phantom{u}}$ the image in A'), we obtain $\psi_1 M_1 = \sum_{j=1}^{n_i} A' \overline{u}_{1,j}$ resp. $\psi_2 M_2 = \sum_{j=1}^{n_2} A' \overline{u}_{2,j}$ and $\{\overline{u}_{1,j}\}$ and $\{\overline{u}_{2,j}\}$ are free bases. Since $\mu$ is an isomorphism, we have $n_1 = n_2 = n$. Moreover, $\mu$ may be described by its action on $\{\overline{u}_{1,j}\}$ say, i.e. $\mu(\overline{u}_{1,i}) = \Sigma\, \mu_{ij}\ \overline{u}_{2,j}$, where $(\mu_{ij}) \in Gl_n(A')$.

The assumptions imply that R is a local ring and that (★) is a Milnor square, since now at least $f_1$ or $f_2$ is surjective. Assume $f_2$ is surjective. The semi local ring $A_2$ is finitely generated over the local ring R and therefore $f_2$ induces an epimorphism $Gl_n(A_2) \to Gl_n(A')$, cf. H. Bass [18 ], hence there exists an automorphism $\pi$ of $M_2$ over $A_2$ such that

$\pi(u_{2,j}) = \Sigma\, \nu_{ij}\ u_{2,j}$ , where $\overline{\nu}_{ij} = \mu_{ij}$,

$$M_2 = \sum_{i=1}^{n} A_2 w_i, \quad \mu(\overline{u}_{1,i}) = w_i \quad (1 \leqslant i \leqslant n)$$

and it follows that $\{(u_{1,i}, w_i);\ 1 \leqslant i \leqslant n\}$ is an A-basis of $S(\underline{M})$. □

In the sequel of this section we assume all rings to be $\kappa$-finitely generated over R. Consequently *all* rings considered are $\kappa$-noetherian.

Let us assume that the cartesian diagram (★) is locally a Milnor square and each of the occuring morphisms has the form $\varphi : U \to V$ with $V = S \otimes_R U$ for some R-algebra S. It then follows in particular that if M is a U-bimodule (over R, as always), then $V \otimes_U M$ is a V-bimodule and that if $u : M \to N$ is a U-bimodule morphism, then so is $V \otimes_U u : V \otimes_U M \to V \otimes_U N$ over V. Note that there is some abuse of notation in the last statement : $V \otimes_U u$ is defined for *left* U-linear morphisms, but not for right U-linear morphisms. We have actually identified $V \otimes_U u$ with $S \otimes_R u$, and then all of the assertions may be verified.

Denote by $Pic_R(A,\kappa)$ the category of $\kappa$-invertible A-bimodules over R and A-bimodule isomorphisms (over R). It is clear that $\varphi : U \to V$ as above induces a functor $Pic_R(U,\kappa) \to Pic_R(V,\kappa)$ by sending $P \in Pic_R(U,\kappa)$ to $Q_\kappa(V \underset{U}{\otimes} P) = Q_\kappa(S \underset{R}{\otimes} P) \in Pic_R(V,\kappa)$. Indeed, since the left and right localizations at $\kappa$ coincide in our context, it is easy to check e.g. that $Q_\kappa(Q_\kappa(V \underset{U}{\otimes} Q) \underset{V}{\otimes} Q_\kappa(V \underset{U}{\otimes} P)) \cong Q_\kappa(V)$ if $Q_\kappa(Q \underset{U}{\otimes} P) \cong Q_\kappa(U)$. It follows that (★) induces a diagram :

$$\begin{array}{ccc} Pic_R(A,\kappa) & \xrightarrow{\eta_1} & Pic_R(A_1,\kappa) \\ {\scriptstyle \eta_2}\downarrow & & \downarrow{\scriptstyle \varphi_1} \\ Pic_R(A_2,\kappa) & \xrightarrow[\varphi_2]{} & Pic_R(A',\kappa) \end{array} \qquad \beta : \varphi_1\eta_1 \xrightarrow{\sim} \varphi_2\eta_2$$

Denote by ${}_R(A,\kappa)$, resp. ... the category of $\kappa$-closed A-bimodules defined over R, resp. ... i.e. ${}_R(A,\kappa) = (A \underset{R}{\otimes} A^{opp},\kappa)$-mod, and construct the fibre product $\Gamma_R(A,\kappa) = {}_R(A_1,\kappa) \underset{{}_R(A',\kappa)}{\times} {}_R(A_2,\kappa)$, which is a *non-full* subcategory of $\Delta(A,\kappa)$. There is a canonical functor $S : \Gamma_R(A,\kappa) \to {}_R(A,\kappa)$, which associates to any $\underline{M} : (M_1,\mu,M_2) \in \Gamma_R(A,\kappa)$ the $S(\underline{M}) \in (A,\kappa)$-mod, defined by the pull-back diagram

$$\begin{array}{ccc} S(\underline{M}) & \longrightarrow & M_2 \\ \downarrow & & \downarrow \\ M_1 \longrightarrow \varphi_1 M_1 & \xrightarrow[\mu]{\sim} & \varphi_2\, M_2 \end{array}$$

and clearly $S(\underline{M}) \in {}_R(A,\kappa)$ in this case. We also have a canonical functor $T : {}_R(A,\kappa) \to \Gamma_R(A,\kappa)$, which sends $M \in {}_R(A,\kappa)$ to $(\eta_1 M, \beta_M, \eta_2 M) \in \Gamma_R(A,\kappa)$. By their very construction, it follows that $S$ and $T$ are adjoint functors, which fit into a commutative diagram

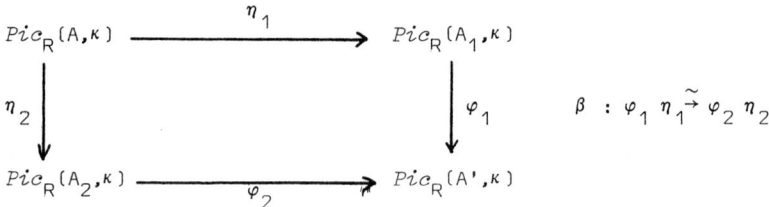

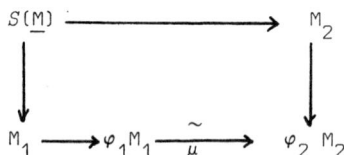

$$\begin{array}{ccccccc}
\Gamma_R(A,\kappa) & \xrightarrow{\mathcal{S}} & {}_R(A,\kappa) & \xrightarrow{\mathcal{T}} & \Gamma_R(A,\kappa) & \xrightarrow{\mathcal{S}} & {}_R(A,\kappa) \\
\downarrow & & \downarrow & & \downarrow & & \downarrow \\
\Delta(A,\kappa) & \xrightarrow[S]{} & (A,\kappa)\text{-mod} & \xrightarrow[T]{} & \Delta(A,\kappa) & \xrightarrow[S]{} & (A,\kappa)\text{-mod},
\end{array}$$

where the vertical arrows are not necessarily full.

7.9. <u>Theorem.</u> The diagram

$$\begin{array}{ccc}
\mathit{Pic}_R(A,\kappa) & \xrightarrow{\eta_1} & \mathit{Pic}_R(A_1,\kappa) \\
{\scriptstyle \eta_2}\downarrow & & \downarrow{\scriptstyle \varphi_1} \\
\mathit{Pic}_R(A_2,\kappa) & \xrightarrow[\varphi_2]{} & \mathit{Pic}_R(A',\kappa)
\end{array} \qquad \beta : \varphi_1\,\eta_1 \xrightarrow{\sim} \varphi_2\,\eta_2$$

is cartesian.

<u>Proof.</u> One may prove this directly, by showing that $\mathcal{S}$ and $\mathcal{T}$ are mutually inverse equivalences, when restricted to the appropriate subcategories of $\Gamma_R(A,\kappa)$ resp. ${}_R(A,\kappa)$. We leave this to the reader and prefer to present an alternative approach.

We have seen before, that if (★) is locally a Milnor square, then S and T induce inverse equivalences between $\mathbb{P}(A,\kappa)$ and $\mathbb{P}(A_1,\kappa) \underset{\mathbb{P}(A',\kappa)}{\times} \mathbb{P}(A_2,\kappa)$. Obviously, if $P \in \mathbb{P}(A,\kappa)$ is an A-bimodule defined over R, then T(P) is in $\Gamma_R(A,\kappa)$. Moreover, if $P \to Q$ is a morphism in ${}_R(A,\kappa)$ between objects in $\mathbb{P}(A,\kappa)$, then $T(P) \to T(Q)$ is a morphism in $\Gamma_R(A,\kappa)$. Denoting by $\Pi_R(A,\kappa)$ the full subcategory of $\Gamma_R(A,\kappa)$ consisting of objects in $\mathbb{P}(A_1,\kappa) \underset{\mathbb{P}(A',\kappa)}{\times} \mathbb{P}(A_2,\kappa)$ and by $P_R(A,\kappa)$ the full subcategory of ${}_R(A,\kappa)$, consisting of objects in $\mathbb{P}(A,\kappa)$, then T and $\mathcal{T}$ coincide on $P_R(A,\kappa)$, and map into $\Pi_R(A,\kappa)$. On the other hand, apply S to $M \in \Pi_R(A,\kappa)$, then we obtain an object in $\mathbb{P}(A,\kappa)$ (by construction !). More precisely, S and $\mathcal{S}$ agree on $\Pi_R(A,\kappa)$ and induce a functor $\Pi_R(A,\kappa) \to P_R(A,\kappa)$. From

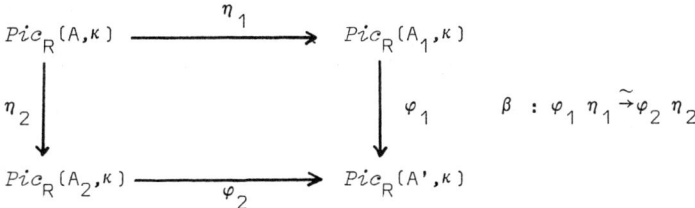

the proof of the fact that S and T define equivalences we infer that $\Pi_R(A,\kappa)$ and $P_R(A,\kappa)$ are equivalent as well. Let us denote (without ambiguity) by S and T the equivalences between $\Pi_R(A,\kappa)$ and $P_R(A,\kappa)$.
Now, $\mathit{Pic}_R(A,\kappa) \subset P_R(A,\kappa)$ resp. $\Pi^1_R(A,\kappa) = \mathit{Pic}_R(A_1,\kappa) \underset{\mathit{Pic}_R(A',\kappa)}{\times} \mathit{Pic}_R(A_2,\kappa)$ $\subset \Pi_R(A,\kappa)$ as full subcategories. It is clear that T maps $\mathit{Pic}_R(A,\kappa)$ into $\Pi^1_R(A,\kappa)$. Conversely, let us show that S maps $\Pi^1_R(A,\kappa) \subset \Pi_R(A,\kappa)$ to $\mathit{Pic}_R(A,\kappa)$.
Since there is an obvious left-right symmetry and since localizing at any $p \in C(\kappa)$ is compatible with all actions involved, the following lemma will suffice to finish the proof.

7.10. Lemma. Let

$$\begin{array}{ccc} A & \xrightarrow{h_1} & A_1 \\ {\scriptstyle h_2}\downarrow & & \downarrow{\scriptstyle f_1} \\ A_2 & \xrightarrow[f_2]{} & A' \end{array}$$

be a cartesian diagram of R-algebras, with properties as above. Let $P_i$ be an $A_i$-invertible bimodule and $\mu : A' \underset{A_1}{\otimes} P_1 \to A' \underset{A_2}{\otimes} P_2$ an isomorphism of A'-bimodules, then $S(P_1,\mu,P_2)$ is an invertible A-bimodule.

Proof. Clearly, $S(P_1,\mu,P_2) = P$ is finitely generated projective on both sides as an A-bimodule. It thus suffices to check that $\mathrm{End}({}_AP) \approx A$, via the action of A on P, the other side will follow similarly.
Take $\varphi \in \mathrm{End}({}_AP)$, then $\varphi$ may be identified with some $\tilde{\varphi} = (u_1,u_2)$, where $u_i \in \mathrm{End}({}_{A_i}P_i)$ such that the following diagram commutes

$$\begin{array}{ccc} A' \underset{A_1}{\otimes} P_1 & \xrightarrow{A' \otimes u_1} & A' \underset{A_1}{\otimes} P_1 \\ {\scriptstyle \mu}\downarrow & & \downarrow{\scriptstyle \mu} \\ A' \underset{A_2}{\otimes} P_2 & \xrightarrow[A' \otimes u_2]{} & A' \underset{A_2}{\otimes} P_2 \end{array}$$

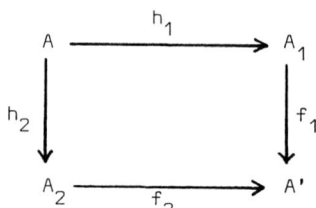

Now $P_i$ is an invertible $A_i$-bimodule, so $\mathrm{End}({}_{A_i}P_i) \approx A_i$, i.e. $u_i$ is just right multiplication by an element of $A_i$, which we also denote by $u_i$. In other words, we may identify $\tilde{\varphi}$ with some $u = (u_1,u_2) \in A_1 \times A_2$. Let us check that $u \in A_1 \underset{\Lambda'}{\times} A_2$. This follows from the commutativity of the above diagram : let $p \in P_1$ and assume $\mu(1 \otimes p) = \Sigma\, \lambda'_\alpha \otimes p_\alpha$, where $\lambda'_\alpha \in A'$ and $p_\alpha \in P_2$, then $\mu(1 \otimes p)f_1(u_1) = \mu(1 \otimes p)(1 \otimes u_1) = \mu(1 \otimes pu_1) = \mu(1 \otimes u_1(p))$
$\mu((A' \otimes u_1)(1 \otimes p)) = (A' \otimes u_2)\mu(1 \otimes p) = (A' \otimes u_2)(\Sigma\, \lambda'_\alpha \otimes p_\alpha) = \Sigma\, \lambda'_\alpha \otimes u_2(p_\alpha)$
$= \Sigma\, \lambda'_\alpha \otimes p_\alpha u_2 = (\Sigma\, \lambda'_\alpha \otimes p_\alpha)(1 \otimes u_2) = (\Sigma\, \lambda' \otimes p_\alpha)f_2(u_2) = \mu(1 \otimes p)f_2(u_2)$.
Since $\mu$ is also left $A'$-linear, it follows that $f_1(u_1)-f_2(u_2)$ annihilates $A' \underset{A_2}{\otimes} P_2$ on the right, hence $f_1(u_1) = f_2(u_2)$ as $A' \underset{A_2}{\otimes} P_2$ is faithful, being an invertible $A'$-bimodule. □

For notational convenience, assume all algebras to be $\kappa$-closed in the sequel of this section (otherwise, insert localization at $\kappa$ in the appropriate places!).For any algebra $A$ denote by $u(A)$ the multiplicative group $U(Z(A)) = Z(A)^*$ of central units. Keep the same notations and assumptions as before. There is a group homomorphism

$$\partial : u(A') \to \mathrm{Pic}_R(A,\kappa)$$

defined as follows. For any $u \in u(A')$, denote by $\mu_u : A' \to A'$ the left (= right) multiplication by $u$. The element $(A_1,\mu_u, A_2)$ then corresponds to an $A$-bimodule in $\mathcal{Pic}_R(A,\kappa)$, which we denote by $Au$. The map $\partial$ then sends $u \in u(A')$ to $[\,Au\,] \in \mathrm{Pic}_R(A,\kappa)$.

7.11. Theorem. With assumptions as above, we obtain an exact sequence of groups

$$1 \to u(A) \xrightarrow{i} u(A_1)\times u(A_2) \xrightarrow{m} u(A') \xrightarrow{\partial} \mathrm{Pic}_R(A,\kappa) \xrightarrow{p} \mathrm{Pic}_R(A_1,\kappa)\times \mathrm{Pic}_R(A_2,\kappa)$$

Proof. The maps are defined as follows : $i(u) = (h_1(u),h_2(u)) \in u(A_1)\times u(A_2)$ resp. $p([\,P\,]) = ([\eta_1 P]\,,\,[\eta_2 P]\,)$. Moreover, $m$ maps $(u_1,u_2) \in u(A_1)\times u(A_2)$

to $f_1(u_1)f_2(u_2)^{-1} \in u(A')$. It is clear that the composition of two successive morphisms is trivial. On the other hand, since $A = A_1 \times_{A'} A_2$, it is obvious that i is injective. If $m((u_1,u_2)) = 1$, then $f_1(u_1) = f_2(u_2)$ and clearly $(u_1,u_2) \in u(A)$. For a $P \in \mathit{Pic}_R(A,\kappa)$ such that $p([P]) = 1$, we have $\eta_1 P \cong A_1$ and $\eta_2 P \cong A_2$. But then, since the bimodule automorphisms of A' stem for u(A') by (1.16.), it follows that $P \cong Au$ for some $u \in u(A')$, so $[P] \in \mathrm{Im}(\partial)$. Finally, $\mathrm{Ker}(\partial) = \mathrm{Im}(m)$ follows from the following lemma. □

7.12. Lemma. The following assertions are equivalent :

1. $Au \cong Au'$ as A-bimodules over R,
2. there exists $(u_1,u_2) \in u(A_1) \times u(A_2)$, such that $f_1(u_1)uf_2(u_2) = u'$.

Proof. Only (1.) ⇒ (2.) needs a proof. If $Au \cong Au'$, then $(A_1,\mu_u,A_2) \cong (A_1,\mu_{u'},A_2)$, hence we may find $A_i$-bimodule automorphisms $\varphi_i$ of $A_i$ such that the following diagram is commutative

$$\begin{array}{ccc} A' \otimes_{A_1} A_1 & \xrightarrow{A' \otimes \varphi_1} & A' \otimes_{A_1} A_1 \\ \downarrow{\scriptstyle \mu_u} & & \downarrow{\scriptstyle \mu_{u'}} \\ A' \otimes_{A_2} A_2 & \xrightarrow[A' \otimes \varphi_2]{} & A' \otimes_{A_2} A_2 \end{array}$$

Since $\varphi_1$ is multiplicative by $u_1^{-1}$ and $\varphi_2$ is multiplication by $u_2$, for some $u_i \in u(A_i)$, we have $u'f_1(u_1^{-1}) = f_2(u_2)u$, i.e. $u' = f_1(u_1)uf_2(u_2)$. □

7.13. Corollary. If $\mathrm{Pic}_R(A',\kappa) = 1$, then p is surjective.

Proof. Pick $P_i \in \mathit{Pic}_R(A_i,\kappa)$ $(i = 1,2)$, then $\varphi_1 P_1 \cong \varphi_2 P_2$, so there exists $P \in \mathit{Pic}_R(A,\kappa)$ with $\eta_i P \cong P_i$. □

7.14. Corollary. If A' is commutative, then the exact sequence above extends to

$$\ldots \to \mathrm{Pic}_R(A_1,\kappa)\times\mathrm{Pic}_R(A_2,\kappa) \xrightarrow{q} \mathrm{Pic}_R(A',\kappa),$$

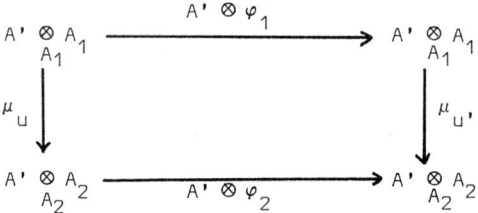

where $q([P_1], [P_2]) = [\varphi_1 P_1][\varphi_2 P_2]^{-1} \in \mathrm{Pic}_R(A',\kappa)$. □

7.15. Note. Earlier we defined the notion of a $\kappa$-quasifree left A-module. Let us now say that M is $\kappa$-quasifree *of rank* n (on the left) if for all $p \in C(\kappa)$ the rank of $M_p$ as a left $A_p$-module is n. If P is a $\kappa$-invertible A-bimodule over R, then P is $\kappa$-quasifree of rank 1 on the left if and only if it is $\kappa$-quasifree of rank 1 on the right. Define $\mathrm{LFP}_R(A,\kappa)$ to consist of all $[P] \in \mathrm{Pic}_R(A,\kappa)$ such that P is $\kappa$-quasi of rank 1. If is easy to see that $\mathrm{LFP}_R(A,\kappa)$ is a subgroup of $\mathrm{Pic}_R(A,\kappa)$. Moreover, applying (7.8.) one may show, mimicking the proof of (7.11.) that there is an exact sequence

$$1 \to u(A) \to u(A_1)\times u(A_2) \to u(A') \to \mathrm{LFP}_R(A,\kappa) \to \mathrm{LFP}_R(A_1,\kappa) \times \mathrm{LFP}_R(A_2,\kappa).$$

For $\kappa = \sigma_1 = \inf\{\sigma_{R-p}; p \in X^{(1)}(R)\}$, this recovers one of the main results of I. Reiner, S. Ullom [123] proved there for RR noetherian integrally closed.

Let us conclude this section by another easy application. Assume R is a noetherian domain with finite normalization $\overline{R}$ and let $I = c(\overline{R}/R)$ be the conductor of $\overline{R}$ in R. Assume that the algebra A is a finitely generated projective module over the central subring R and let $\Sigma = A \otimes_R K$ be a central simple algebra, then $\overline{A} = A \otimes_R \overline{R}$ is an R-order in $\Sigma$. Let $\sigma_1$ be as before and denote by $\overline{\sigma}_1$ the induced idempotent kernel functor in $\overline{R}$-mod. We claim that $\overline{\sigma}_1 = \inf\{\kappa_{\overline{R}-P}; P \in X^{(1)}(\overline{R})\}$. Indeed, let $P \in X(\overline{\sigma}_1)$ i.e. (by definition) P is prime and $\sigma_1(\overline{R}/P) = \overline{\sigma}_1(\overline{R}/P) = 0$. If $p = P \cap R$, then this implies that $\sigma_1(R/P) = 0$, so $\mathrm{ht}(p) \leqslant 1$. The incomparability-property then implies that $\mathrm{ht}(P) \leqslant 1$. Conversely, if ht $P \leqslant 1$, then $\sigma_1(\overline{R}/P) = 0$, indeed, if not, then $\sigma_1(\overline{R}/P) \neq 0$, so $P \in L(\overline{\sigma}_1)$, hence $p = P \cap R \in L(\sigma_1)$, i.e. ht $p > 1$. Since $\overline{R}/R$ is finite, we may apply "going down" and "incomparability" to derive a contradiction.

Let $J = A \otimes_R I = AI \subset A$, then we obtain a Milnor square

$$\begin{array}{ccc} A & \longrightarrow & \overline{A} \\ \downarrow & & \downarrow \\ A/J & \longrightarrow & \overline{A}/J \end{array}$$

hence a long exact sequence :

$$1 \to u(Q_\sigma(A)) \to u(Q_{\sigma_1}(\overline{A})) \otimes u(Q_{\sigma_1}(A/J)) \to u(Q_{\sigma_1}(\overline{A}/J)) \to \mathrm{Pic}_R(A,\sigma_1) \to$$

$$\to \mathrm{Pic}_R(\overline{A},\overline{\sigma}_1) \times \mathrm{Pic}_R(A/J,\sigma_1)$$

Let us take a concrete example : assume R to be an irreducible algebraic variety having only one nonnormal point of codimension 1, say $p \in X^{(1)}(R)$ and let $P_1$, $P_2$ be the points of the normalization lying over it. We have $I = p = P_1 \cap P_2$. Let $\sigma_1'$ be the idempotent kernel functor induced by $\sigma_1$ in R/p-mod, then $\sigma_1' = \sigma_{R/p-0}$. Hence $Q_{\sigma_1}(A/J) = Q_{R/p-0}(A \otimes_R R/p) = A \otimes_R \mathbb{k}(p) =: A(p)$. Similarly $\mathrm{Pic}_R(A/J,\sigma) = \mathrm{Pic}_R(A(p))$. Now, in order to calculate the remaining terms, denote by T the ideal $P_1 + P_2$ of $\overline{R}$ and let

$$R/p \longrightarrow \overline{R}/P_1 \oplus \overline{R}/P_2$$

denote the canonical inclusion. It is finite, and T/p is a common ideal of both rings. We thus obtain a Milnor square

$$\begin{array}{ccc} \overline{A}/p\overline{A} & \longrightarrow & \overline{A}/P_1\overline{A} \oplus \overline{A}/P_2\overline{A} \\ \downarrow & & \downarrow \\ \overline{A}/T\overline{A} & \longrightarrow & \overline{A}/T\overline{A} \oplus \overline{A}/T\overline{A} \end{array}$$

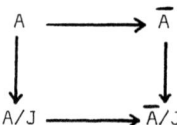

One easily checks that $T \in \mathcal{L}(\overline{\sigma}_1)$, hence from the Mayer-Vietoris sequence one obtains that $u(Q_\sigma(\overline{A}/p\overline{A}) \cong u(Q_\sigma(\overline{A}/P_1\overline{A})) \oplus u(Q_\sigma(\overline{A}/P_2\overline{A}))$ - this may also be derived directly from the above cartesian diagram, of course. Now, arguing as before, one easily sees that $u(Q_{\sigma_1}(\overline{A}/p\overline{A})) = u(\overline{A}(P_1)) \oplus u(\overline{A}(P_2))$, and the sequence reduces to :

$$1 \to u(Q_{\sigma_1}(A)) \to u(Q_{\sigma_1}(\overline{A})) \oplus u(A(P)) \to u(\overline{A}(P_1)) \oplus u(\overline{A}(P_2)) \to \mathrm{Pic}_R(A,\sigma_1) \to$$

$$\to \mathrm{Pic}_R(\overline{A},\sigma) \oplus \mathrm{Pic}_R(A(p)).$$

Now in many special cases this sequence will simplify considerably, e.g. : if $\overline{A}$ happens to be a *maximal* order over $\overline{R}$, then $Q_{\sigma_1}(\overline{A}) = \overline{A}$ (since $\overline{A}$ is then reflexive) and $\mathrm{Pic}_R(\overline{A},\sigma_1)$ reduces to Cl(A), cf. J. Riley [124]. We will come back to this in chapter E.

# C RELATIVE AZUMAYA ALGEBRAS REVISITED

## C.1. Relative Azumaya Algebras and Generalized Crossed Products.

Some of the results and techniques developed in [162] Chapter III will be prerequisites for this section. Since relative Azumaya algebras will be about the nicest maximal orders we will encounter, we aim to make their description as complete as possible. Galois cohomology and related crossed product theorems present a first approach to the problem. Here however we proceed to derive directly certain structural properties of relative Azumaya algebras which turn out to be equivalent to the existence of the Chase-Rosenberg sequence in Galois cohomology. This method works because the crossed product theory may be freed from assumptions about the vanishing of certain Picard groups at the cost (actually we consider it a gain) of utilizing generalized crossed products instead of the common crossed products. We refer to T. Kanzaki [ 79 ] for certain results in the absolute case while the relative case appeared in [ 26 ] and [ 27 ] . At the end of the section we include a few remarks concerning graded Azumaya algebras. Apart from putting right some imprecise statements in Chapter VI, Section VI.5. in [ 162] (due to the non-vanishing of the graded Picard group), these remarks are also useful in the constructive methods of Chapter D.

Let A be a graded ring of type G, say $A = \bigoplus_{\sigma \in G} A_\sigma$. We consider an idempotent kernel functor $\kappa$ on $A_e$-mod. We say that A is <u>relatively strongly graded with respect to $\kappa$</u> or simply <u>$\kappa$-graded</u> if $Q_\kappa(A_\sigma A_\tau) = A_{\sigma\tau}$ for all $\sigma, \tau \in G$. In case $\kappa$ is trivial, i.e. $\mathcal{L}(\kappa) = \{A_e\}$; then A is $\kappa$-graded if and only if it is strongly graded. A less trivial example may be obtained as follows. Let $\Lambda$ be a tame order over a Krull domain and let $\sigma_1$ be the idempotent kernel functor $\inf\{\kappa_{Z(\Lambda)-p}; p \in X^{(1)}(Z(\Lambda))\}$. A $\sigma_1$-graded ring A with $A_e = \Lambda$ is nothing but a divisorially graded ring in the sense of [26,97 ]; these will reappear in subsequent chapters. In the sequel of this section we always assume $\kappa$ to be a noetherian idempotent kernel functor; this condition is equivalent to the fact that $Q_\kappa(-)$ commutes with direct sums and it is clearly a very mild restriction. If $A_e$ is a maximal order over a Krull domain, then $A_e$ certainly has the property that $\sigma$ is noetherian. We will also assume for simplicity's sake $\kappa$ to be symmetric. Actually, in most applications, $\kappa$ will be a *central* kernel functor. With these notations and conventions we have :

1.1. Lemma. The ring A is $\kappa$-graded by G if and only if $Q_\kappa(AA_{\bar\tau}) = A$ for all $\tau \in G$. In this case every graded left ideal L of A has the property that $Q_\kappa(L) = Q_\kappa(AL_e)$.

Proof. Suppose that A is $\kappa$-graded, then $A_\sigma$ is $\kappa$-closed for all $\sigma \in G$ and since $\kappa$ is noetherian, $A = \bigoplus_{\sigma \in G} A_\sigma$ is also $\kappa$-closed. Moreover, for $x \in A_\sigma$ we may calculate that $A_{\sigma\tau^{-1}} A_{\tau\sigma^{-1}} x \subset AA_\tau$. As $A_{\sigma\tau^{-1}} A_{\tau\sigma^{-1}} \in \mathcal{L}(\kappa)$, we obtain that $Q_\kappa(AA_\tau) = A$ for all $\tau \in G$. Conversely, from $Q_\kappa(AA_\tau) = A$ for all $\tau \in G$, it follows that for every $x \in A_e$, there is an $I \in \mathcal{L}(\kappa)$ such that $Ix \subset AA_\tau$. Hence $Ix \subset (AA_\tau)_e$ or $Ix \subset A_{\tau^{-1}}A_\tau$, and thus $x \in Q_\kappa(A_{\tau^{-1}}A_\tau)$ or $A_e \subset Q_\kappa(A_{\tau^{-1}}A_\tau)$. On the other hand, as $A_{\tau^{-1}}A_\tau \subset A_e$ and $A_e$ is $\kappa$-closed (since A is $\kappa$-closed), it follows that $A_e = Q_\kappa(A_{\tau^{-1}}A_\tau)$ for all $\tau \in G$. We now first check the second statement in the lemma. If L is a graded left ideal and $x \in L_\tau$ for some $\tau \in G$, then $A_\tau A_{\tau^{-1}} x \subset AL_e$, hence $L \subset Q_\kappa(AL_e)$ and also $Q_\kappa(L) = Q_\kappa(AL_e)$. Now, for all $\gamma, \tau \in G$ we have $A_\gamma A_{\gamma^{-1}} A_{\gamma\tau} \subset A_\gamma A_\tau$ and since $A_\gamma A_{\gamma^{-1}} \in \mathcal{L}(\kappa)$, we finally obtain $A_{\gamma\tau} = Q_\kappa(A_\gamma A_\tau)$ for all $\gamma, \tau \in G$. □

We say that $\kappa$ is G-*invariant* if for all $I \in \mathcal{L}(\kappa)$ and all $\sigma \in G$ we have $A_\sigma I A_{\sigma^{-1}} \in \mathcal{L}(\kappa)$. Obviously, if $\kappa$ central, i.e. $\mathcal{L}(\kappa)$ has a cofinal subset consisting of ideals which are centrally generated, then $\kappa$ is G-invariant.

1.2. Proposition. Let A be $\kappa$-graded by G, where $\kappa$ is noetherian, symmetric and G-invariant. If M is a graded $\kappa$-closed left A-module, then $Q_\kappa(A \otimes_{A_e} M_e) \cong M$ as graded A-modules.

Proof. Consider the exact sequence $0 \to K \to A \otimes_{A_e} M_e \xrightarrow{f} M$, where f is the canonical graded morphism of degree e. From $\kappa\, M = 0$, we derive that $\kappa(A \otimes_{A_e} M_e) \subset K$. On the other hand, writing an $x \in K_\tau$ as $x = \sum_i \lambda_\tau^{(i)} \otimes m_e^{(i)}$ with $\lambda_\tau^{(i)} \in A_\tau$ and $m_e^{(i)} \in M_e$, we derive from $f(x) = 0$ that $A_\tau A_{\tau^{-1}} x \subset A_\tau \otimes_{A_e} A_{\tau^{-1}} \sum_i \lambda_\tau^{(i)} m_e^{(i)} = 0$, hence $\kappa \in \kappa(A \otimes_{A_e} M_e)$ or $K = \kappa(A \otimes_{A_e} M_e)$. Therefore, the canonical map $i : Q_\kappa(A \otimes_{A_e} M_e) \to Q_\kappa(M) = M$ is monomorphic. Now Im(i) is a $\kappa$-closed module and for $m_\tau \in M_\tau$ we have $A_\tau A_{\tau^{-1}} m_\tau \subset A_\tau M_e$ with $A_\tau A_{\tau^{-1}} \in \mathcal{L}(\kappa)$, thus $m_\tau \in i(Q_\kappa(A \otimes_{A_e} M_e))$ or i is onto. Finally, since $M_\tau / A_\tau M_e$ is $\kappa$-torsion, it follows that $Q_\kappa(A_\tau \otimes_{A_e} M_e) = M_\tau$ and i is graded of degree e. □

Let $\kappa$ be as before, then

1.3. Theorem. Let A be $\kappa$-graded by G. The $A_e$-bimodule isomorphism class $[A_\sigma]$ of $A_\sigma$ is in $\mathrm{Pic}(A_e, \kappa)$ for all $\sigma \in G$. Actually $\sigma \mapsto [A_\sigma]$ defines a group homomorphism $G \to \mathrm{Pic}(A_e, \kappa)$.

Proof. We first prove that $Q_\kappa(M) \in$ A-gr. Since $\kappa\, M$ is a graded submodule of M, we may assume that M is $\kappa$-torsion free, i.e. $M \subset Q_\kappa(M)$. Put $M'_\tau = Q_\kappa(M_\tau)$ in $Q_\kappa(M)$. For $\gamma \in G$ and $x \in M'_\tau$ we obtain $R_\gamma I R_{\gamma^{-1}} X R_\gamma x \subset M_{\gamma\tau}$, if $I \in \mathcal{L}(\kappa)$ has been chosen such that $I x \subset M_\tau$. Since $\kappa$

G-invariant, $R_{\gamma\tau} \subset M'_{\gamma\tau}$ and $M' = \bigoplus_{\gamma \in G} M'_\gamma$ is a graded R-module, containing M. Because $\kappa$ is noetherian, $Q_\kappa(-)$ commutes with direct sums and thus $M' = Q_\kappa(M)$. So we have established our claim that $Q_\kappa(M)$ is graded with $Q_\kappa(M)_\gamma = Q_\kappa(M_\gamma)$ for all $\gamma \in G$. Secondly, if we take for M in Proposition (1.2.) the $\tau$-shifted A-module $A(\tau)$ defined by $A(\tau)_\sigma = A_{\sigma\tau}$ for all $\sigma \in G$, then $A(\tau) = Q_\kappa(A \otimes_{A_e} A_\tau)$ and $A_{\sigma\tau} = Q_\kappa(A_\sigma \otimes_{A_e} A_\tau)$ for all $\sigma \in G$. In particular, for $\tau = \sigma^{-1}$ we obtain $A = Q_\kappa(A_\sigma \otimes_{A_e} A_{\sigma^{-1}}) = Q_\kappa(A_{\sigma^{-1}} \otimes_A A_\sigma)$.

As a consequence of this, it will suffice to establish that $A_\sigma$ is $\kappa$-flat in order to prove that $A_\sigma$ is $\kappa$-invertible. Consider a $\kappa$-torsion left $A_e$-module M" and put $M = Q_\kappa(A \otimes_{A_e} M")$, hence M is G-graded such that $M_e \cong Q_\kappa(M")$. Therefore we have that $M = Q_\kappa(A \otimes_{A_e} M") = Q_\kappa(A \otimes_{A_e} M_e) = Q_\kappa(A \otimes_{A_e} Q_\kappa(M"))$. From $Q_\kappa(M") = 0$ we then derive that $M = 0$ and $A \otimes_{A_e} M"$ is $\kappa$-torsion. We are left to establish the following property : if $i : M' \to M$ is a monomorphism in $R_e$-mod, then $\mathrm{Ker}(A_\sigma \otimes i)$ is $\kappa$-torsion. We start from the following exact sequence :

$$0 \to K_1 \to A_{\sigma^{-1}} \otimes_{A_e} K \to A_{\sigma^{-1}} \otimes_{A_e} A_\sigma \otimes_{A_e} M' \to A_{\sigma^{-1}} \otimes_{A_e} A_\sigma \otimes_{A_e} M.$$

By localization at $\kappa$ we obtain :

$$0 \to Q_\kappa(K_1) \to Q_\kappa(A_{\sigma^{-1}} \otimes_{A_e} K) \to Q_\kappa(M') \cong Q_\kappa(A_{\sigma^{-1}} \otimes_{A_e} A_\sigma \otimes_{A_e} M') \to$$

$$\to Q_\kappa(M) \cong Q_\kappa(A_{\sigma^{-1}} \otimes_{A_e} A_\sigma \otimes_{A_e} M).$$

Since $Q_\kappa(M') \subset Q_\kappa(M)$ we have $Q_\kappa(A_{\sigma^{-1}} \otimes_{A_e} K) = 0$ and thus $A_{\sigma^{-1}} \otimes_{A_e} K$ is $\kappa$-torsion. By the first part then, $A_\sigma \otimes_{A_e} A_{\sigma^{-1}} \otimes_{A_e} K$ is also $\kappa$-torsion and this finally entails that K is $\kappa$-torsion, because $Q_\kappa(K) =$ $= Q_\kappa(A_\sigma \otimes_{A_e} A_{\sigma^{-1}} \otimes_{A_e} K) = 0$. □

Let us recall, for completeness' sake, some basic definitions and properties from the commutative theory.

Let R be an arbitrary commutative ring, let A be an R-algebra and let $\kappa$ be an idempotent kernel functor in R-mod. Recall that if A is commutative, then we call A a $\kappa$-Galois extension of R if :

1. A is a $\kappa$-separable extension of R and a $\kappa$-progenerator, where A is $\kappa$-separable if and only if it is a $\kappa$-quasiprojective $Q_\kappa(A^e) = A^f$-algebra.
2. there is a finite group G of R-automorphisms of R such that $R = A^G$.
3. If $\overline{\kappa}$ is the idempotent kernel functor induced on A-mod by $\kappa$, then for each $P \in \mathcal{C}(\kappa)$ and each $\sigma \neq e$ in G there exists an $a \in A$ such that $\sigma(a) - a \notin P$.

From the commutative theory we recall :

1.4. Lemma 1. 1. Let R and $\kappa$ be as above, assume that R is $\kappa$-closed and let A be a $\kappa$-closed R-algebra which is $\kappa$-finitely presented as an R-module. Then A is a $\kappa$-Azumaya algebra over R if and only if $A_p$ is an Azumaya algebra over $R_p$ for all $p \in \mathcal{C}(\kappa)$.
2. If A is a commutative $\kappa$-closed R-algebra, which is a $\kappa$-progenerator and $\kappa$-separable over R and if G is a finite group of R-automophisms of A such that $R = A^G$, then A is a $\kappa$-Galois extension of R if and only if $A_p$ is a Galois extension of $R_p$ with Galois group G for every $p \in \mathcal{C}(\kappa)$.

Proof. 1. cf Proposition III.3.8., p. 55 in [162] ; 2. cf. Proposition III.2.6. , p. 51 in [162]. □

We are now ready to state the main result of this section:

1.5. Theorem. Let A be a $\kappa$-Azumaya algebra over R containing as a maximal commutative subring a $\kappa$-Galois extension S of R with Galois group G, then A is $\kappa$-graded by G such that $A_e = S$.

Proof. For each $\sigma \in G$ we consider $A_{(\sigma)} = \{u \in A;\ u\,s = \sigma(s)\,u$ for all $s \in S\}$. Since $0 \in A_{(\sigma)}$ for each $\sigma \in G$, we have that each $A_{(\sigma)}$ is nonempty and one easily verifies that each $A_{(\sigma)}$ is an S-bimodule.

It is also clear that $A_{(\sigma)} A_{(\tau)} \subset A_{(\sigma\tau)}$ for any $\sigma, \tau \in G$. We claim that the $A_{(\sigma)}$ are $\kappa$-closed. As A itself is $\kappa$-closed, it will suffice to check that $I\,a \subset A_{(\sigma)}$ with $I \in \mathcal{L}(\kappa)$ and $a \in A$ implies that $a \in A_{(\sigma)}$. Now, $I\,a \subset A_{(\sigma)}$ entails that $i\,a\,s - \sigma(s)i\,a = 0$ for all $s \in S$ and $i \in I$, hence $I(as - \sigma(s)a) = 0$, so $(as - \sigma(s)a) \in \kappa A = 0$, and $a \in A_{(\sigma)}$ follows. Take a prime ideal $p \in \mathcal{C}(\kappa)$, then the foregoing lemma yields that after localizing at p we obtain a situation $R_p \subset S_p \subset A_p$, where $A_p$ is an Azumaya algebra over the local ring $R_p$ and containing a Galois extension $S_p$ over $R_p$ with Galois group G as a maximal commutative subring. Since $R_p$ and $S_p$ have trivial Picard group, it follows that $A_p$ is a crossed product : $A_p = \bigoplus_{\sigma \in G} S_p u_\sigma$, where each $u_\sigma$ is a unit of $A_p$ inducing $\sigma$ on $S_p$ by conjugation. Now, we may find $t \in R\text{-}p$ such that $tu_\sigma \in j_{R-p}(A) = A/\kappa_{R-p}A$. Pick any $y \in A$ such that $j_{R-p}: A \to A/\kappa_{R-p}A$ maps y to $tu_\sigma$, then $ys - \sigma(s)y \in \kappa_{R-p}A$ for any $s \in S$. Since S is $\kappa$-finitely generated, we may choose a finitely generated R-module $S' = Rx_1 + \ldots + Rx_m$ such that $S/S'$ is $\kappa$-torsion. For each $1 \leqslant i \leqslant m$ we may select $\lambda_i \in R\text{-}p$ such that $\lambda_i(yx_i - \sigma(x_i)y) = 0$. Put $\lambda = \lambda_1 \ldots \lambda_m$, then $\lambda \in R\text{-}p$ and $\lambda(ys' - \sigma(s')y) = 0$ for all $s' \in S'$. If $s \in S$, then $J\,s \subset S'$ for some $J \in \mathcal{L}(\kappa)$, consequently $\lambda(yjs - \sigma(js)y) = 0$ for all $j \in J$. Now, $j \in R$ is left fixed under $\sigma$ and it is a central element, hence $j\lambda(ys - \sigma(s)y) = 0$ for all $j \in J$, i.e. $\lambda(ys - \sigma(s)y) \in \kappa\,A = 0$ and $\lambda\,y \in A_{(\sigma)}$ or $\lambda\,t\,u_\sigma \in S_pA_{(\sigma)} = (A_{(\sigma)})_p$, but $\lambda t \in R\text{-}p$ then yields that $u_\sigma \in S_pA_{(\sigma)} = (A_{(\sigma)})_p$. Consequently, $A_p = \bigoplus_{\sigma \in G} (A_{(\sigma)})_p$ and $(A_{(\sigma)})_p = S_p\,u_\sigma$.

Now, for every $\sigma \neq \tau$ in G, we have that $A_{(\sigma)} \cap A_{(\tau)}$ is $\kappa$-closed, whereas on the other hand, this module localizes to zero at each $p \in \mathcal{C}(\kappa)$, since $(A_{(\sigma)})_p \cap (A_{(\tau)})_p = S_p\,u_\sigma \cap S_p\,u_\tau = 0$.

Therefore $A_{(\sigma)} \cap A_{(\tau)} = 0$ for every $\sigma \neq \tau$ in G.

Consider $B = \bigoplus_{\sigma \in G} A_{(\sigma)} \subset A$. This is a $\kappa$-closed R-module, and A/B is $\kappa$-torsion as $B_p = A_p$ for each $p \in \mathcal{C}(\kappa)$, hence B = A. Finally, $A_{(\sigma)} A_{(\tau)} \subset A_{(\sigma\tau)}$ and both coincide locally at each $p \in \mathcal{C}(\kappa)$, hence $Q_\kappa(A_{(\sigma)} A_{(\tau)}) = Q_\kappa(A_{(\sigma\tau)}) = A_{(\sigma\tau)}$. Putting $A_\sigma = A_{(\sigma)}$ for $\sigma \in G$ proves that A is $\kappa$-graded over $S = A_e$. □

1.6. Special Case : The absolute case.

If $\mathcal{L}(\kappa)$ is trivial in the foregoing theorem then the result states that every Azumaya algebra containing a Galois extension S of its center R (with Galois group G) as a maximal commutative subring can be viewed as a strongly graded ring (of type G) with $A_e = S$. In this way we recover results of T. Kanzaki [ 79 ]. Let us recall how strongly graded rings of the type described above relate to $H^2(G,U(S))$, cf. C. Nastasescu, F. Van Oystaeyen [107 ], Chapter A.I.3., Theorem I.3.16 and Corollary I.3.18. For $\sigma,\tau \in G$ we have $A_e$-bimodule isomorphisms $g_{\sigma,\tau} : A_\sigma \otimes_{A_e} A_\tau \to A_{\sigma\tau}$. The morphism $\Phi : G \to \text{Pic}(A_e)$ gives rise to a morphism $G \to \text{Pic}(A_e) \to \text{Aut}(Z(A_e))$ and this defines an action G on $U(Z(A_e))$.
In the case we are studying here, this action corresponds to the Galois action of G on U(S). The family $\{g_{\sigma,\tau} ; \sigma, \tau \in G\}$ forms a factor set with respect to $\Phi$. If $\{f_{\sigma,\tau} ; \sigma,\tau \in G\}$ and $\{g_{\sigma,\tau} ; \sigma,\tau \in G\}$ are two factor sets associated to the same $\Phi : G \to \text{Pic}(A_e)$, then the $q_{\sigma,\tau} = g_{\sigma,\tau}\, i\, f^{-1}_{\sigma,\tau}$ define $A_e$- bimodule automorphisms of $A_\sigma \otimes_{A_e} A_\tau$ and then $q_{\sigma,\tau}$ is just multiplication by some $\overline{q}_{\sigma,\tau} \in U(Z(A_e))$. The map $\overline{q} : G \times G \to U(Z(A_e)) = S$ defines a 2-cocycle and A corresponds upto graded isomorphisms over S to an element of $H^2(G,U(S))$ once $\Phi$ has been fixed. The explicit structure of an Azumaya algebra representing an $\alpha \in \text{Br}(S/R)$ in terms of $\Phi : G \to \text{Pic}(S)$ and some $\overline{q} \in H^2(S/R,U)$ entails exactness of the following sequence :

$$H^0(S/R,\text{Pic}) \to H^2(S/R,U) \to \text{Br}(S/R) \to H^1(S/R,\text{Pic}).$$

By purely cohomological methods, this sequence may be completed to the well-known Chase-Rosenberg exact sequence :

$$1 \to H^1(S/R,U) \to Pic(R) \to H^0(S/R,Pic) \to H^2(S/R,U) \to Br(S/R) \to H^1(S/R,Pic) \to$$
$$\to H^3(S/R,U).$$

So, we have obtained a rather "direct" proof of this sequence.

1.7. Special case : the divisorial case.

We assume R to be a Krull domain and $\sigma_1$ associated to the set $X^{(1)}(R) \subset$ Spec(R) as before. In this case theorem 1.5. states that a reflexive Azumaya algebra, i.e. a representative of an element in the "reflexive Brauer group" $\beta(R)$ of R, cf. M. Orzech [ 112] or F. Van Oystaeyen, A. Verschoren [ 162], containing a reflexive Galois extension S of R with Galois group G is a divisorial ring. In C. Năstăsescu , F. Van Oystaeyen [109 ] it was shown that divisorially graded rings A relate in a similar way to $H^2(G,U(Z(A_e)))$, i.e. $H^2(G,U(S))$ in our case, with $Pic(S,\kappa)$ replacing Pic(S) everywhere. Now, $Pic(R,\sigma_1) = Cl(R)$, the class-group of R. The explicit structure of a reflexive Azumaya representing $\alpha \in \beta(S/R)$ in terms of $\Phi : G \to Pic(S,\sigma_1)$ and some $\bar{q} \in H^2(S/R,U)$ entails exactness of the following sequence :

$$H^0(G,Pic(S,\sigma_1)) \to H^2(G,U(S)) \to \beta(S/R) \to H^1(G,Pic(S,\sigma_1))$$

Clearly, one of the useful consequences of Theorem 1.5. is that $\kappa$-Azumaya algebras split by a $\kappa$-Galois extension may be studied by using relative techniques on one hand and graded techniques on the other. In the commutative theory, $\mathbb{Z}$-graded relative Azumaya algebras played an interesting role in describing Br((Proj(R)) and $Br^g(R)$ for a $\mathbb{Z}$-graded ring R. For completeness' sake, we verify that the above results are compatible with $\mathbb{Z}$-gradations, whenever there are being considered. We start with a few additions to the theory of $\mathbb{Z}$-graded Galois extensions, which have not been included in [162] .

Throughout R is a $\mathbb{Z}$-graded commutative ring and S is a graded R-algebra.

1.8. Lemma. The idempotents of a $\mathbb{Z}$-graded commutative ring R are homogeneous of degree zero.

Proof. If R is gr-local, i.e. R has a unique maximal graded ideal, then the result is easily verified. In general, suppose e is an idempotent in R and let $e_0$ be the part of degree zero in the homogeneous decomposition of e in R. For every general prime ideal p of R, the element $e-e_0$ maps to zero in $Q_p^g(R) = \Sigma^{-1}R$, where $\Sigma = (R-p) \cap h(R)$. Hence $e-e_0 \in \kappa^g_{R-p}(R)$ for all graded prime ideals p of R, consequently $e-e_0 = 0!$ □

A graded Azumaya algebra over a graded ring R is just an Azumaya algebra over R, which happens to be graded, cf. [151 ]. A similar definition holds for Galois extensions.

1.9. Theorem. Let R be a commutative $\mathbb{Z}$-graded ring and let S be a $\mathbb{Z}$-graded Galois extension of R with Galois group G. All R-isomorphisms of S in G are homogeneous of degree zero.

Proof. We first introduce some terminology, following F. De Meyer, E. Ingraham [42 ]. Let $\Delta = \Delta(S:G)$ be the S-algebra defined by taking the free S-module generated by $\{u_\sigma; \sigma \in G\}$ and introducing a multiplication law by $(au_\sigma)(bu_\tau) = a\sigma(b)u_{\sigma\tau}$ for all $\sigma, \tau \in G$. Let $\nabla = \nabla(S:G)$ be the S-algebra defined by taking the free S-module generated by $\{v_\sigma; \sigma \in G\}$ with multiplication $(av_\sigma)(bv_\tau) = ab\,\delta_{\sigma,\tau}\, v_\sigma$. Both $\Delta$ and $\nabla$ are graded S-algebras if we put $\deg(u_\sigma) = \deg(v_\sigma) = 0$ for all $\sigma \in G$. If S is a Galois extension of R, then it follows from [ 42 ], prop. III 1.2. that there is an algebra isomorphism $f: S \otimes_R S \to \nabla(S:G)$ given by $f(a \otimes b) = \sum_{\sigma \in G} a\,\sigma(b)v_\sigma$.

Write $e_\sigma = f^{-1}(v_\sigma)$, then $e_\sigma$ is an idempotent, hence it has degree zero. Now, consider an element $s \in S$, which is homogeneous of degree $r \in \mathbb{Z}$. Let $p_\sigma\ \nabla(S:G) \to S$ be the projection on the component $Sv_\sigma$, then $p_\sigma f(1 \otimes s) = p_\sigma(\sum_{\sigma \in G} \sigma(b)v_\sigma) = \sigma(b)$ actually has degree r, and so $\sigma$ has degree 0. □

1.10. Corollary. If S is a graded Galois extension of R with Galois group G, then the canonical isomorphisms $g:\Delta(S:G) \to \mathrm{Hom}_R(S,S)$ and $f : S \otimes_R S \to \nabla(S:G)$ are graded of degree zero.

Proof. It has been established above that f has degree zero. To see that g has degree zero, it suffices to note that each $\sigma \in G$ is a graded morphism of degree zero and that g is defined by $g(au_\sigma)(x) = a\sigma(x)$, for $a,x \in S$. □

As a further corollary we obtain the equivalence between the notion of graded Galois extension and the notion of gr-Galois extensions, which are defined completely in the framework of the intrinsic graded theory.

1.11. Corollary. Let S be a $\mathbb{Z}$-graded extension of R and let G be a finite group of R-automorphisms of S, then the following statements are equivalent :

1. S is a graded Galois extension of R;
2. a. $S^G = R$;
   b. $\sigma$ is a graded morphism of degree zero, for every $\sigma \in G$;
   c. for each gr-maximal ideal $m$ of S and for each $\sigma \neq e$ in G, there is an $x \in S$ such that $\sigma(x)-x \notin m$.

Proof. 1. ⇒ 2. is straightforward. To prove 2. ⇒ 1., let $\sigma \neq e$ and consider the graded ideal I of S generated by the elements $\sum_{j=1}^{n} x_j(y_j-\sigma(y_j))$, where $x_j$ and $y_j$ are homogeneous in S. Of course, I cannot be contained in any gr-maximal ideal and so I = S. The sequel of the proof is now identical to the proof of 5 ⇒ 2 in proposition III.1.2. in [42], so we do not repeat it here. □

1.12. Proposition. (Embedding property). Let S be a $\mathbb{Z}$-graded extension of R, which is a finitely generated projective separable extension with no nontrivial idempotents, then there exists a graded extension S of S' which is a graded Galois extension of R and which still has no nontrivial idempotents.

Proof. An easy modification of the corresponding ungraded statement, cf. [42], Theorem III.2.9. □

1.13. Corollary. If S' is as in (1.12.), then every separable R-subalgebra T of S' is graded.

Proof. Consider a Galois extension S as in (1.12.) above. By the fundamental theorem of Galois theory (III.1.1. in [42]), there is a subgroup H of the Galois group of S over R such that $T = S^H$. Since the automorphisms in H are graded morphisms of degree zero, writing $x = x_{i_1} + \dots + x_{i_n}$ with $x_{i_r} \in h(S)$, we obtain that $x = \sigma(x) =$ $= \sigma(x_{i_1}) + \dots + \sigma(x_{i_n})$ and then the uniqueness of the homogeneous decomposition of x in S yields $x_{i_r} = \sigma(x_{i_r})$, i.e. $x_{i_r} \in S^H = T$ for all r, proving that T is graded. □

Note that some of the above results have analogues in the case of graded $\kappa$-Galois extensions. This is not surprising, since upto some coherence conditions, the relative theory is just a "partial globalization" of the local data obtained from the common theory. We did not go into details here, in order to avoid unnecessary abstraction. Let us just mention the following useful corollary of Theorem 1.5. :

1.14. Proposition. Let R be $\mathbb{Z}$-graded commutative ring and let $\kappa^g$ be a graded idempotent kernel functor in the sense of [108] (i.e. $L(\kappa^g)$ has a cofinal subset consisting of graded R-ideals). Let A be a $\mathbb{Z}$-graded Azumaya algebra containing as a maximal (graded) commutative subring a $\mathbb{Z}$-graded $\kappa^g$-Galois extension S of R with Galois group G. Then A is $\kappa^g$-graded by G such that $A_e = S$. Moreover, the G-gradation is compatible with the $\mathbb{Z}$-gradation of A, i.e. each of the $A_\sigma$ is a $\mathbb{Z}$-graded S-module.

Proof. The definition of $A_{(\sigma)}$ in Theorem 1.5. makes it into a $\mathbb{Z}$-graded S-module; the other statements follow directly from Theorem 1.5. and its proof. □

The above result has some corollaries concerning the relative version of the Chase-Rosenberg sequence in the graded context. We leave this as an exercise to the zealous reader interested in the cohomology of graded rings.

## C.2. Relative Cohomology and Crossed Products.

The crossed-product results in the foregoing Section go back essentially to T. Kanzaki [ 79] for the absolute case and to F. Van Oystaeyen [155] for the relative case. In [ 26], S. Caenepeel, M. Van den Bergh and F. Van Oystaeyen showed that the generalized crossed product theorems were actually equivalent to (the relative version of) the Chase-Rosenberg sequence in Galois cohomology. A complete work-out in terms of Amitsur- and Galois cohomology has been developed by S. Caenepeel, A. Verschoren in [ 27]. Using $\kappa$-faithfully flat descent as the main tool we stay close to the set-up used in [27 ], upto some minor modifications and simplifications.

Throughout this section, R is a commutative ring and $\kappa$ will be an idempotent kernel functor in R-mod. The quotient category $(R,\kappa)$-mod is aGrothendieck category. For any morphism is $(R,\kappa)$-mod, we may identify its kernel with its kernel in R-mod. However, the cokernel of $f : N \to M$ in $(R,\kappa)$-mod is just $Q_\kappa(N/\mathrm{Im}(f))$ and the image of f in $(R,\kappa)$-mod is $Q_\kappa(\mathrm{Im}(f))$, where $\mathrm{Im}(f)$ is the image of f in R-mod. Whereas $Q_\kappa(-)$ is not necessarily exact in R-mod. it is true that the *reflector* $a_\kappa$ : R-mod $\to$ $(R,\kappa)$-mod : $M \mapsto Q_\kappa(M)$ is an exact functor. Therefore, if M is $\kappa$-closed, then $Q_\kappa(M \otimes_R -) = M \perp_\kappa -$ is a right exact endofunctor of $(R,\kappa)$-mod. The definition of $\kappa$-flatness may then be expressed by saying that a $\kappa$-closed M is $\kappa$-flat if $M \perp -$ is exact in $(R,\kappa)$-mod. We now say that M is $\kappa$-faithfully flat if it satisfies one of the equivalent conditions in the following proposition :

2.1. Proposition. Assume that $\kappa$ is noetherian. The following statements are equivalent for any $\kappa$-closed R-module N :

1. A sequence $M' \xrightarrow{\alpha} M \xrightarrow{\beta} M''$ is exact in $(R,\kappa)$-mod if and only if $N \perp_\kappa M' \xrightarrow{N \perp \alpha} N \perp_\kappa M \xrightarrow{N \perp \beta} N \perp_\kappa M''$ is exact in $(R,\kappa)$-mod;
2. N is $\kappa$-flat and for all $\kappa$-closed R-modules M we have that $N \perp_\kappa M = 0$ if and only if $M = 0$;
3. N is $\kappa$-flat and $Q_\kappa(N/pN) \neq 0$ for all $p \in C(\kappa)$.

Proof. Straightforward verification using the correct definitions of "Im" and "Coker" in $(R,\kappa)$-mod. We content ourselves to point out how to prove that 3. implies 2.. Let $H \neq 0$ be $\kappa$-closed. If $x \neq 0$ in M, then $I = \mathrm{Ann}_R(x) \notin L(\kappa)$ and Rx is isomorphic to R/I in R-mod. Since $\kappa$ is noetherian, there exists a $p \in C(\kappa)$ containing I. Then $Q_\kappa(N/IN) = (R/I) \perp N \neq 0$, because otherwise $Q_\kappa(N/pN) = (R/p) \perp N$ would vanish too, contradicting the hypotheses. From the $\kappa$-flatness of N, it follows that $(R/I) \perp N \cong Rx \perp N \subset M \perp N$ and therefore $M \perp N \neq 0$.

2.2. Proposition. Assume that $\kappa$ is noetherian and let N be a $\kappa$-closed R-module, then N is $\kappa$-faithfully flat if and only if $N_p$ is a faithfully flat $R_p$-module for all $p \in \mathcal{C}(\kappa)$.

Proof. The result is an easy exercise in local-global techniques, by localization at the $p \in \mathcal{C}(\kappa)$, since an R-linear $f : M' \to M''$ gives rise to an isomorphism $Q_\kappa(f) : Q_\kappa(M') \to Q_\kappa(M'')$ if and only if the local morphisms $f_p = M'_p \to M''_p$ are isomorphisms for all $p \in \mathcal{C}(\kappa)$. □

2.3. Proposition. Let $\kappa$ be noetherian and let S be a commutative $\kappa$-closed R-algebra. The following statements are equivalent :

1. S is a $\kappa$-faithfully flat R-algebra;
2. S is $\kappa$-flat and for all $p \in X(\kappa)$, there is a $q \in \mathrm{Spec}(S)$ such that $q \cap R = p$;
3. S is $\kappa$-flat and for all $p \in \mathcal{C}(\kappa)$, there is a $q \in \mathrm{Spec}(S)$ such that $q \cap R = p$.

Proof. 1. ⇒ 2. If $p \in X(\kappa)$, then $S_p$ is faithfully flat over $R_p$, hence there exists a $Q \in \mathrm{Spec}(S_p)$ lying over $pR_p$. If $j_p : S \to S_p$ is the canonical localization morphism, then $q = j_p^{-1}(Q)$ has the property $q \cap R = p$.

2. ⇒ 3. Obvious, since $\mathcal{C}(\kappa) \subset X(\kappa)$.

3. ⇒ 1. If $Q_\kappa(S/pS) = 0$ for some $p \in \mathcal{C}(\kappa)$, then we find $I \in \mathcal{L}(\kappa)$, $q \in \mathrm{Spec}(S)$ with $q \cap R = p$, such that $I \subset pS \subset q$. Hence $I \in q \cap R = p$, but this contradicts $p \in \mathcal{C}(\kappa)$. Proposition 2.1. then finishes the proof. □

Note that if S is an arbitrary R-algebra and if $\bar{\kappa}$ denotes the idempotent kernel functor in S-mod induced by $\kappa$, then $\bar{\kappa}$ is noetherian, if $\kappa$ is noetherian, and if S is $\kappa$-faithfully flat over R, then $X(\bar{\kappa}) = \{q \in \mathrm{Spec}(S);\ q \cap R \in X(\kappa)\}$.

2.4. Proposition. Let $\kappa$ be noetherian and assume that the R-algebra S is $\kappa$-faithfully flat over R, then for a $\kappa$-closed R-module M the following statements are equivalent :

1. M is $\kappa$-finitely generated over R;
2. $M \perp_{\kappa} S$ is $\overline{\kappa}$ -finitely generated over S.

Proof. 1.⇒ 2. Obvious.

2. ⇒ 1. By (B.4.7.), it is clear that $M \otimes_R S$ is $\overline{\kappa}$-finitely generated so we may find a finitely generated R-module M' ⊂ M such that the induced map $f : M' \otimes_R S \to M \otimes_R S$ has $\kappa$-torsion cokernel, hence $Q_{\kappa}(f)$ is epimorphic in $(R,\kappa)$-mod. Since S is $\kappa$-faithfully flat, the inclusion $M' \to M$ is epimorphic in $(R,\kappa)$-mod too, hence M is $\kappa$-finitely generated. □

2.5. Lemma. Assume that S is a $\kappa$-faithfully flat R-algebra and consider $P \in X(\overline{\kappa})$ and $p = P \cap R$. Then the induced ring morphism $R_p \to S_P$ is faithfully flat.

Proof. We have seen that $p \in X(\kappa)$. If $Q \in \mathrm{Spec}(R_p)$, then it suffices to show that $Q S_P \neq S_P$ and it will be sufficient to establish this for $Q = p R_p$. Now $QS_P = pS_P \subset PS_P \neq S_P$. □

2.6. Proposition. If S is a $\kappa$-faithfully flat R-algebra and M a $\kappa$-closed R-module, then the following statements are equivalent.

1. M is $\kappa$-quasiprojective;
2. $M \perp_{\kappa} S$ is $\overline{\kappa}$-quasiprojective.

Proof. If $P \in X(\overline{\kappa})$, then $p = P \cap R \in X(\kappa)$ and we have $(M \perp_{\kappa} S)_P =$ $((M \otimes_R S)_p)_P = (M_p \otimes_{R_p} S_p)_P = M_p \otimes_{R_p} S_P)$. If we assume 1., then $M_p$ in $C_p$-projective for any $P \in X(\overline{\kappa})$ and $p = P \cap R$, hence $M_p \otimes_{R_p} S_P$ is $S_P$-projective, proving 2.

If we assume 2., then for every $p \in X(\kappa)$ there is a $P \in X(\overline{\kappa})$ such that $p = P \cap R$. Since $S_P$ is a faithfully flat $R_p$-algebra, it follows from the $S_P$-projectivity of $(M \underset{\kappa}{\perp} S)_P = M_P \underset{R_P}{\otimes} S_P = M_p \underset{R_p}{\otimes} S_P$ that $M_p$ is $R_p$ free. □

2.7. <u>Corollary</u>. a. Let R be a $\kappa$-noetherian ring and let S be a $\kappa$-closed, $\kappa$-finitely generated and $\kappa$-faithfully flat R-algebra, then for any $\kappa$-closed R-module M, the following statements are equivalent :

1. M is a $\kappa$-progenerator;
2. $M \underset{\kappa}{\perp} S$ is a $\overline{\kappa}$-progenerator.

b. Suppose that R is $\kappa$-noetherian and let S be a commutative R-algebra which is a $\kappa$-progenerator as an R-module, then S is $\kappa$-faithfully flat.

<u>Proof</u>. a. Obvious from the foregoing.

b. For every $p \in X(\kappa)$ we know that $S_p$ is an $R_p$-progenerator, hence $S_p$ is a faithfully flat $R_p$-algebra, whence the result. □

We now obtain the final "descent"-results.

2.8. <u>Theorem</u>. Assume that R is a $\kappa$-noetherain ring and S is a commutative R-algebra which is a $\kappa$-progenerator. If A is a $\kappa$-closed R-algebra, with the property that $A \underset{\kappa}{\perp} S \cong \mathrm{End}_S(P)$ for some $\overline{\kappa}$-progenerator P over S, then A is a $\kappa$-Azumaya algebra.

<u>Proof</u>. From the foregoing it follows that A is a $\kappa$-progenerator, so we only have to establish that $A \perp A^{opp} \cong \mathrm{End}_R(A)$. We have : $(A \perp A^{opp}) \perp S \cong \mathrm{End}_S(P) \underset{S}{\perp} (\mathrm{End}_S(P))^{opp} \cong \mathrm{End}_S(\mathrm{End}_S(P))$ and $\mathrm{End}_R(A) \perp S \cong \mathrm{End}_S(A \perp S) \cong \mathrm{End}_S(\mathrm{End}_S(P))$. In the following commutative diagrams :

$$\begin{array}{ccc} S \perp (A \perp A^{opp}) \perp S & \longrightarrow & S \perp \mathrm{End}_S(\mathrm{End}_S(P)) \\ \downarrow & & \downarrow \varphi \\ (A \perp A^{opp}) \perp S \perp S & \longrightarrow & \mathrm{End}_S(\mathrm{End}_S(P)) \perp S \end{array}$$

resp.

$$\begin{array}{ccc} S \perp \mathrm{End}_R(A) \perp S & \longrightarrow & S \perp \mathrm{End}_S(\mathrm{End}_S(P)) \\ \downarrow & & \downarrow \psi \\ \mathrm{End}_R(A) \perp S \perp S & \longrightarrow & \mathrm{End}_S(\mathrm{End}_S(P)) \perp S \end{array}$$

one easily checks that $\varphi = \psi$ and that $A \perp A^{opp} \cong \mathrm{End}_R(A)$ follows. □

2.9. Proposition. Let R be $\kappa$-noetherian and let S be an R-algebra which is also a $\kappa$-progenerator. For a $\kappa$-closed R-module M, the following statements are equivalent:

1. M is a $\kappa$-invertible R-module ;
2. $M \perp S$ is a $\bar{\kappa}$-invertible S-module.

Proof. 1 ⇒ 2. is obvious and 2. ⇒ 1. follows from a descent argument similar to the one in the above theorem. □

At this point it is necessary to introduce some notation and terminology used in the cohomological theory :

Let S be a commutative R-algebra and consider $M_1,\dots,M_n \in$ R-mod. Define $e_i : M_1 \underset{R}{\otimes} \dots \underset{R}{\otimes} M_n \to M_1 \underset{R}{\otimes} \dots \underset{R}{\otimes} M_{i-1} \underset{R}{\otimes} S \underset{R}{\otimes} \dots \underset{R}{\otimes} M_n$ by $e_i(m_1 \otimes \dots \otimes m_n) = m_1 \otimes \dots \otimes m_{i-1} \otimes 1 \otimes \dots \otimes m_n$ and $\epsilon_i = Q_\kappa(e_i)$. Write $S^{\otimes n}$ resp. $S^{\perp n}$ for $S \otimes \dots \otimes S$ resp. $S \perp \dots \perp S$. Define $\tau_{ij} = Q_\kappa(t_{ij})$ :
$M_1 \perp \dots \perp M_i \perp \dots \perp M_j \perp \dots \perp M_n \to M_1 \perp \dots \perp M_j \perp \dots \perp M_i \perp \dots \perp M_n$ where $t_{ij}$ is the usual switch-map. Given $M_1',\dots,M_n' \in$ R-mod and a map $\gamma : M_1 \perp \dots \perp M_n \to M_1' \perp \dots \perp M_n'$, we define
$\gamma_i : M_1 \perp \dots \perp M_{i-1} \perp S \perp \dots \perp M_n \to M_1' \perp \dots \perp M_{i-1}' \perp S \perp \dots \perp M_n'$
by $\gamma_i = \tau_{i,n+1} \circ (\gamma \perp S) \circ \tau_{i,n+1}$.

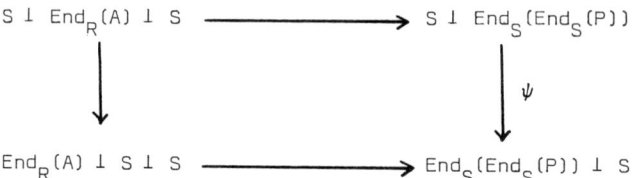

The relative Amitsur complex $C_\kappa^+(S/R,Q)$ is the complex obtained from the usual Amitsur complex $C^+(S/R)$, cf. [ 32] or [42 ], by letting $Q_\kappa(-)$ act, i.e. we have

$$0 \to Q_\kappa(R) \xrightarrow[\Delta_{-1}]{} Q_\kappa(S) \xrightarrow[\Delta_0]{} S \perp S \xrightarrow[\Delta_1]{} \dots \to \overbrace{S \perp \dots \perp S}^{n} \xrightarrow[\Delta_{n-1}]{} \overbrace{S \perp \dots \perp S}^{n+1} \to \dots,$$

where $\Delta_n = \sum_{i=1}^{n+2} (-)^{i+1} \epsilon_i$.

2.10. __Proposition.__ Let S be a $\kappa$-closed $\kappa$-faithfully flat commutative R-algebra and let N be a $\kappa$-closed R-module, then de complex $N \perp C^+(S/R,Q_\kappa)$ is exact in $(R,\kappa)$-mod.

__Proof.__ That $C^+(S/R,Q_\kappa)$ and $N \perp C^+(S/R,Q_\kappa)$ are complexes is easily verified. Furthermore, $N \otimes_R C^+(S/R) \otimes_R S$ is an exact sequence, hence $N \perp C^+ (S/R,Q_\kappa) \perp S$ is exact in $(R,\kappa)$-mod and the result follows from the fact that S is $\kappa$-faithfully flat. □

2.11. __Corollaries.__

1. (Descent of elements). If S is as above and $x \in S$, then $x \in Q_\kappa(R)$ if and only if $1 \perp x = x \perp 1$.

2. (Descent of homomorphisms). If M and M' are $\kappa$-closed R-modules, then $\gamma : M \perp S \to M' \perp S$ is of the form $\varphi \perp S$ for some $\varphi : M \to M'$ in R-mod if and only if $\gamma_2 = \gamma_3$.

__Proof.__ 1. From $1 \perp x = x \perp 1$ it follows that $1 \otimes x - x \otimes 1 \in \kappa(S \otimes_R S)$, so $x \in \text{Ker}(\Delta_0) = Q_\kappa(\text{Im}(\Delta_1)) = \text{Im}(\Delta_1)$ and $x \in Q_\kappa(R)$.

2. Restricting $\gamma$ to M, we obtain $\varphi : M \to M' \perp S$. If $\gamma_2 = \gamma_3$, then $(M' \perp \Delta_0)\,\varphi(x) = (M' \perp (\epsilon_1 - \epsilon_0))\,\gamma(x \perp 1) = (\gamma_2 - \gamma_3)(x \perp 1) = 0$, so $\varphi(x) \in Q_\kappa(\text{Im}(M' \perp \Delta_{-1})) = M'$. Hence we may consider $\varphi$ as a map $\varphi : M \to M'$ The converse is obvious. □

2.12. __Proposition.__ (Descent of modules and algebras).

Suppose that R is $\kappa$-closed and let S be a $\kappa$-closed $\kappa$-faithfully flat commutative R-algebra. Let M be a $\overline{\kappa}$-closed S-module and let there be given an $S^{\perp 2}$-isomorphism $\gamma : S \perp M \to M \perp S$ with $\gamma_2 = \gamma_3\gamma_1$. Then there exists a $\kappa$-closed R-module N and an S-isomorphism $\eta : N \perp S \xrightarrow{\sim} M$ such that the following diagram of $S^{\perp 2}$-isomorphisms is commutative.

$$\begin{array}{ccc} S \perp N \perp S & \xrightarrow{S \perp \eta} & S \perp M \\ \Big\downarrow \tau_{12} & & \Big\downarrow \gamma \\ N \perp S \perp S & \xrightarrow[\eta \perp S]{} & M \perp S \end{array}$$

The pair $(N,\eta)$ is determined upto isomorphism and one may take $N = \{x \in M; \epsilon_2(x) = \gamma_1(x)\} = \mathrm{Ker}(\epsilon_2 - \gamma_1)$ and $\eta$ the map induced by multiplication. Furthermore, if M has the structure of an S-algebra and $\gamma$ is an $S \perp S$-algebra isomorphism, then N may be given the structure of an R-algebra in an essentially unique way such that $\eta$ is an S-algebra isomorphism.

<u>Proof.</u> Let N be the kernel (or equalizer) of $M \underset{\gamma_1}{\overset{\epsilon_2}{\rightrightarrows}} M \perp S$, then N is $\kappa$-closed. Consider the following diagram of S-homomorphisms :

$$\begin{array}{ccccccc} 0 & \longrightarrow & S \perp N & \longrightarrow & S \perp M & \underset{S \perp \gamma\epsilon_1}{\overset{S \perp \epsilon_2}{\rightrightarrows}} & S \perp M \perp S \\ & & & & \Big\downarrow \gamma & & \Big\downarrow \gamma_3 \\ 0 & \longrightarrow & M & \longrightarrow & M \perp S & \underset{\epsilon_2}{\overset{\epsilon_3}{\rightrightarrows}} & M \perp S \perp S \end{array}$$

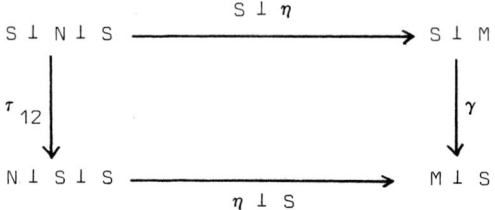

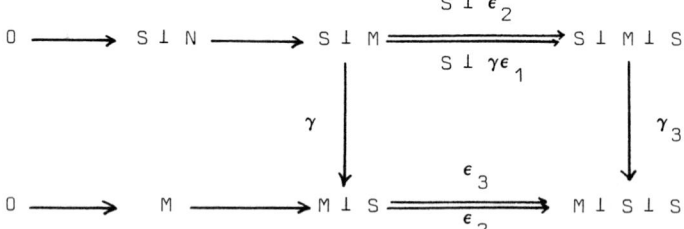

From the flatness hypotheses it follows that $S \perp N$ is the equalizer of the first row, and also M is the equalizer of the second row. The two squares in the diagram are commutative. Indeed, $\gamma_3 \circ (S \perp \epsilon_2) = \gamma_3 \circ \epsilon_3 = \epsilon_3 \circ \gamma$ and $\gamma_3 \circ (S \perp (\gamma \circ \epsilon_1)) = \gamma_3 \circ \gamma_1 \circ \epsilon_2 = \gamma_2 \circ \epsilon_2 = \epsilon_2 \circ \gamma$. By restriction, we obtain an isomorphism $\varphi : S \perp N \to M$. As $\varphi \circ (S \perp \eta) = \epsilon_2^{-1} \circ \gamma \circ (S \perp \eta) = \epsilon_2^{-1} \circ (\eta \perp S) = \eta$, we obtain $\varphi = \eta \circ \tau$, where $\eta$ is induced by multiplication. This proves that $\eta$ is an isomorphism. Observing that $S\perp M$ and $M\perp S$ are also $S^{\perp 2}$-modules, we may write down the following commutative diagram :

$$\begin{array}{ccc} S\perp N\perp S & \xrightarrow{\;S\perp\eta\;} & S\perp M \\ {\scriptstyle \eta\perp S}\downarrow & & \downarrow{\scriptstyle \gamma} \\ M\perp S & \longrightarrow & M\perp S \end{array}$$

If $(P,\pi)$ and $(N,\eta)$ are two solutions to the problem, put $\rho = \eta^{-1}\circ\pi$, then we have a commutative diagram

$$\begin{array}{ccccc} & & N\perp S\perp S & & \\ & {\scriptstyle \eta_2^{-1}}\nearrow & & \searrow{\scriptstyle \eta_3} & \\ S\perp M & & \xrightarrow{\qquad\gamma\qquad} & & M\perp S \\ & {\scriptstyle \pi_2^{-1}}\searrow & & \nearrow{\scriptstyle \pi_3} & \\ & & N\perp S\perp S & & \end{array}$$

Then $\pi_3\pi_2^{-1} = \eta_3\eta_2^{-1}$ and thus $\rho_2 = \eta_2^{-1}\pi_2 = \eta_3^{-1}\pi_3 = \rho_3$ and $\rho = \alpha\perp S$ for $\alpha : P \to N$. The final part of the proof is a slight modification of the classical proof, which goes by considering the diagram :

$$\begin{array}{ccc} P\perp S & \xrightarrow{\;\alpha\perp S\;} & N\perp S \\ & {\scriptstyle \rho}\searrow \quad \swarrow{\scriptstyle \eta} & \\ & M & \end{array}$$

□

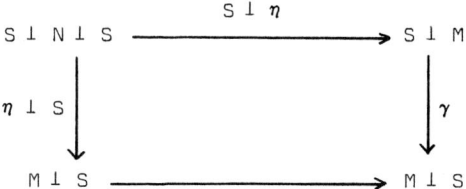

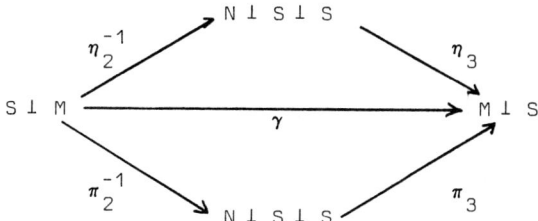

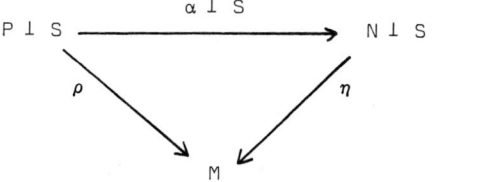

In the first section we have hinted at the relative version of the Chase-Rosenberg sequence and how it could be derived from the relative crossed product theorem, Theorem 1.5. Now, we derive the long exact sequence from the results on relative cohomology we have at hand. We then relate the relative Amitsur complex and the relative Galois complex, proving that these are isomorphic over a $\kappa$-noetherian ring R, and then close the circle by rediscovering Theorem 1.5. Basic properties about $Br(R,\kappa)$ may be found in [162], the theory of $Pic(R,\kappa)$ that we need may also be found in [162], but all we need may also be derived from the noncommutative theory expounded in Chapter B.

2.13. Lemma. Let P and Q be $\kappa$-progenerators over R. Each isomorphism $\alpha : End_R(P) \tilde{\to} End_R(Q)$ is induced by an isomorphism $f : P \perp I \tilde{\to} Q$, for some $\kappa$-invertible R-module I. The pair (I,f) is unique upto isomorphism.

Proof. Let $\kappa'$ be the central idempotent kernel functor induced by $\kappa$ on $End_R(P)$-mod. We may view Q as a $\kappa'$-closed $End_R(P)$-module, with action defined by $b.q = \alpha(b)q$ for any $b \in End_R(P)$ and $q \in Q$. The relative Morita theory in [162], III.1.3. entails that $P \perp I \cong Q$, where $I = P^{\star} \perp_B Q$, with $P^{\star}$ the dual of P and $B = End_R(P)$. The rest of the proof is now reduced to an easy verification. □

For any commutative R-algebra S, put $U_\kappa(S) = U(Q_\kappa(S))$, the multiplicative group of units in $Q_\kappa(S)$.

2.14. Theorem. Let R be a $\kappa$-noetherian commutative ring. Let S be a commutative R-algebra, which is a $\kappa$-progenerator as an R-module, then the following sequence is exact :

$$1 \to H^\circ(S/R,U_\kappa) \xrightarrow{\alpha_\circ} U_\kappa(R) \to 1 \to H^1(S/R,U_\kappa) \xrightarrow{\alpha_1} Pic(R,\kappa) \xrightarrow{\beta_1}$$
$$\to H^\circ(S/R,Pic(-,\kappa)) \xrightarrow{\gamma_1} H^2(S/R,U_\kappa) \xrightarrow{\alpha_2} Br(S/R,\kappa) \xrightarrow{\beta_2} H^1(S/R,Pic(-,\kappa))$$
$$\xrightarrow{\gamma_2} H^3(S/R,U_\kappa),$$

where the $H^i(S/S,F)$ stand for the Amitsur cohomology groups associated to the functor F and $Br(S/R,\kappa)$ is the kernel of $Br(R,\kappa) \longrightarrow Br(S,\bar{\kappa})$.

Proof. Having deduced all of the necessary tools above, the proof becomes an easy adaptation of the proof of the classical Chase-Rosenberg theorem e.g. as given by M.A. Knus in [81]. Here we only present a description of the connecting homomorphisms in the sequence, leaving the verification of the details (concerning exactness) to the reader.

a. $H^\circ(S/R,U_\kappa) = U_\kappa(S)$;

b. *Definition of* $\alpha_1$. If t represents $[t] \in H^1(S/R,U_\kappa)$, then multiplication by t defines a descent datum $S \perp S \to S \perp S$ which determines a $\kappa$-invertible module I such that $I \perp S = S$. Put $\alpha_1([t]) = [I]$.

c. We define $\beta_1$ by $\beta_1([I]) = I \perp S$.

d. *Definition of* $\gamma_1$. If $[I] \in H^\circ(S/R,\mathrm{Pic}(-,\kappa))$, then we have an $S \perp S$-isomorphism $\varphi : S \perp I \to I \perp S$ and $\varphi_2^{-1}\ \varphi_3\ \varphi_1$ determines an $S^{\perp 3}$-automorphism of $S \perp S \perp I$, so it is multiplication by a unit t in $U(S^{\perp 3})$; put $\gamma_1([I]) = [t]$.

e. *Definition of* $\alpha_2$. If P denotes the S-module $S \perp S$, where the action is defined by $a.s = (a \perp S)s$, then for each $[t] \in H^2(S/R,U_\kappa)$, we define $f(t) : P_1 \to P_2$ by $f(t)y = \tau_{23}\ t\ y$. Clearly $f(t)$ induces a map $\varphi(t) : S \perp \mathrm{End}_S P \to \mathrm{End}_S P \perp S$, which defines an Azumaya algebra $A(t)$ such that $A(t) \perp S \cong \mathrm{End}_S P$. Put $\alpha_2([t]) = [A(t)]$.

f. *Definition of* $\beta_2$. Consider $[A] \in \mathrm{Br}(S/R,\kappa)$. We have an isomorphism $\sigma : S \perp A \to \mathrm{End}_{Q_\kappa(R)}(Q)$ for some $\kappa$-progenerator Q. Put $\varphi = \sigma_3\tau_{23}\sigma_1^{-1}$ : $\mathrm{End}_{S \perp S}(S \perp Q) \to \mathrm{End}_{S \perp S}(Q \perp S)$. By Lemma 2.13., $\varphi$ is induced by some map $f : (S \perp Q) \underset{S \perp S}{\perp} I \to Q \perp S$. From $\varphi_2 = \varphi_3\ \varphi_1$, it follows that $I_2 = I_3 \perp I_1$ and therefore, we may put $\beta_2([A]) = [I]$.

g. *Definition of* $\gamma_2$. Pick $[I] \in H^1(S/R,\mathrm{Pic}(-,\kappa))$. From the existence of the isomorphism $f : I_1 \underset{S^{\perp 3}}{\perp} I_3 \to I_2$, it follows that $f_4^{-1}\ f_2^{-1}\ f_3 f_1$ is an automorphism of $I_{11} \perp I_{31} \perp I_{33}$, tensorproducts being taken over $S^{\perp 3}$, therefore, it is multiplication by a unit u in $S^{\perp 3}$. It is not hard to verify that u is a cocycle and that we may define $\gamma_2$ by putting $\gamma_2$ by putting $\gamma_2([I]) = [u]$. □

2.15. Corollary. Let R be a Krull domain and let $\sigma_1$ be the idempotent kernel in R-mod associated to the height one primes of R. A $\sigma_1$-Azumaya algebra is just an iv-algebra in the sense of S. Yuan [178] or M. Orzech [112]. Furthermore, $\mathrm{Pic}(R,\sigma_1) = \mathrm{Cl}(R)$ and $\mathrm{Br}(R,\sigma_1) = \beta(R)$, the latter being the algebra class group defined in [12,13,112,178]. Let S be a commutative R-algebra which is also a divisorial R-lattice, then we have an exact sequence :

$$1 \to H^1(S/R,U_{\sigma_1}) \to \mathrm{Cl}(R) \to H^\circ(S/R,\mathrm{Cl}) \to H^2(S/R,U_{\sigma_1}) \to \beta(S/R)$$
$$\to H^1(S/R,\mathrm{Cl}) \to H^3(S/R,U_{\sigma_1}). \qquad \square$$

This is a direct translation of the sequence obtained above. We thus have proved a conjecture of S. Yuan's [178], but note that we had to introduce $H^i(S/R,U_{\sigma_1})$ instead of $H^i(S/R,U)$.

It is possible to derive the Galois cohomological equivalent of the above exact sequences. One may develop the theory of $\kappa$-Galois extensions and prove results similar to the absolute case. Let us first mention the following strengthening of (1.4.2.).

2.16. Lemma. Let S be a $\kappa$-closed R-algebra and assume that R is $\kappa$-noetherian. If G is a finite group of R-automorphisms of S, then the following assertions are equivalent :

1. S is a $\kappa$-Galois extension of R with $\mathrm{Gal}_\kappa(S/R) = G$;
2. $S_p$ is Galois over $R_p$, with $\mathrm{Gal}(S_p/R_p) = G$ for all $p \in \mathcal{C}(\kappa)$. $\square$

2.17. Proposition. (Fundamental Theorem of Galois theory)
Let R be a $\kappa$-closed $\kappa$-noetherian commutative ring and let S be a $\kappa$-Galois extension of R with Galois group G. There is a bijective correspondence between normal subgroups of G and subrings of S which are $\kappa$-Galois extensions of R; in this correspondence, a normal subgroup H of G corresponds

to $S^H$ and a $\kappa$-Galois extension T of R in S corresponds to $\{g \in G; g(x) = x \text{ for all } x \in T\}$.

Proof. Cf. [ 27 ]. □

2.18. Theorem. Let S be a $\kappa$-Galois extension of the $\kappa$-noetherian ring R. The Amitsur complex $0 \to S \to S \perp S \to \ldots$ and the Galois complex $0 \to K^0(G,S) \to K^1(G,S) \to K^2(G,S) \to \ldots$ are isomorphic : hence $H^i(G,U(S)) = H^i(S/R,U_\kappa)$ and $H^i(G,Pic(S,\kappa)) = H^i(S/R,Pic(-,\kappa))$. Therefore, we obtain an exact sequence :

$$1 \longrightarrow H^0(G,U(S)) \xrightarrow{\alpha_0} U(S) \longrightarrow 1 \longrightarrow H^1(G,U(S)) \xrightarrow{\alpha_1} Pic(R,\kappa) \xrightarrow{\beta_1}$$

$$\to H^0(G,Pic(S,\kappa)) \xrightarrow{\gamma_1} H^2(G,U(S)) \xrightarrow{\alpha_2} Br(S/R,\kappa) \xrightarrow{\beta_2} H^1(G,Pic(S,\kappa))$$

$$\xrightarrow{\gamma_2} H^3(G,U(S)).$$

Proof. By the classical techniques, using Lemma 2.16. □

2.19. Corollary. The statement in Theorem 1.5. follows from the exact sequence above (in case R is $\kappa$-noetherian). So, over a $\kappa$-noetherian commutative ring R, the relative version of the Chase-Rosenberg sequence is equivalent to the generalized crossed-product theorem proved by F. Van Oystaeyen in [154].

# D CONSTRUCTION OF ORDERS

## D.1. Relative Maximal Orders.

Torsion-theoretic notations are the same as in Chapter B.
If $L^2(\sigma)$ is a multiplicatively closed set of (two-sided) ideals of a ring $\Lambda$, then we denote by $L^1(\sigma)$ resp. $L^r(\sigma)$ the filter of left resp. right ideals of $\Lambda$ generated by $L^2(\sigma)$. If $L^1(\sigma)$ resp. $L^r(\sigma)$ is an idempotent filter in $\Lambda$-mod resp. mod-$\Lambda$, then we denote by $Q^1_\sigma(-)$ resp. $Q^r_\sigma(-)$ the associated localization functor.

Throughout, $\Sigma$ denotes a fixed ring and all rings considered are subrings of $\Sigma$. Usually, $\Sigma$ will be a (classical symmetric or maximal) ring of quotients of a given $\Lambda$. Note that at this point we do not assume $\Sigma$ to be a prime ring.

If $\Lambda$ is any ring, then a torsion couple $(\lambda,\rho)$ consists of a filter $L(\lambda)$ of left ideals of $\Lambda$ and a filter $L(\rho)$ of right ideals of $\Lambda$ satisfying the condition $L(\lambda)L(\rho) \subset L^2(\lambda,\rho)$, where $L^2(\lambda,\rho)$ consists of all *ideals* I of $\Lambda$ lying in $L(\lambda)$ *and* $L(\rho)$.
If $\Lambda$ is a subring of $\Sigma$ and if $(\lambda,\rho)$ is a torsion couple of $\Lambda$, then we say that $\Lambda$ is a $(\lambda,\rho)$-order in $\Sigma$ if o is the only element x in $\Lambda$ such that I x = o or x I = o for some $I \in L^2(\lambda,\rho)$. Alternatively, if $\mathrm{Ann}^r(I) = \mathrm{Ann}^1(I) = o$ for all $I \in L^2(\lambda,\rho)$. A $(\lambda,\rho)$-order $\Lambda$ in $\Sigma$ is said to be a maximal $(\lambda,\rho)$-order if there exists no intermediate ring $\Lambda \subsetneq \Gamma \subsetneq \Sigma$ satisfying $I\,\Gamma\,J \in L^2(\lambda,\rho)$ for some $I \in L(\lambda)$ and $J \in L(\rho)$.

Example. Let $\Lambda$ be a prime Goldie ring and let $\Sigma$ be its classical ring of fractions, which exists by the Goldie theorems. If $\lambda$ is the left Goldie kernel functor, i.e. $L(\lambda)$ consists of all left ideals of $\Lambda$

containing a regular element, and if $\rho$ is the right Goldie kernel functor, then one easily verifies that $\Lambda$ is a $(\lambda,\rho)$-order in $\Sigma$. Moreover, $\Lambda$ is a maximal order in the classical sense if and only if $\Lambda$ is a maximal $(\lambda,\rho)$-order.

We will now provide a relative version of the notion of a fractional $\Lambda$-ideal. Let $\Lambda$ be a $(\lambda,\rho)$-order in $\Sigma$, then a *left fractional* $\Lambda$-ideal is by definition a left $\Lambda$-submodule A of $\Sigma$ such that $I \subset A$ for some $I \in L(\lambda)$ and $A\,J \in L^2(\lambda,\rho)$ for some $J \in L(\rho)$ . Similarly, a *right fractional* $\Lambda$-*ideal* is a right $\Lambda$-submodule B of $\Sigma$ such that $J \subset B$ for some $J \in L(\rho)$ and $I\,B \in L^2(\lambda,\rho)$ for some $I \in L(\lambda)$. Finally, A is a *fractional* $\Lambda$-*ideal* if it is both a right and a left fractional $\Lambda$-ideal. Our first objective is to generalize some of the usual properties of maximal orders, as developed in e.g. I. Reiner [122] or the original paper by M. Auslander and O. Goldman [15 ]. For any two subsets A and B of $\Sigma$ we write $B \;{}^{\bullet}\!.\; A = \{x \in \Sigma;\ x\,A \subset B\}$ and $B \;.^{\bullet}\; A = \{x \in \Sigma; Ax \subset B\}$.

1.1. *Proposition*. If $\Lambda$ is a $(\lambda,\rho)$-order in $\Sigma$, then the following assertions are equivalent :

1. $\Lambda$ is a maximal $(\lambda,\rho)$-order;
2. if A is a left fractional $\Lambda$-ideal such that $A^2 \subset A$ and $\Lambda \subset A$, then $\Lambda = A$; if B is a right fractional $\Lambda$-ideal such that $B^2 \subset B$ and $\Lambda \subset B$, then $\Lambda = B$;
3. if A is a left fractional $\Lambda$-ideal, then $A \;.^{\bullet}\; A = \Lambda$; if B is a right fractional $\Lambda$-ideal, then $B \;{}^{\bullet}\!.\; B = \Lambda$;
4. if $I \in L^2(\lambda,\rho)$, then $I \;{}^{\bullet}\!.\; I = I \;.^{\bullet}\; I = \Lambda$.
5. if A is a fractional $\Lambda$-ideal, then $A \;{}^{\bullet}\!.\; A = A \;.^{\bullet}\; A = \Lambda$.

*Proof*. 1. $\Rightarrow$ 2. Clearly, $\Lambda \subset A \;{}^{\bullet}\!.\; A \subset \Sigma$. There exists a right ideal $I \in L(\rho)$ such that $A\,I \in L^2(\lambda,\rho)$, hence $(A \;{}^{\bullet}\!.\; A)\;A\,I = A\,I \in L^2(\lambda,\rho)$. Because $\Lambda$ is a maximal $(\lambda,\rho)$-orders, this implies that $\Lambda = A \;{}^{\bullet}\!.\; A$, whence $A \subset \Lambda$, because $A \subset A \cdot. A$. The second statement follows from left-right symmetry.

2. ⇒ 3. If A is a left fractional $\Lambda$-ideal, so is $A \cdot. A$, for it is clearly a left $\Lambda$-submodule of $\Sigma$ and $\Lambda \subset A \cdot. A$. Furthermore, if $A J \in L^2(\lambda,\rho)$ for some $J \in L(\rho)$ and if $I \subset A$ for some $I \in L(\lambda)$, then $I J \subset (A \cdot. A) I J \subset A J \in L^2(\lambda,\rho)$, whence $(A \cdot. A) I J \in L^2(\lambda,\rho)$. Clearly $(A \cdot. A)^2 \subset A \cdot. A$ and by 2. this entails that $A \cdot. A = \Lambda$. Left-right symmetry yields the result.

3. ⇒ 4. Trivial.

4. ⇒ 5. There exists a right ideal $J \in L(\rho)$ such that $A J \in L^2(\lambda,\rho)$ and therefore, $\Lambda \subset A \cdot. A \subset A J \cdot. AJ \subset \Lambda$.

5. ⇒ 1. Suppose that $\Gamma$ is an overring of $\Lambda$ in $\Sigma$ such that $I \Gamma J \in L^2(\lambda,\rho)$ for some $I \in L(\lambda)$ and $J \in L(\rho)$, then $I \Lambda \supset I \Gamma J I \Lambda \in L^2(\lambda,\rho)$ whence $\Gamma J I \Lambda \subset I \Lambda \cdot. I \Lambda = \Lambda$, entailing that $\Gamma J I \Lambda \in L^2(\lambda,\rho)$. This proves that $\Gamma$ is a left fractional $\Lambda$-ideal. Similarly, one proves that $\Gamma$ is a right fractional $\Lambda$-ideal, so $\Gamma \subset \Gamma \cdot. \Gamma \subset \Lambda$, finishing the proof. □

An important consequence of the foregoing proposition is that a $(\lambda,\rho)$-order $\Lambda$ is maximal if and only if $\Lambda$ is maximal as a $(\sigma^l,\sigma^r)$-order, where $L^2(\sigma) = L^2(\lambda,\rho)$ and $\sigma^l$ and $\sigma^r$ are defined by $L(\sigma^l) = L^l(\sigma)$ resp. $L(\sigma^r) = L^r(\sigma)$. For simplicity's sake, we call $\Lambda$ a <u>$\sigma$-maximal order</u> if and only if $L^2(\sigma)$ is a multiplicatively closed filter of ideals of $\Lambda$ such that $\Lambda$ is a maximal $(\sigma^l,\sigma^r)$-order in $\Sigma$.

Let $L^2(\text{Reg})$ be the (multiplicatively closed) filter of ideals I of $\Lambda$ such that $\text{Ann}^l(I) = \text{Ann}^r(I) = 0$ and define $Q^l(\Lambda) = \varinjlim \text{Hom}_\Lambda(I,\Lambda)$, where the limit is taken over all $I \in L^2(\text{Reg})$. The generalized Martindale ring of quotients of $\Lambda$ is then defined as

$$Q_{Mar}(\Lambda) = \{x \in Q^l(\Lambda);\ \exists\ I \in L^2(\text{Reg}),\ x I \subset \Lambda \text{ and } I x \subset \Lambda\}.$$

In particular, if $\Lambda$ is prime Goldie with classical ring of quotients $\Sigma$, then $Q_{Mar}(\Lambda)$ consists of all $x \in \Sigma$ such that $I\,x \subset \Lambda$ and $x\,I \subset \Lambda$ for some ideal I of $\Lambda$ and $\Lambda$ is a maximal $(\sigma^1,\sigma^r)$-order in $\Sigma$ if and only if it is maximal as a $(\sigma^1,\sigma^r)$-order in $Q_{Mar}(\Lambda)$. Therefore, from now on $\Lambda$ will be a $\sigma$-maximal order in $\Sigma$, the generalized Martindale ring of quotients of $\Lambda$. Recall that we denote by $F_\sigma(\Lambda)$ the set of all fractional $\Lambda$-ideals.

1.2. Lemma. If $\Lambda$ is a $\sigma$-order in $\Sigma$, then $F_\sigma(\Lambda)$ is closed under (ordinary) multiplication.

Proof. Suppose that A and B are fractional $\Lambda$-ideals and let $I,J,K,L \in L^2(\sigma)$ such that $I \subset A$, $J \subset B$ and $KA$ , $AK$ , $LB$, $BL$ $\in L^2(\sigma)$, then $L(KA)\ (BL) \subset (LK)AB \subset LB \subset \Lambda$ and $L(KA)(BL) \in L^2(\sigma)$, yielding that $(LK)(AB) \in L^2(\sigma)$. One proves in a similar way, that $(AB)(LK) \in L^2(\sigma)$. □

Usually, $F_\sigma(\Lambda)$ turns out to be too large for proper description and one has to restrict attention to so-called relative divisorial $\Lambda$-ideals. Thus, for any $\sigma$-order $\Lambda$ in $\Sigma$, we call a fractional $\Lambda$-ideal A divisorial if and only if $\Lambda .^{\cdot} (\Lambda \,^{\cdot}. A) = A$ and $\Lambda \,^{\cdot}. (\Lambda .^{\cdot} A) = A$. The set of divisorial $\Lambda$-ideals is denoted by $D_\sigma(\Lambda)$. We then have the following result :

1.3. Theorem. If $\Lambda$ is a $\sigma$-maximal order , then $D_\sigma(\Lambda)$ is an abelian group, when endowed with the multiplication $A \star B = \Lambda .^{\cdot} (\Lambda \,^{\cdot}. AB)$. In order to prove this, let us recall a construction due to E. Artin, cf. G. Maury, J. Raynaud [ 98 ].

Let T be an ordered set endowed with a multiplication law, then T is called an Artin setting if it is subject to the following conditions :

1. T is a semigroup with unit element e;
2. T is a lattice, i.e. sup(a,b) and inf(a,b) exist for any $a,b \in T$;
3. $a \leqslant b$ implies $ac \leqslant bc$ and $ca \leqslant cb$ for all $a,b,c \in T$;
4. c. sup(a,b) = sup(ca,cb) and sup(a,b).c = sup(a.c,b.c) for all $a,b,c \in T$;

5. for every bounded subset $\{a_i;\ i \in I\}$ of T every $c \in T$, we have $\sup_i\{a_i\}.c = \sup_i\{a_i.c\}$ and $c.\sup_i\{a_i\} = \sup_i\{c.a_i\}$

6. there exists a mapping $(-)^{-1}: T \to T$ such that $a.a^{-1}.a \leqslant a$ for every $a \in T$ and if $a.x.a \leqslant a$, then $x \leqslant a^{-1}$;

7. if $a^2 \leqslant a$ and $e \leqslant a$, then $a = e$.

We call two elements $a,b \in T$ quasi-equal if $a^{-1} = b^{-1}$. This defines an equivalence relation on T, and the set of equivalence classes is an abelian group, when endowed with the multiplication $[a] . [b] = [((a.b)^{-1})^{-1}]$, cf. loc. cit. Using this, (1.3.) follows from :

1.4. Proposition. $F_\sigma(\Lambda)$ is an Artin-setting.

Proof. $F_\sigma(\Lambda)$ is ordered by inclusion, the unit element will be $\Lambda$ and for any $A \in F_\sigma(\Lambda)$ we put $A^{-1} = \Lambda \,\dot{.}\, A$. Let us check the conditions :

1. follows from 1.2. ;

2. Let $A, B \in F_\sigma(\Lambda)$, then $A \cap B \in F_\sigma(\Lambda)$, for if $I,J,K,L \in L^2(\sigma)$ such that $I \subset A$, $J \subset B$ and $KA$, $AK$, $LB$, $BL \in L^2(\sigma)$, then $IJ \subset A \cap B$. Moreover, $KL(AKLB) \subset KL(A \cap B) \subset KLA \cap KLB \subset \Lambda$ and $KL(AK)(LB) \in L^2(\sigma)$, hence so is $KL(A \cap B)$. Also, $A + B \in F_\sigma(\Lambda)$, as $I \subset A+B$ and $KL(A+B)K \subset KL(A+B) \subset KLA + KLB \subset \Lambda$, whence $KL(A+B) \in L^2(\sigma)$.

3. and 4. are trivially satisfied.

5. Let $A_i \subset \sum_i A_i \subset B \in F_\sigma(\Lambda)$ such that $I_iA$, $A_iI$, $IB$ and $BI \in L^2(\sigma)$, then $IA_iI \subset IA_i \subset I(\sum_i A_i) \subset \Lambda$ and $IA_iI \in L^2(\sigma)$, hence so is $I(\sum_i A_i)$. Similarly $(\sum_i A_i)I \in L^2(\sigma)$. If $J_i \in L^2(\sigma)$ with $J_i \subset A_i$, then of course $\sum_i J_i \subset \sum_i A_i$.

6. If A is a fractional $\Lambda$-ideal, then we claim that $\Lambda \,\dot{.}\, A$ is a left $\Lambda$-ideal. Indeed, if $I,J \in L^2(\sigma)$ are such that $I \subset A$, and $JA$, $AJ \in L^2(\sigma)$, then $\Lambda \subset \Lambda \,\dot{.}\, A$ and $I \subset (\Lambda \,\dot{.}\, A)I \subset \Lambda$. Similarly, $\Lambda \,.\dot{}\, A$ is a right $\Lambda$-ideal. Now, $A(\Lambda \,\dot{.}\, A)A \subset A$, hence $A(\Lambda \,\dot{.}\, A) \subset (A \,\dot{.}\, A) = \Lambda$, and therefore, $\Lambda \,\dot{.}\, A \subset \Lambda \,.\dot{}\, A$. By the symmetry we obtain that $\Lambda \,\dot{.}\, A = \Lambda \,.\dot{}\, A$ is a fractional $\Lambda$-ideal satisfying the requirement.

7. easy ! □

For any fractional $\Lambda$-ideal I, we will write $I^d = \Lambda:(\Lambda : I)$ and $I^{-1} = \{x \in \Sigma;\ Ix\ I \subset I\}$.

1.5. Lemma. Let $\Lambda$ be a $\sigma$-maximal oreder in $\Sigma$. Let $A \in D_\sigma(\Lambda)$ and $I \in F_\sigma(\Lambda)$, then $A \mathrel{.^{\cdot}} I$ and $A \mathrel{^{\cdot}.} I$ are divisorial $\Lambda$-ideals and $A \mathrel{.^{\cdot}} I = A \mathrel{.^{\cdot}} I^d = A \mathrel{^{\cdot}.} I^d = A \mathrel{^{\cdot}.} I$.

Proof. One easily verifies that both $(A^{-1}\ I)^{-1}$ and $I^{-1} \star A$ are divisorial $\Lambda$-ideals and that $(A^{-1}\ I)^{-1} = I^{-1} \star A$. Now, suppose that $x \in \Sigma$ has the property that $Ix \subset A$, then $A^{-1}I\ x \subset A^{-1}A \subset \Lambda$, hence $x \in (A^{-1}I)^{-1}$. Conversely, if $x \in I^{-1} \star A$, then $Ix \subset I(I^{-1}A)^d \subset A$, and therefore $A \mathrel{.^{\cdot}} I = I^{-1} \star A$. Similarly, one may prove that $A \star I^{-1} = A \mathrel{^{\cdot}.} I$. Finally, since $D_\sigma(\Lambda)$ is abelian and $(I^d)^{-1} = I^{-1}$, we obtain $A \mathrel{.^{\cdot}} I = A \mathrel{.^{\cdot}} I^d = I^{-1} \star A = A \star I^{-1} = A \mathrel{^{\cdot}.} I^d = A \mathrel{^{\cdot}.} I$. □

Definition. We call $\Lambda$ a $\sigma$-Krull order in $\Sigma$ if it is a $\sigma$-maximal order and if it satisfies the ascending chain condition (A.C.C.) on divisorial $\Lambda$-ideals contained in $\Lambda$.

1.6. Proposition. If $\Lambda$ is a $\sigma$-Krull order in $\Sigma$, then $D_\sigma(\Lambda) \cong \mathbb{Z}^{(I)}$ for some index set I, and this isomorphism is order preserving.

Proof. Reverse the ordering on $D_\sigma(\Lambda)$, i.e. $A \leqslant B$ if $B \subset A$. It follows that every finite subset $\{A_1,\dots,A_n\}$ of $D_\sigma(\Lambda)$ has supremum $A_1 \cap \dots \cap A_n$ and infimum $\Lambda:(\Lambda:A_1 + \dots + A_n)$. Moreover, by the A.C.C. any nonempty subset of positive elements of $D_\sigma(\Lambda)$, i.e. of divisorial $\Lambda$-ideals contained in $\Lambda$, has a minimal element, A well-known theorem of N. Bourbaki [21] on commutative ordered groups satisfying these properties yields that $D_\sigma(\Lambda) \cong \mathbb{Z}^{(I)}$ for some index set I and this group isomorphism is order preserving. Of course, the order on $\mathbb{Z}^{(I)}$ here is given by

$(a_i)_{i \in I} \leq (b_i)_{i \in I}$ if and only if $a_i \leq b_i$ for every $i \in I$. □

1.7. Proposition. If $\Lambda$ is a $\sigma$-Krull order in $\Sigma$, then $D_\sigma(\Lambda)$ is generated by $P(\sigma)$, a set of prime ideals contained in $L^2(\sigma)$.

Proof. Since the group isomorphism of (1.6.) is order preserving, it is easy to verify that the maximal divisorial $\Lambda$-ideals (with respect to the inclusion !) contained in $\Lambda$ form a set of generators of $D_\sigma(\Lambda)$. We claim that these generators are prime ideals of $\Lambda$. Indeed, for any two elements a, b in an Artin setting T, we have $(a.b)^d = (a^d.b)^d = (a.b^d)^d = (a^d.b^d)^d$, where $a^d = (a^{-1})^{-1}$, cf. loc. cit. Now if P is a maximal divisorial $\Lambda$-ideal contained in $\Lambda$ and if I and J are ideals of $\Lambda$ such that P is properly contained in I and J, then $IJ \subset P$ implies $(IJ)^d \subset P^d = P$. But, $(IJ)^d = (I^dJ^d)^d = \Lambda$, which yields a contradiction. □

1.8. Lemma. If $\Lambda$ is a $\sigma$-Krull order in $\Sigma$ and if P is a divisorial prime ideal which is minimal in $L^2(\sigma)$, then $P \in P(\sigma)$.

Proof. By the foregoing proposition, there exist prime ideals $P_1,\dots,P_n$ in $L^2(\sigma)$, such that $P_1^{k_1} \cdot \dots \cdot P_n^{k_n} \subset P_1^{k_1} \star \dots \star P_n^{k_k} = P$, hence $P = P_i$ for some $1 \leq i \leq n$. □

Unless otherwise indicated, assume that $\Lambda$ is a $\sigma$-Krull order in $\Sigma$.
Let $L^2(\tau) = \cap\, \{L^2(\Lambda\text{-}P) \cap L^2(\sigma); P \in P(\sigma)\}$, where $L^2(\Lambda\text{-}P)$ is the multiplicatively closed set of ideals not contained in P. By abuse of notation, for any $\Lambda$-ideal A, we define $Q^l_\tau(A) = \{x \in \Sigma;\ \exists\, I \in L(\tau^l),\ I\,x \subset A\}$ resp. $Q^r_\tau(A) = \{x \in \Sigma;\ \exists\, I \in L(\tau^r),\ x\,I \subset A\}$. In all examples, both $L(\tau^r)$ and $L(\tau^l)$ happen to be idempotent filters, hence $Q^l_\tau(A)$ and $Q^r_\tau(A)$ are just the usual localizations.

1.9. Proposition. For any pair $A,B \in D_\sigma(\Lambda)$ we have $Q^l_\tau(AB) = A \star B = Q^r_\tau(AB)$.

Proof. If $x \in Q^r_\tau(AB)$, then $x\, I \subset AB \subset (AB)^d$ for some $I \in L^2(\tau)$. Now $I^d = \Lambda$, since $I \in F_\sigma(\Lambda)$ and $I \not\subset P$ for all $P \in P(\sigma)$. It follows that $x\, I^d = x\, \Lambda \subset (AB)^d = A \star B$. Conversely, if $x \in (AB)^d$, then $x(\Lambda:AB) \subset \Lambda$, whence $x(\Lambda:AB)AB \subset AB$. Here $(\Lambda : AB)AB$ is an ideal of $\Lambda$ not contained in any $P \in P(\sigma)$, as $((\Lambda : AB)AB)^d = \Lambda$ and $(\Lambda : AB)AB \in L^2(\sigma)$, $x \in Q^r_\tau(AB)$. Similarly, $Q^1_\tau(AB) = A \star B$, which finishes the proof. □

1.10. Proposition. If $\Lambda$ is a $\sigma$-Krull order in $\Sigma$ and if $A \in D_\sigma(\Lambda)$, then A is $\tau$-invertible whenever $\tau = \tau^1$ is idempotent.

Proof. Let us first check that $A \otimes -$ is inner in the class of $\tau$-torsion modules. For, let N" be a $\tau$-torsion left $\Lambda$-module and assume that $\Sigma\, a_i \otimes m_i \notin \tau(A \otimes M")$. There exists an ideal $I \in L^2(\tau)$ such that $Im_i = 0$ for all $m_i$. As $D_\sigma(\Lambda)$ is a commutative group, $Q_\tau(AI) = Q_\tau(IA)$, hence for any $i \in I$, there exists an ideal $J \in L^2(\tau)$ with $Jia_i \in AI$. So $Ji(\Sigma\, a_i \otimes m_i) = 0$ hence $I(\Sigma a_i \otimes m_i) \subset \tau(A \otimes M")$ and as $\tau^1$ is idempotent, $\Sigma\, a_i \otimes m_i \in \tau(A \otimes M")$, a contradiction. On the other hand, let us establish more generally that for $A,B \in D_\sigma(\Lambda)$, we have $Q_\tau(A \otimes B) \cong Q_\tau(AB)$ as $\Lambda$-bimodules. Consider the multiplication map $\mu : A \otimes B \to AB$. One easily checks that $Q_\tau(AA^{-1}) = Q_\tau(A^{-1}A) = \Lambda$ implies that $\mathrm{Ker}(\mu)$ is $\tau$-torsion, exactly as in (B.6.10), proving the assertion. □

This result may be strengthened as follows :

1.11. Proposition. If $\kappa$ is a symmetric kernel functor in $\Lambda$-mod such that $L^2(\tau) \subset L^2(\kappa) \subset L^2(\sigma)$ and if $A \in D_\sigma(\Lambda)$, then A is $\kappa$-invertible.

Proof. By the foregoing, A is $\tau$-invertible. If $\alpha : M' \to M$ is a monomorphism then $\mathrm{Ker}(A \otimes \alpha)$ is $\tau$-torsion, hence certainly $\kappa$-torsion. As in (1.10), one proves that $A \otimes M"$ is $\kappa$-torsion, if M" is. Hence A is $\kappa$-flat. Finally there is an isomorphism of $\Lambda$-bimodules $Q_\kappa(A \otimes A^{-1}) \cong Q_\kappa(Q_\tau(A \otimes A^{-1})) \cong Q_\kappa(\Lambda)$. □

From G. Maury, J. Raynaud [ 98 ] we retain that maximal orders in the usual sense are a nice generalization in the noncommutative case of completely integrally closed domains. Indeed, R is a commutative maximal order if and only if R is completely integrally closed in its field of fractions. We will prove below a relative version of this result.

Let R be a commutative ring and denote by K its generalized Martindale ring of quotients. Let $\mathcal{L}(\kappa)$ be a multiplicatively closed filter of nonzero ideals of R, then we say that R is $\kappa$-completely integrally closed in K if R is $\kappa$-torsionfree and satisfies the property : if for $x \in K$ there exists an ideal $I \in \mathcal{L}(\kappa)$ for which $I x^n \subset R$ for all positive integer n, then $x \in R$.

1.12. Lemma. The following statements are equivalent :

1. R is a $\kappa$-maximal order in K;
2. R is $\kappa$-completely integrally closed.

Proof. 1. ⇒ 2. If $x \in K$ has the property that $I x^n \in R$ for some $I \in \mathcal{L}(\kappa)$ and all $n \in \mathbb{N}$, then $I R[x] \subset R$. Because R is a $\kappa$-maximal order, this implies that $R = R[x]$, i.e. $x \in R$.

2. ⇒ 1. Let R' be an overring of R in K, such that $I R' \subset R$ for some $I \in \mathcal{L}(\kappa)$. If $x \in R'$, then $I x^n \in R$ for all $n \in \mathbb{N}$. Since R is $\kappa$-completely closed, this entails that $x \in R$. So $R = R'$, which proves the assertion. □

If $\Lambda$ is a $\sigma$-order in $\Sigma$ with center R, then we let $\mathcal{L}(c(\sigma))$ consist of all ideals I of R such that $\Lambda I \in \mathcal{L}^2(\sigma)$, then $\mathcal{L}(c(\sigma))$ is a multiplicatively closed filter of nonzero ideals of R such that R is $c(\sigma)$-torsion free.

1.13. Proposition.

1. If $\Lambda$ is a $\sigma$-maximal order in $\Sigma$, then R is $c(\sigma)$-completely integrally closed.
2. If $\Lambda$ is a $\sigma$-Krull order in $\Sigma$, then R is a $c(\sigma)$-Krull order.

Proof. 1. Suppose there exists an element x such that $I x^n \subset R$ for some $I \in L(c(\sigma))$ and all $n \in \mathbb{N}$, then $\Lambda[x]$ is an overring of $\Lambda$ in $\Sigma$ such that $(\Lambda I)\Lambda[x] \subset \Lambda$, hence $x \in \Lambda \cap K = R$, where K consists of all $x \in \Sigma$ such that $I x \subset R$ for some $I \in L(c(\sigma))$.

2. In view of 1. we are left to show that R satisfies the A.C.C. on divisorial R-ideals contained in I. For any $I \in L(c(\sigma))$, write $(\Lambda I)^{-1} = \{x \in \Sigma; x\Lambda I \subset \Lambda\}$ resp. $I^{-1} = \{k \in K; k I \subset R\}$, and $\overline{\Lambda I} = \{x \in \Sigma; (\Lambda I)^{-1}x \subset \Lambda\}$ resp. $\overline{I} = \{k \in K; I^{-1}k \subset R\}$. Clearly, $\overline{\Lambda I} \cap R \supset I$ if I is a divisorial ideal. Moreover, $(\overline{\Lambda I} \cap R)I^{-1} \subset \overline{\Lambda I}(\Lambda I)^{-1} \cap K \subset \Lambda \cap K = R$, so $I = \overline{\Lambda I} \cap R$ and R satisfies the A.C.C. on divisorial ideals if $\Lambda$ does. □

Remark. It may happen that $L(c(\sigma))$ is trivial, even if $L^2(\sigma)$ is not. Let us give an example. If L is a field and $\alpha$ is an automorphism of L of infinite rank, then $\Lambda = L[X,\alpha]$ is a ($\kappa_{\Lambda-o}$-) maximal order, and for $L^2(\sigma) = L^2(\Lambda\text{-}o)$ we have $L(c(\sigma)) = \{L^\alpha\} = \{Z(\Lambda)\}$. We will prove below that such a situation cannot occur if $\Lambda$ is a PI-ring.

We will now briefly consider "one-dimensional" relative Krull orders. Call a $\sigma$-order $\Lambda$ in $\Sigma$ a $\sigma$-Asano order if and only if $F_\sigma(\Lambda)$ is a group under multiplication. We then have :

1.14. Proposition. If $\Lambda$ is a $\sigma$-Asano-order, then any fractional $\Lambda$-ideal is finitely generated as a left and right projective $\Lambda$-module and $\Lambda$ is a $\sigma$-Krull order in $\Sigma$.

Proof. Let A be a fractional $\Lambda$-ideal, then $BA = \Lambda = AB$ for some $\Lambda$-ideal B, hence $B \subset \Lambda \,^{\cdot}.\, A$ and $B \subset \Lambda \,.^{\cdot}\, A$. Moreover $\sum_{i=1}^{n} b_i a_i = 1$ for some $b_i \in B$ and $a_i \in A$, hence for any $a \in A$ we have $\sum_{i=1}^{n}(ab_i)a_i = a$ and $ab_i \in \Lambda$, yielding that $A = \sum_{i=1}^{n} \Lambda a_i$. Finally, $\Lambda \,.^{\cdot}\, A$ is contained in

$\mathrm{Hom}_\Lambda(A,\Lambda)$ and the dual basis theorem yields that A is a finitely generated projective left $\Lambda$-module. A similar argument shows that A is finitely generated projective on the right also. Let us show that $\Lambda \overset{\cdot}{.} I = \Lambda \,.\!\!\overset{\cdot}{}\, I = I^{-1}$ for any $I \in L^2(\sigma)$. Since $(\Lambda \overset{\cdot}{.} I)I = \Lambda \in L^2(\sigma)$, it follows that $I(\Lambda \overset{\cdot}{.} I)I = I, I(\Lambda \overset{\cdot}{.} I) = I(\Lambda \;. I)II^{-1} = II^{-1} = \Lambda$. Hence $\Lambda \overset{\cdot}{.} I$ and $\Lambda \,.\!\overset{\cdot}{}\, I$ are elements of $F_\sigma(\Lambda)$, which proves the assertion. Now, $I^{-1}(I \overset{\cdot}{.} I)I = \Lambda$ entails that $I^{-1}(I \overset{\cdot}{.} I) = I^{-1}$, hence $I \overset{\cdot}{.} I = II^{-1} = \Lambda$. Similarly, $I \,.\!\overset{\cdot}{}\, I = \Lambda$, hence by (1.1.) it follows that $\Lambda$ is a $\sigma$-maximal order. Finally, $\Lambda$ satisfies the A.C.C. on divisorial ideals, since every fractional ideal is finitely generated. □

1.15. Proposition. If $\Lambda$ is a $\sigma$-Asano order, then $F_\sigma(\Lambda)$ is the free abelian group generated by the prime ideals of $\Lambda$ contained in $L^2(\sigma)$.

Proof. Since $A^{-1} = \Lambda : A$ for any $\Lambda$-ideal A, we obtain that $F_\sigma(\Lambda) = D_\sigma(\Lambda)$, hence there exists an order preserving isomorphism $\psi : F_\sigma(\Lambda) \to \mathbb{Z}^{(I)}$ for some index set I. Put $1_i = (\delta_{ij})_{j \in I}$ and let $P_i = \psi^{-1}(1_i)$, then any $\Lambda$-ideal A may be written uniquely as $A = P_1^{n_1} \cdot \ldots \cdot P_k^{n_k}$. We claim that any prime ideal $P \in L^2(\sigma)$ equals $P_i$ for some i. Indeed, if P is a maximal ideal contained in $L^2(\sigma)$, then $P = P_1^{n_1} \cdot \ldots \cdot P_k^{n_k} \subset P_i$, hence $P = P_i$. Now, let P be any prime ideal in $L^2(\sigma)$ and let Q be a maximal ideal containing P, then $\psi(Q) \leqslant \psi(P)$, so $P = Q^n P_1^{l_1} \cdot \ldots \cdot P_k^{l_k}$ where $n > 0$. Thus, either $P = Q$ or $P_i \subset P \subset Q = P_j$ for some $1 \leqslant i \leqslant k$, contradicting $\psi(P_j) \not\leqslant \psi(P_i)$ for $i \neq j$. Proposition 1.7. finishes the proof. □

The proof of (1.15) also shows that any $P \in P(\sigma)$ is a maximal ideal of $\Lambda$. Moreover, $L^2(\sigma) = \{P_1^{n_1} \cdot \ldots \cdot P_k^{n_k};\ P_i \in P(\sigma)\}$.

1.16. Proposition. Let $\Lambda$ be a $\sigma$-Asano order and let $L^2(\rho) \subset L^2(\sigma)$ be a multiplicative closed filter of ideals, then $\rho^r$ and and $\rho^l$ have property (T).

Proof. Let us first check that $\rho^l$ and $\rho^r$ are idempotent. In view of (1.14), $L(\rho^l)$ has a cofinal set of finitely generated $\Lambda$-ideals. Let J be a left ideal of $\Lambda$ such that $\rho^l(I/J) = I/J$ for some $I \in L(\rho^l)$, which we may assume to be an ideal of $\Lambda$, finitely generated on the left, say $I = \Lambda i_1 + \dots + \Lambda i_k$. There exists an ideal $K \in L^2(\rho)$ such that $Ki_j \subset J$ for all $1 \leqslant j \leqslant k$, hence, $K I \subset J$ and because $L^2(\rho)$ is multiplicatively closed, $J \in L(\rho^l)$. As $\rho^l$ is idempotent, $Q^l_\rho(\Lambda) = \varinjlim \mathrm{Hom}_\Lambda(I,\Lambda)$, where the limit is taken over all $I \in L^2(\rho)$. It follows that $Q^l_\rho(\Lambda)$ consists of all $x \in \Sigma$ such that $s\, I \subset \Lambda$ for some $I \in L^2(\rho)$. Clearly, $Q^l_\rho(\Lambda) = Q^r_\rho(\Lambda)$ because $\Lambda \,^{\cdot}\!. \, I = \Lambda \,.^{\cdot}\, I$ for any $I \in L^2(\sigma)$. Finally, if I is an ideal in $L^2(\rho)$, then $I\, I^{-1} = \Lambda$, so $I^{-1} \subset Q^l_\rho(\Lambda)$, which shows that $Q^l_\rho(\Lambda)\; j_\rho(I) = Q^l_\rho(\Lambda)$, proving that $\rho^l$ has property (T) indeed. □

If $\sigma$ and $\tau$ are idempotent kernel functors in $\Lambda$-mod, then we say that $\sigma$ is $\tau$-geometrical if $Q_\tau(\Lambda)\; j_\tau(I)$ is an ideal of $Q_\tau(\Lambda)$ for every ideal in $L(\sigma)$. We then have :

1.17. Proposition. If $\Lambda$ is a $\sigma$-Asano order and if $L^2(\rho) \subset L^2(\sigma)$ is a multiplicatively closed filter of ideals of $\Lambda$, then :

1. $\rho$ is $\sigma$-geometrical;
2. $\Lambda$ is a $\rho$-Asano order;
3. $Q^l_\rho(\Lambda)$ is a $\sigma(\rho)$-Asano order, where $L^2(\sigma(\rho))$ is generated by the $Q^l_\rho(\Lambda)I$, where $I \in L^2(\sigma)$.

Proof. 1. Let $A = \Lambda a_1 + \dots + \Lambda a_n$ be an ideal in $L^2(\sigma)$ and let $x \in Q^l_\rho(\Lambda)$ be such that $I\, x \subset \Lambda$ for some $I \in L^2(\rho) \subset L^2(\sigma)$. Since $F_\sigma(\Lambda)$ is an abelian group, $I(Ax) = AIx \subset A\,\Lambda \subset A$, hence $I(Ax) \subset Q^l_\rho(\Lambda)A$, finishing the proof. Moreover, 2. Is trivial and 3. follows from some compatibility results in F. Van Oystaeyen [148]. □

Let $\Lambda$ be any ring and let $\Sigma$ be its generalized Martindale ring of quotients and let $A$ be the set of all ideals of $\Lambda$ which are not invertible in $\Sigma$. It is clear that $A$ is inductively ordered (by inclusion) and that the maximal elements in $A$ are prime. Denote by $A(\Lambda)$ this collection of maximal elements of A and by $L^2(\sigma)$ the intersection of all $L^2(\Lambda\text{-P})$, where $P \in A(\Lambda)$. Since $L^2(\sigma)$ consists of invertible ideals, the associated kernelfunctor $\sigma$ is idempotent and has property (T). It is easily verified that $\sigma$ is the largest symmetric kernel functor such that $\Lambda$ is a $\sigma$-Asano order.

1.18. Examples. Let us conclude this section by briefly mentioning some examples of $\sigma$-Asano orders.

a. Let A be a prime Goldie ring and let $\sigma^l$ resp. $\sigma^r$ be the left resp. right Goldie kernel functor, then we have pointed out before that A is a $(\sigma^l,\sigma^r)$-order, which is a maximal $(\sigma^l,\sigma^r)$-order exactly when it is a maximal order in the usual sense. Since $Q^l_\sigma(A) = Q^r_\sigma(A)$ is the classical simple artinian ring of fractions of A, the set $F_\sigma(A)$ consists of all fractional A-ideals in the usual sense. In this setting, Proposition 1.15. is the classical result that the set $F(A)$ of fractional ideal of A is the free abelian group generated by the non-zero prime ideals of A.

b. Krull P I rings may be viewed as $\sigma$-Krull orders. Indeed, for a P I-ring A the various definitions of noncommutative Krull rings given by M. Chamarie [29 ], R.M. Fossum [50 ], E. Jespers, L. Le Bruyn and P. Wauters [75 ] and H. Marubayashi [92 ]coincide and therefore a Krull P I ring may be defined as being a maximal order over a Krull domain. In this case, we let $\sigma = \kappa_{A-0}$, which is central by Posner's theorem, cf. [116] and $P(\sigma) = X^{(1)}(A)$, the set of prime ideals of height one of A. These Krull P I rings also have the property that the $\kappa_{A-p}$ are central for $p \in X^{(1)}(A)$, by a result of M. Chamarie's [29 ] for example. Let us also mention that $\Omega$-Krull rings, cf. [75 ], yield examples of relative

Krull orders, even in the absence of the P I hypothesis (see also E.1.).

c. Recall that an HNP ring is a left and right noetherian prime ring A with the property that every left resp. right ideal of A is a left resp. right projective A-module. We claim that every HNP P I-ring A is a relative maximal order with respect to a central idempotent kernel functor $\sigma$ which has property (T). Indeed, let $\sigma = \inf\{\kappa_{A-P}; P \in Id(A)\}$, where Id(A) consists of all idempotent prime ideals of A and recall that a prime ideal of A is necessarily either invertible or idempotent. It follows that $L^2(\sigma)$ consists of all ideals I of A not contained in any $P \in Id(A)$. Now, an arbitrary ideal of A may be written as the product of an invertible ideal and an eventually idempotent ideal. Therefore, $L^2(\sigma)$ consists of invertible ideals only, since an eventually idempotent ideal cannot be contained in an invertible prime ideal P, since $\bigcap_n P^n = 0$. Now, A being a noetherian P I ring, each idempotent filter is completely characterized by the *prime* ideals contained in it and here this is exactly the set $P(\sigma) = Inv(A)$, the collection of invertible prime ideals of A. Finally, for every invertible ideal I of A we have that I : I = A, so A is a $\sigma$-maximal order, indeed. A result due to E. Nauwelaerts, F. Van Oystaeyen [110] states that $\sigma$ is a central idempotent kernel functor. Of course, in this case we also have that A is a $\sigma$-Krull order.

d. Combining examples b. and c. it is easy to see that tame orders over a Krull domain may also be viewed as relative maximal orders.

## D.2. Graded Rings over Goldie Rings.

Let $A = \bigoplus_{\sigma \in G} A_\sigma$ be a graded ring, graded by an arbitrary group G. We call a graded left ideal L of A gr-essential if L is an essential subobject of A in the category A-gr. From Lemma I.2.8. in [109] we retain that L is gr-essential if and only if $\underline{L}$ is an essential left ideal of A(in A-mod) where, for any $M \in A$-gr we denote by $\underline{M}$ the underlying A-module, forgetting the grading. It follows that if A is strongly graded of type G, then L is gr-essential in A if and only if $L_e$ is an essential left ideal of $A_e$, in view of (A.4.2.). These observations are useful in studying graded rings over (left) Goldie rings and (left) gr-Goldie rings, where a graded ring A is said to be a gr-Goldie ring if it has finite Goldie dimension in A-gr and satisfies the ascending chain condition on graded left annihilators. We do not intend to investigate in full generality the graded equivalents of Goldie's theorems (cf. [108] for the $\mathbb{Z}$-graded case) but we content ourselves to derive some fundamental properties for graded rings of type G, where G is an ordered abelian group.

2.1. Lemma. Let A be a graded ring of type G, where G is a finitely generated abelian group. If M is a graded left A-module, then M is gr-noetherian (in A-gr) if and only if $\underline{M}$ is noetherian (in A-mod).

Proof. cf. [109]. □

2.2. Proposition. Let A be a semiprime ring which is strongly graded by a torsionfree, finitely generated abelian group G, then A is a (left) Goldie ring if and only if A is a (left) gr-Goldie ring if and only if $A_e$ is a (left) Goldie ring.

Proof. It is evident from Theorem A.4.2. that A is a gr-Goldie ring if and only if $A_e$ is a Goldie ring. The fact that A is a gr-Goldie ring if it is a Goldie ring is obvious. So, let us assume that $A_e$ is a Goldie ring. By the first part A is then a gr-Goldie ring and hence it satisfies the ascending chain condition on graded left annihilators. Since G is an ordered group, we may define for every left ideal L of A the left A-ideal $\tilde{L}$ generated by the homogeneous components of highest degree appearing in the decompositions of elements of L. If L is an essential left ideal of A, then $\tilde{L}$ is a gr-essential left ideal of A and thus $\tilde{L}$ is an essential left ideal of A too. It follows that the left singular radical J of A is a graded ideal, hence it is generated by its part of degree e, and since $J_e = 0$ it follows that $J = 0$. Therefore, we may consider the maximal ring of quotients $Q_{max}(A)$ and the graded maximal ring of quotients $Q^g_{max}(A)$ (defined in a similar way, using graded objects only). Since the left singular radical of A is zero, we have $A \subset Q^g_{max}(A) \subset Q_{max}(A)$. Now, $Q^g_{max}(A)_e = Q(A_e)$ and $Q^g_{max}(A)$ is strongly graded of type G. From Corollary I.5.3. in [109] we retain that $Q^g_{max}(A)$ is Von Neumann gr-regular. Since A has finite Goldie dimension in A-gr, it follows that $Q^g_{max}(A)$ has finite Goldie dimension in $Q^g_{max}(A)$-gr and thus $Q^g_{max}(A)$ is a gr-semisimple gr-Artinian ring. In particular, $Q^g_{max}(A)$ is also left gr-noetherian and, by the lemma, it follows that $Q^g_{max}(A)$ is a left noetherian ring. Thus, $Q^g_{max}(A)$ has finite Goldie dimension and since $Q_{max}(A)$ is a left essential extension of $Q^g_{max}(A)$ in $Q^g_{max}(A)$-mod, it follows that $Q_{max}(A)$ has finite Goldie dimension. Consequently, A has finite Goldie dimension too. The statement of the Proposition now follows (note that it also follows from the fact that $Q_{max}(A)$ is a left noetherian regular ring !). □

2.3. Corollary. If $A_e$ is a (semi)prime left Goldie ring, then the set of regular elements of $A_e$ (which are also regular in A, because A is strongly graded) is a left Ore set of A and the left ring of fractions $Q^g$ at $S_e$ has the property that $(Q^g)_e = Q(A_e)$. Since $Q(A_e)$ is simple artinian, Lemma I.3.15 in [109] implies that $Q^g$ is a crossed product of $Q(A_e)$ and G. Every graded left ideal of $Q^g$ is of the form $Q^g u$ for some $u \in Q(A_e)$. □

2.4. Proposition. Let A be strongly graded of type G, where G is a torsion free, finitely generated abelian group.

1. If P is a (semi)prime ideal of $A_e$ such that AP is an ideal of A, then AP is (semi)prime.

2. Suppose that $A_e$ satisfies the ascending chain condition on ideals. If P is a semiprime ideal of A, then $P \cap A_e$ is a semiprime ideal of $A_e$. For a prime ideal P of A, the ideal $P \cap A_e$ is of the form $M \cap A_{\tau_1} M A_{\tau_1^{-1}} \cap \ldots \cap A_{\tau_n} M A_{\tau_n^{-1}}$ for some positive integer n and some prime ideal M of $A_e$, which is minimal amongst the prime ideals of $A_e$ containing $P \cap A_e$.

<u>Proof</u>. 1. Easy.

2. For every $\tau \in G$, associating $A_\tau P A_{\tau^{-1}}$ to P defines a bijection of $\mathrm{Spec}(A_e)$ onto itself, which takes a closed set V(I) associated to an ideal I of $A_e$ to the closed set $V(A_\tau I A_{\tau^{-1}})$. Therefore, if $\mathrm{rad}(I) = \bigcap_{i=1}^n M_i$, then $\mathrm{rad}(A_\tau I A_{\tau^{-1}}) = \bigcap_{i=1}^n A_\tau M_i A_{\tau^{-1}}$. Now, from $A_{\tau^{-1}}(\bigcap_{i=1}^n A_\tau M_i A_{\tau^{-1}})A_\tau \subset \bigcap_{i=1}^n A_{\tau^{-1}} A_\tau M_i A_{\tau^{-1}} A_\tau = \bigcap_{i=1}^n M_i$, it follows that $\bigcap_{i=1}^n A_\tau M_i A_{\tau^{-1}} \subset A_\tau(\bigcap_{i=1}^n M_i) A_{\tau^{-1}}$. Since the converse inclusion is obvious, equality holds. Consequently for all $\tau \in G$, we have $\mathrm{rad}(A_\tau I A_{\tau^{-1}}) = A_\tau \mathrm{rad}(I) A_{\tau^{-1}}$. For all $\tau \in G$, we also have $A_\tau(P \cap A_e) = P \cap A_\tau = (P \cap A_e)A_\tau$, hence $A_\tau \mathrm{rad}(P \cap A_e)A_{\tau^{-1}} = \mathrm{rad}(P \cap A_e)$ and $A\, \mathrm{rad}(P \cap A_e) = \mathrm{rad}(P \cap A_e)A$. Because $A_e$ satisfies the A.C.C., we know that $(\mathrm{rad}(P \cap A_e))^m \subset P \cap A_e$ for some $m \in \mathbb{N}$. Therefore, $(A\, \mathrm{rad}(P \cap A_e))^m \subset A(P \cap A_e) \subset P$ and this yields that $A\, \mathrm{rad}(P \cap A_e) \subset P$ or $\mathrm{rad}(P \cap A_e) = P \cap A_e$. The A.C.C. on ideals of $A_e$ and the above remarks entail that $P \cap A_e = (\bigcap_i) \bigcap_{\tau \in G} A_\tau M_i A_{\tau^{-1}}$, where the $M_i$ are minimal amongst prime ideals containing $P \cap A_e$ and the intersection is finite. For each $M_i$ one easily checks that $A(\bigcap_{\tau \in G} A_\tau M_i A_{\tau^{-1}}) = (\bigcap_{\tau \in G} A_\tau M_i A_{\tau^{-1}})A$, so if P is a prime ideal of A, then we obtain, upto a renumbering of the $M_i$, that $A(\bigcap_{\tau \in G} A_\tau M_1 A_{\tau^{-1}}) = P$ and therefore $P \cap A_e = \bigcap_{\tau \in G} A_\tau M_1 A_{\tau^{-1}}$. Finally, we may take a finite intersection, i.e.

$P \cap A_e = A_{\tau_1} M_1 A_{\tau_1^{-1}} \cap \ldots \cap A_{\tau_n} M_1 A_{\tau_n^{-1}}$. □

2.5. <u>Remarks</u>. 1. If $P$ is a (semi)prime ideal of $A$ such that $P \cap A_e = 0$, then $A$ is (semi) prime. Indeed, $P \supset P_g$ and $P_g$ is (semi) prime, since $G$ is ordered, so $P_g = 0$ or else $P_g = A(P_g \cap A_e)$. But, $P \cap A_e = 0$ entails $P_g \cap A_e = 0$, hence $P_g = 0$ and $A$ is a (semi)prime ring.

2. If $A_e$ is a prime Goldie ring, then an ideal $I$ of $A$ contains a regular element of degree zero, exactly when $I \cap A_e \neq 0$.

3. If $I$ is an ideal of $A$ such that $A/I$ satisfies the ascending chain condition on left annihilators, then $Q^g I$ is an ideal of $Q^g$. Indeed, let $\pi : A \to A/I$ denote the canonical projection and let $S_e$ be the set of regular elements of $A_e$. Clearly, $\pi(S_e)$ is a left Ore set in $A/I$ and by our assumption, $\pi(S_e)$ is left reversible in $A/I$. If $a \in A$ and $s \in S_e$ have the property that $as \in I$, then $s'a \in I$ for some $s' \in A_e$. If $i \in I$ and $s \in I$ and $s \in S_e$, then $ti = as$ for some $a \in A$ and $t \in S_e$. But $as \in I$ yields $s'a \in I$ for an $s'$ as above, hence $s'ti = s'as$ with $s't \in S_e$ and $s'a \in I$. Consequently, $IQ^g \subset Q^g I$. □

At this point we specialize to the case $G = \mathbb{Z}$.

2.6. <u>Proposition</u>. Let $A$ be strongly graded of type $\mathbb{Z}$ and assume that $A_e$ is a semiprime left and right Goldie ring. If $I$ is an ideal of $A$, then $Q^g I = IQ^g$ is an ideal of $Q^g$.

<u>Proof</u>. Since $(Q^g)_0 = Q(A_0)$ is a semisimple artinian ring and $\mathrm{Pic}(Q(A_0)) = 1$ it follows that $Q^g = Q(A_0)[X, X^{-1}, \varphi]$, where $\varphi$ is an automorphism of $Q(A_0) = Q_0$. Hence $Q^g$ is a left and right principal ideal ring and therefore $Q^g I Q^g = Q^g a = aQ^g$ for some $a \in Q^g$. We claim that we may choose $a$ in $I$. Indeed, since $A_0$ is a left and right Goldie ring, there exist $s, t \in S_0$ such that $sat \in I$ and we obtain that $Q^g sat = Q^g at = Q^g a Q^g t = Q^g a Q^g = Q^g a$, and similarly $sat\, Q^g = aQ^g$. Thus : $Q^g I = IQ^g$. □

2.7. Proposition. Let $A$ be strongly graded of type $\mathbb{Z}$ over a prime left Goldie ring $A_0$.

1. If $P$ is a prime ideal of $A$ such that $P \cap A_0 = 0$ and $A/P$ is a left Goldie ring, then $Q^g P \cap A = P$.

2. Suppose that $A/P$ is a left Goldie ring for each prime ideal $P$ of $A$ with $P \cap A_0 = 0$, then there is a bijective correspondence between proper prime ideals of $Q^g$ and prime ideals $P$ of $A$ with $P \cap A_0 = 0$. Moreover, for each of these, $ht(P) \leqslant 1$.

Proof. 1. Put $I = Q^g P \cap A$. If $I \neq P$, then $I/P$ is a nonzero ideal of the prime left Goldie ring $A/P$, so $I/P$ contains a regular elements $\bar{a}$ of $A/P$. If $a \in I$ represents $\bar{a}$, then $sa \in P$ for some $s \in P$, a contradiction !

2. If $P$ is a prime ideal of $A$, such that $P \cap A_0 = 0$, then $Q^g P$ is a proper prime ideal of $Q^g$, because $Q^g P \cap A = P$. Conversely, if $M \neq 0$ is a proper prime ideal of $Q^g$, then $M \cap A \subset P$ for some ideal $P$ of $A$, which is maximal with the property of lying over 0. It is clear that $P$ is a prime ideal and that $M = Q^g(M \cap A) \subset Q^g P$. Since $Q^g$ is a prime principal left and right ideal ring, $M = Q^g P$ follows. In conclusion, this yields $M \cap A = Q^g P \cap A = P$. The statement about $ht(P)$ follows immediately from the observation that if $P' \subset P$ is another prime ideal of $A$, then $P' \subset A_0 = 0$, and so, since $Q^g P' \subset Q^g P$ are prime ideals of $Q^g$, we obtain $P' = 0$ or $P = P'$. □

The condition : $A/P$ is a left Goldie ring for every prime ideal $P$ of $A$ such that $P \cap A_e = 0$ is satisfied e.g. in case $A$ is left noetherian or a PI ring and in all cases considered in subsequent sections, so it is to be considered a very mild condition here.

2.8. Proposition. Let $A$ be strongly graded of type $\mathbb{Z}$ over a left Goldie ring $A_0$ and assume that for all $P \in \mathrm{Spec}(A)$ with $P \cap A_0 = 0$, we have that $A/P$ is a left Goldie ring, then for any $P$ lying over 0, $\mathcal{C}(P)$ is a regular left Ore set of A and $A \subset Q_P^1(A) = Q_M(Q^g)$, where $M = Q^gP$. Therefore, $Q_P^1(A)$ is a bounded prime principal left and right ideal ring, with unique maximal ideal $Q_P^1(A)P$. If $\mathcal{P}$ denotes the set of all $P \in \mathrm{Spec}(A)$ with $P \cap A_0 = 0$, then $Q^g = \bigcap_{P \in \mathcal{P}} Q_P^1(A) \cap S(Q^g)$, where $S(Q^g)$ is a simple noetherian ring.

Proof. First we establish that $\mathcal{C}(P) \subset \mathcal{C}(Q^gP)$. If $c \in \mathcal{C}(P)$ and $q \in Q^g$ are such that $qc \in Q^gP$, then we may write $q = s^{-1}a$ for some $s \in A$, so $ac \in Q^gP \cap A = P$. Hence $a \in P$ and $q \in Q^gP$, establishing the claim. Also if $s \in S_0$ and $c \in A$ are such that $ac \in P$ with $a \in A$, then $ass^{-1}c \in Q^gP$ yields $as \in Q^gP$ and $a \in Q^gP \cap A = P$. Next, $\mathcal{C}(M)$ is a regular left and right Ore set of $Q^g$, as $Q^g$ is a prime principal left and right ideal ring. It remains to verify that $Q_M(Q^g)$ is the left ring of fractions of A with respect to $\mathcal{C}(P)$. It is evident that $A \subset Q_M(Q^g)$. Any $c \in \mathcal{C}(P) \subset \mathcal{C}(M)$ is invertible in $Q_M(Q^g)$. On the other hand, any $q \in Q_M(Q^g)$ may be written as $q = u^{-1}v$ with $u \in \mathcal{C}(M)$, $v \in Q^g$. Put $u = s^{-1}c$ and $v = s^{-1}d$ for some $s \in S_0$ and $c,d \in A$, then $u \in \mathcal{C}(M)$ entails $c \in \mathcal{C}(P)$, hence c is invertible in $Q_M(Q^g)$. This yields that $q = (s^{-1}c)^{-1}s^{-1} = c^{-1}d$ and it follows that $Q_M(Q^g) = Q_P^1(A)$ or $\mathcal{C}(P)$ is a regular left Ore set of A. The last statement of the Proposition easily follows from the foregoing. □

## D.3. Generalized Rees Rings.

We will introduce a class of graded rings that will be very useful for the study of certain aspects of the theory of class groups of (maximal) orders. The construction of certain generalized Rees rings over the ring which one is studying often allows to infer a property of the ring from graded properties of the constructed Rees rings; this makes for a good number of applications. There is one main problem : how general should our Rees rings be ? We include some very general definitions, but afterwards we turn to the study of Rees rings over arithmetical orders.

If A is a left noetherian ring an M is an A-bimodule, then the Rees ring A(M) is defined to be $A \oplus M \oplus M^{(2)} \oplus \dots \oplus M^{(n)} \oplus \dots$, where $M^{(n)} = M \otimes_A \dots \otimes_A M$ is the n-fold tensorproduct. In case M is an ideal of A, the Rees ring and modules over it may be used to express properties of the M-adic filtration on A, e.g. properties related to the Artin-Rees property. More generally, if M is an invertible A-module, one may define the $\mathbb{Z}$-graded ring $\check{A}(M)$ to be $\check{A}(M) = \bigoplus_{n \in \mathbb{Z}} M^{(n)}$, where $M^{(o)} = A$ and $M^{(n)} = (M^{-1})^{(-n)}$ if n is negative. Multiplication in $\check{A}(M)$ is induced by the tensorproduct maps $M^{(n)} \otimes M^{(m)} \to M^{(n+m)}$. Such constructions, in the commutative case, have been studied in [155,146] and there A is usually assumed to be a Dedekind ring or a Krull domain. In the context of this book, it is clear that we aim to study such rings over (maximal) orders over Dedekind rings or Krull domains, but first we have to introduce some more general constructions.

Let A be a strongly graded ring of type G and let $\varphi : G \to \mathrm{Pic}(A_e)$ be the corresponding group morphism, cf. Section A.4. . Assume that $A_e$ is a semi prime left Goldie ring, so it has a semisimple artinian left ring of quotients $Q_e$. The canonical map $\varphi : \mathrm{Pic}(A_e) \to \mathrm{Aut}(Z(A_e))$ associates to the class of $A_\sigma$, with $\sigma \in G$, an automorphism $\varphi_\sigma$ of $Z(A_e)$. The automorphism $\varphi_\sigma$ extends to the so-called Miyashita automorphisms of $A^{A_e}$ the centralizer of $A_e$ in A. Only in some particular cases do the $\varphi_\sigma$ extend to automorphisms of A. Since A is strongly graded, $Q^g \cong Q_e \otimes_{A_e} A$ is a gr-Goldie ring by Theorem A.4.2. and it is actually a gr-simple gr-artinian ring. In is clear that one cannot expect much from this construction if $Q^g$ does not have nice properties, in particular, it should be a left Goldie ring, or in the most general case some relative version of this fact. So, here we restrict attention to groups G which are torsion free abelian (hence ordered) and depending on the situation we may add a finiteness condition like being finitely generated or satisfying the A.C.C. on cyclic subgroups, i.e. G is a Krull group in the terminology of Krull semigroups, cf. [ 3, 34 ]

We have the well-known exact diagram :

$$\begin{array}{ccccccccc}
0 & \longrightarrow & \mathrm{Inn}(A_e) & \longrightarrow & \mathrm{Aut}(A_e) & \xrightarrow{\pi} & \mathrm{Pic}(A_e) & & \\
 & & \downarrow & & \downarrow & & \downarrow j & & \\
0 & \longrightarrow & \mathrm{Inn}(Q_e) & \longrightarrow & \mathrm{Aut}(Q_e) & \xrightarrow[\pi]{} & \mathrm{Pic}(Q_e) & \longrightarrow & 1
\end{array}$$

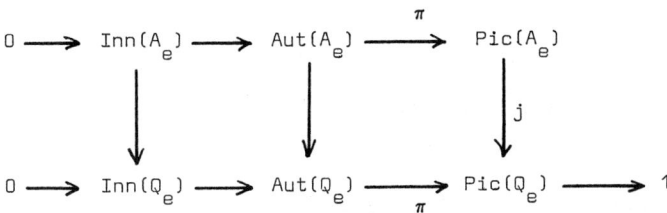

where the vertical arrows derive from $Q_e \otimes_{A_e}$ - and the extension of (inner) automorphism of $A_e$ to (inner) automorphisms of $Q_e$. From Lemma I.3.25 of [109] (or also from the surjectivity of $\overline{\pi}$, taking into account the structure theory for strongly graded rings), it follows that $Q^g$ is a crossed product over $Q_e$, say $Q^g \cong Q_e \star G$. We say that A is a <u>generalized Rees ring of type</u> G, if for all $\sigma \in G$ we have $j([A_\sigma]) \in j\,\pi(\mathrm{Aut}(A_e))$. Considering A as a subring of $Q_e \star G$, we may write $A = \sum_{\sigma \in G} I_\sigma X_\sigma$, with $[I_\sigma] \in \mathrm{Pic}(A_e)$, satisfying the relations $I_\sigma \varphi_\sigma(I_\tau) = I_{\sigma\tau}$, for $\sigma, \tau \in G$, where $\varphi_\sigma \in \mathrm{Aut}(A_e)$ is such that $\pi(\varphi_\sigma) = [A_\sigma]$. Note that $\varphi_\sigma$ is only determined upto an inner automorphism by the latter description, but when the $X_\sigma$ are being chosen to establish the isomorphism $Q^g \cong Q_e \star G$, then also the $\varphi_\sigma$ are fixed. As a final step in the generalization procedure, we may consider a central idempotent kernel functor $\kappa$ in $A_e$-mod and define a $\kappa$-<u>Rees ring</u> in the obvious way, i.e. replace $\mathrm{Pic}(A_e)$ by $\mathrm{Pic}(A_e,\kappa)$ and $\mathrm{Pic}(Q_e)$ by $\mathrm{Pic}(Q_e,\kappa)$, where necessary and write $A = \sum_{\sigma \in G} I_\sigma X_\sigma$ with $Q_\kappa(I_\sigma \varphi_\sigma(I_\tau)) = I_{\sigma\tau}$ for any $\sigma, \tau \in G$, i.e. A is a $\kappa$-graded ring of type G over $A_e$. We do not study $\kappa$-Rees rings in this generality since our aim with these constructions is to obtain a tool for studying certain orders over Krull rings. Hence we will be interested in rings $A_e$ with an arithmetical ideal theory. Therefore, we restrict to rings $A_e$ which are $\sigma$-maximal orders or $\sigma$-Krull orders in the sense of Section D.1. and for $\kappa$ we take $\tau$ as in loc. cit. . Note that in all examples mentioned at the end of loc. cit. $\kappa$ is indeed a central kernel functor; in some results we only need the fact that $\kappa$ is G-invariant in the sense of Section - C.1. We say that A is <u>divisorially graded of type G</u> if A is a graded ring of type G satisfying the conditions :

DG.1. $A_e$ is a $\sigma$-maximal order in the sense of D.I.;

DG.2. for all $\gamma \in G$ we have $Q_\tau(AA_\gamma) = A$, where $\tau$ is the kernel functor derived from $\sigma$.

A $\tau$-Rees ring which is divisorially graded of type G will just be called a <u>generalized Rees ring</u> again and here the term has reached the most general meaning we consider in this book.

For any divisorially graded ring A of type G, the properties established in Lemma C.1.1. and Proposition C.1.2. are valid, with $\kappa = \tau$. By Theorem C.1.3.,it follows that the map $\gamma \to [A_\gamma]$ defines a group homomorphism $G \to \mathrm{Pic}(A_e,\tau)$. Using the results of $\tau$-graded rings referred to above, one may derive a generalized crossed product theorem for divisorially graded rings.

Let A be any $\sigma$-maximal order, G any group and let there be given a group homomorphism $\Phi : G \to \mathrm{Pic}(A,\tau) : \gamma \to [A_\gamma]$ . A <u>factor set</u> associated to $\Phi$ is a set $f = \{f_{\gamma,\lambda};\gamma,\lambda \in G\}$ of A-bimodule isomorphisms $f_{\gamma,\lambda} : Q_\tau(P_\gamma \otimes P_\lambda) \tilde{\to} P_{\gamma\lambda}$ together with another A-bimodule isomorphism $\alpha : P_e \tilde{\to} A$ such that the following diagrams commute (compare to (A.4.3.) and the diagrams preceding it) :

$$\begin{array}{ccc}
Q_\tau(P_\gamma \otimes P_\lambda \otimes P_\mu) & \xrightarrow{Q_\tau(P_\gamma \otimes f_{\lambda,\mu})} & Q_\tau(P_\gamma \otimes P_{\lambda\mu}) \\
{\scriptstyle Q_\tau(f_{\gamma,\lambda} \otimes P_\mu)}\downarrow & & \downarrow{\scriptstyle f_{\gamma,\lambda\mu}} \\
Q_\tau(P_{\gamma\lambda} \otimes P_\mu) & \xrightarrow[f_{\gamma\lambda,\mu}]{} & P_{\gamma\lambda\mu}
\end{array}$$

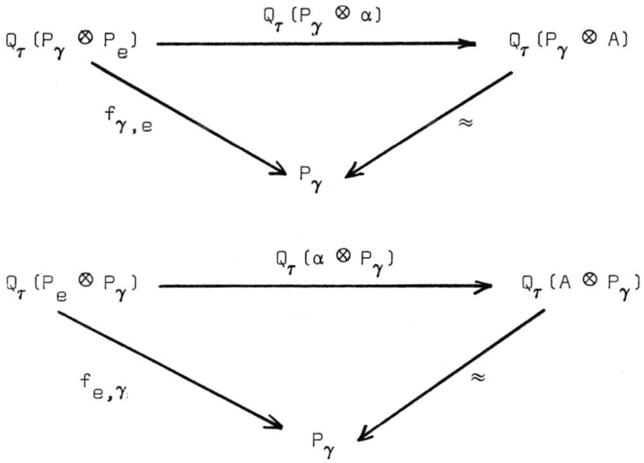


where all tensorproducts are over A. In writing down these diagrams we used the fact that for any $[P] \in \mathrm{Pic}(A,\tau)$ we have that $Q_\tau(P \otimes M) = Q_\tau(P \otimes Q_\tau(M))$ for any $M \in$ A-mod and $Q_\tau(N \otimes P) = Q_\tau(Q_\tau(N) \otimes P)$ for any A-bimodule N. Extending the terminology of Section A.4. to the divisorial case, we write $F_s(\Phi)$ for the collection of factor sets of the above type associated to the given $\Phi$. We define the ring $A < f,\Phi,G >$ with additive group $\bigoplus_{\gamma \in G} P_\gamma$ and multiplication given by $x.y = f_{\gamma\,\lambda}(x \otimes y)$ for any $x \in P_\gamma$ and $y \in P_\lambda$.

3.1. Theorem. The ring $B = A < f,\Phi,G >$ is divisorially graded of type G, containing a subring isomorphic to A, i.e. $B_e = P_e \cong A$. Conversely, if B is a divisorially graded ring of type G such that $\tau$ is G-invariant and noetherian, then there exists a group homomorphism $\Phi : G \to \mathrm{Pic}(B_e,\tau)$ and a factor set $f \in F_s(\Phi)$ such that $B \cong B_e < f,\Phi,G >$ as graded rings.

Proof. The first statement is easy, exactly the same as the proof of this statement in the strongly graded case, cf. Proposition I.3.13. in [109]. For the converse, let B be divisorially graded by G. The canonical $B_e$-bimodule morphism $B_\gamma \otimes_{B_e} B_{\gamma^{-1}} \to B_e$ extends to a surjective morphism of $B_e$-bimodules $Q_\tau(B_\gamma \otimes_{B_e} B_{\gamma^{-1}}) \to B_e$. As a consequence, one obtains that $Q_\tau(B_\gamma \otimes_{B_e} B_{\gamma^{-1}}) \cong B_e \cong Q_\tau(B_{\gamma^{-1}} \otimes_{B_e} B_\gamma)$ as $B_e$-bimodules. To conclude that $B_\gamma$ is $\tau$-invertible one applies Theorem C.1.3.. It is then clear that $\gamma \to [B_\gamma]$ defines a group homomorphism $\Phi : G \to \mathrm{Pic}(B_e,\tau)$. The multiplication map $B_\gamma \otimes_{R_e} B_\lambda \to B_{\gamma\lambda}$ extends to a $B_e$-bimodule isomorphism $f_{\gamma,\lambda}: Q_\tau(B_\tau \otimes_{B_e} B_\lambda) \to B_{\gamma\lambda}$ and so we obtain a factor set f consisting of the $f_{\gamma,\lambda}$ together with $\alpha = 1_{R_e}$. It is obvious that there is an isomorphism of graded rings $B \cong B_e < f,\Phi,G >$. □

3.2. Corollary. Let A be a $\sigma$-maximal order, such that $\tau$ is G-invariant and let B be divisorially graded by G such that $B_e = A$. If $\mathrm{Pic}(A,\tau) = \{[A]\}$, then B is a crossed product of A and G described by a group morphism $G \to \mathrm{Aut}(A)$ and a 2-cocycle $c: G\times G \to U(Z(A))$.

Proof. A straightforward consequence of the theorem. □

3.3. Lemma. Let A be a $\sigma$-maximal order such that $\sigma$ and $\tau$ are G-invariant central localizations, then $L(\sigma)$ is invariant under $\alpha_P$, the automorphism of Z(A) induced by P, for any $[P] \in \mathrm{Pic}(A,\tau)$.

Proof. Put $G = \mathrm{Pic}(A,\tau)$ and let B be the divisorially graded ring $A < f,\Phi,G >$, where $\Phi$ is the identity on $\mathrm{Pic}(A,\tau)$ and f is the trivial factor set associated to $\Phi$ - this allows the use of the graded terminology and simplifies notation a lot. If $J \in L(\sigma)$, then for every $\gamma \in G$ we know that $J' = B_\gamma J B_{\gamma^{-1}}$ is also in $L(\sigma)$. Since $\sigma$ is central we may assume that $J = A(J \cap Z(A))$. The definition of the canonical morphism $\alpha : \mathrm{Pic}(A,\tau) \to \mathrm{Aut}(A(A))$ entails that $JB_\gamma = B_\gamma \alpha_\gamma(J)$. Consequently we obtain : $B_\gamma J B_{\gamma^{-1}} = B_\gamma B_{\gamma^{-1}} \alpha_{\gamma^{-1}}(J) \in L(\sigma)$, since $B_\gamma B_{\gamma^{-1}} \in L(\sigma)$ for every $\gamma \in G$. □

3.4. Lemma. Let A be a $\sigma$-maximal order such that $\sigma$ and $\tau$ are as in Lemma 3.3. and assume also that $\sigma$ induces a perfect localization, then $Q_\sigma(P)$ is an invertible $Q_\sigma(A)$-bimodule for all $[P] \in \mathrm{Pic}(A,\tau)$.

Proof. Since $Q_\sigma^r(A) = Q_\sigma^l(A)$ and $Q_\sigma^l(P) = Q_\sigma^l(A) \otimes_A P$ resp. $Q_\sigma^r(P) = P \otimes_A Q_\sigma^r(A)$ it will suffice to establish that $P \otimes_A Q_\sigma(A) = Q_\sigma(A) \otimes_A P$.

Note that for the A-bimodule P, the $\sigma^l$-torsion and $\sigma^r$ torsion coincide, since $pI = 0$ with $I \in L(\sigma)$ is equivalent to $\alpha_P(I)p = 0$ with $\alpha_P(I) \in L(\sigma)$ (by the foregoing lemma). If $x \in Q_\sigma^r(P)$, put $x = \sum_i p_i \otimes q_i$ for certain $p_i \in P$ and $q_i \in Q_\sigma(A)$. For some $J \in L(\sigma)$ we have $J\, q_i \subset A$ for all $q_i$. Let Q represent $[P]^{-1}$ in $\mathrm{Pic}(A,\tau)$, then $\alpha_Q(J)x \subset \sum_i p_i J \otimes q_i \subset \sum_i p_i \otimes Jq_i \subset P$, using the canonical identifications $A \otimes_A P = P = P \otimes_A A$. Consequently, $x \in Q_\sigma(A) \otimes_A P = Q_\sigma^l(P)$. Conversely, if $y \in Q_\sigma^l(P)$, then $y \in P \otimes_A Q_\sigma(A)$ (with identifications as before) follows in the same way. Therefore, it follows from $Q_\tau(P \otimes_A Q) = Q_\tau(Q \otimes_A P) = A$ that $Q_\sigma(P \otimes_A Q) = Q_\sigma(A) \otimes_A P \otimes_A Q =$ $Q_\sigma(P) \otimes_A Q_\sigma(Q) = Q_\sigma(P) \otimes_{Q_\sigma(A)} Q_\sigma(Q)$. Finally, we obtain $Q_\sigma(A) = Q_\sigma(P \otimes_A Q) =$ $Q_\sigma(P) \otimes_{Q_\sigma(A)} Q_\sigma(Q) = Q_\sigma(P \otimes_A Q) = Q_\sigma(P) \otimes_{Q_\sigma(A)} Q_\sigma(Q)$, hence $Q_\sigma(P)$ is an invertible $Q_\sigma(A)$-bimodule indeed. □

3.5. Proposition. Let A be a $\sigma$-maximal PI-order, where $\sigma$ is central localization at $Z(A)-\{0\}$. Let G be a group, $\Phi: G \to \mathrm{Pic}(A,\tau)$ a group-homomorphism and f a factor set associated to $\Phi$. Consider the divisorially graded ring $B = A < \Phi, f, G >$, then $Q_\sigma(B)$ is a crossed product of $Q_\sigma(A)$ and G.

Proof. Since $Q_\sigma(A) = Q(A)$ is simple artinian by Posner's theorem, the lemma and the fact that $Q_\sigma(B)$ is strongly graded over $Q(A)$ yield the result by the generalized crossed product theorem. □

3.6. Remarks. a. In the examples given in (1.19.) the conditions on $\sigma$ and $\tau$ in the foregoing results are fulfilled indeed.

b. In (3.5.) we implicitly used that B is $\sigma$-torsion free. To see this assume that $Jx_\gamma = 0$ for some $x_\gamma \in B_\gamma$ and $J \in \mathcal{L}(\sigma)$. Since we may assume that $J = A(J \cap Z(A))$, it follows that $J(Bx_\gamma B) = 0$. Now $Q_\tau(Bx_\gamma B) = 0$ if and only if $(Bx_\gamma B)_e$ is $\tau$-torsion if and only if $(Bx_\gamma B)_e = 0$ (since $A = B_e$ is a prime ring), but this is impossible because $B_\gamma$ is $\tau$-torsion free. Therefore we may write $B = \sum_\gamma I_\gamma u_\gamma \subset Q_\sigma(A)[u_\gamma, \varphi, c]$, where each $I_\gamma$ is an A-bimodule in $Q_\sigma(A)$.

c. If we replace Pic$(A,\tau)$ by Picent$(A,\tau)$, then the units $u_\gamma$ commute with the center of $Q_\sigma(A)$ and the automorphism $\varphi_\gamma$ associated to $\gamma$ is inner in $Q_\sigma(A)$, say $\varphi_\gamma(a) = a_\gamma a a_\gamma^{-1}$. A change of generators $u_\gamma \to a_\gamma^{-1} u_\gamma$ and correspondingly changing $\varphi$ to the trivial map $G \to 1_{Q_\sigma(A)}$ and replacing c by some equivalent cocycle c', learns that in the case of Picent $(A,\tau)$, we may suppose that $u_\gamma$ commutes with $Q_\sigma(A)$ and then $B = \sum_{\gamma \in G} I_\gamma u_\gamma$, where the map $\gamma \mapsto I_\gamma$ defines a group homomorphism $G \to \mathcal{D}_\sigma(A)$.

It is precisely the situation of Remark c. above, which we generalize in the following definition, which picks out a special kind of generalized Rees ring.

3.7. Definition. Let there be given a $\sigma$-maximal order A, a group G a group homomorphism $\varphi : G \to \mathrm{Aut}(A)$ such that $\mathcal{L}(\sigma)$ is invariant under each $\varphi(\gamma)$ for $\gamma \in G$, and a 2-cocycle $c: G \times G \to U(Z(A)): (\gamma,\lambda) \mapsto c_{\gamma,\lambda}$. Construct the crossed product $Q_\sigma(A)[u_\gamma, \gamma \in G; \varphi, c]$ in the usual way. Consider a map $\Phi : G \to \mathcal{D}_\sigma(A)$ and the abelian group $\sum_{\gamma \in G} \Phi(\gamma)u_\gamma \subset Q_\sigma(A)[u_\gamma, \gamma \in G; \varphi, c]$. For $\sum_{\gamma \in G} \Phi(\gamma)u_\gamma$ to be a subring of the crossed product, it is necessary and sufficient that $\Phi(\gamma)\varphi(\gamma)(\Phi(\lambda)) \subset \Phi(\gamma\lambda)$ In order to have that this ring is divisorially graded, one has to impose $Q_\tau(\Phi(\gamma)\varphi(\gamma)(\Phi(\lambda))) = \Phi(\gamma\lambda)$, for any $\gamma, \lambda \in G$. If both conditions hold, then we write $\check{A}(\Phi,\varphi,c) = \sum_{\gamma \in G} \Phi(\gamma)u_\gamma$ and we call $\check{A}(\Phi,\varphi,c)$ the generalized Rees ring associated to $\Phi,\varphi$ and c over A.

3.8. Proposition. In the situation of Proposition 3.5. and under the assumption that B is a PI ring and that A is globally $\varphi$-invariant, we have that $B = \sum_{\gamma \in G} I_\gamma u_\gamma \subset Q_\sigma(A)[u_\gamma, \gamma \in G; \varphi, c]$ with $I_\gamma \in \mathcal{D}_\sigma(A)$ for all $\gamma \in G$. Consequently, a ring $A < \Phi, f, G >$ which is a PI ring is of the form $\check{A}(\Phi,\varphi,c)$ exactly when $\Phi$ maps G into $\mathcal{D}_\sigma(A)$.

Proof. By remark 3.6.b. we may write $B = \sum_{\gamma \in G} I_\gamma u_\gamma$ in $S = Q_\sigma(A)[u_\gamma, \gamma \in G; \varphi, c]$. Consider a fixed $\lambda \in G$ and the subring $B_\lambda$ generated by $u_\lambda$ and $u_{\lambda^{-1}}$ over $Q_\sigma(A)$ in S. We claim that $\varphi(\lambda)^{n_\lambda}$ is inner in $Q_\sigma(A)$ for some positive integer $n_\lambda$, say $\varphi(\lambda)^{n_\lambda}(x) = a_\lambda x a_\lambda^{-1}$ for a certain invertible $a_\lambda \in Q_\sigma(A)$. If $\varphi(\lambda)$ is in the torsion part of $\mathrm{Aut}(Q_\sigma(A))$, then the claim obviously holds . If $\varphi(\lambda)$ is not periodic, then $\lambda$ cannot be periodic in G. Then, consider the restriction $c_\lambda$ of c to $<\lambda> \times <\lambda>$, i.e. $c_\lambda : <\lambda> \times <\lambda> \to U(Z(A))$. Since $<\lambda> \cong \mathbb{Z}$, it is clear that $c_\lambda \sim 1$ and therefore, $B_\lambda \cong Q_\sigma(A)[u_\lambda, u_\lambda^{-1}, \varphi(\lambda)]$. But, being a subring of a PI-ring, $B_\lambda$ is again a PI ring and therefore $\varphi(\lambda)$ has finite order in $\mathrm{Aut}(Q_\sigma(A))/\mathrm{Inn}(Q_\sigma(A))$. Now, for any $\gamma,\lambda \in G$, define $I_\gamma^{(\lambda)} = I_\gamma \cap \varphi(\lambda)(I_\gamma) \cap \ldots \cap \varphi(\lambda)^{n_\lambda}(I_\gamma) \cap Z(Q_\sigma(A))$. Clearly $I_\gamma^{(\lambda)}$ is $\varphi(\lambda)$ invariant. From $I_\gamma u_\gamma . I_{\gamma^{-1}} u_{\gamma^{-1}} = I_\gamma \varphi(\gamma)(I_{\gamma^{-1}}) c_{\gamma,\gamma^{-1}} u_e$, it follows that $I_\gamma \varphi(\gamma)(I_{\gamma^{-1}}) \subset A = B_e$. But $I_{\gamma^{-1}}^{(\gamma)} \cap A \subset Z(A)$ implies that $I_{\gamma^{-1}}^{(\gamma)} \cap A$ commutes with $I_\gamma$ and hence $JI_\gamma = I_\gamma J \subset A$, where $J = A(I_{\gamma^{-1}}^{(\gamma)} \cap A)$. Note that $\varphi(\gamma)^r(I_{\gamma^{-1}}) \cap A \neq 0$ and thus also $\bigcap_{r=1}^{n_\gamma} \varphi(\gamma)^r(I_{\gamma^{-1}}) \cap A \neq 0$. Since A is a semiprime PI ring, nontrivial ideals of A intersect Z(A) nontrivially, hence $I_{\gamma^{-1}}^{(\gamma)} \cap A \neq 0$ or $J \in \mathcal{L}(\sigma)$. This shows that $JI_\gamma \in \mathcal{L}(\sigma)$ and $I_\gamma \in \mathcal{F}_\sigma(A)$. Suppose $Q_\tau(I_\gamma) \supsetneq I_\gamma$, then pick $y \in Q_\tau(I_\gamma) - I_\gamma$ in $Q_\sigma(A)$ and it is clear that this y is in $Q_\tau(B) \subset Q_\sigma(A)[u_\gamma, \gamma \in G; \varphi, c] = Q_\sigma(B)$. Since B is $\tau$-closed, it follows that

$Q_\tau(I_\gamma) = I_\gamma$ and thus $I_\gamma \in D_\sigma(A)$. The final statement follows directly from the above and Proposition 3.5. □

3.9. Remark. If $\check{A}(\Phi,\varphi,c)$ is a generalized Rees ring associated to $\Phi,\varphi$ and c over A, then we may view $\Phi$ as a special map $\Phi : G \to \mathrm{Pic}(A,\tau)$, but also as a *group* homomorphism $\Phi : G \to \mathrm{Pic}(A,\tau)$. Since $D_\sigma(A)$ has a canonical group structure it is natural to ask when $\Phi : G \to D_\sigma(A)$ is a group homomorphism. A necessary and sufficient condition for $\Phi : G \to D_\sigma(A)$ to be a group homomorphism is given by $\varphi(\gamma)(I_\lambda) \subset I_\lambda$ for all $\gamma,\lambda \in G$. Indeed, from $Q_\tau(I_\lambda u_\lambda I_\gamma u_\gamma) = I_{\lambda\gamma}\, u_{\lambda\gamma}$ it follows that $Q_\tau(I_\lambda \varphi(\lambda)(I_\gamma) c_{\lambda,\gamma} u_{\lambda\gamma}) = Q_\tau(I_\lambda \varphi(\lambda)(I_\gamma)) u_{\lambda\gamma}$. If $\Phi : G \to D_\sigma(A)$ is a group homomorphism, then the foregoing reduces to $\varphi(\lambda)(I_\gamma) = I_\lambda^{-1} \star I_{\lambda\gamma} = I_{\lambda^{-1}} \star I_{\lambda\gamma} = I_\gamma$. Conversely, if $\varphi(\lambda)(I_\gamma) \subset I_\gamma$, then we obtain $Q_\tau(I_\lambda I_\gamma) = I_{\lambda\gamma}$, i.e. $I_\lambda \star I_\gamma = I_{\lambda\gamma}$ in $D_\sigma(A)$ i.e. $\Phi : G \to D_\sigma(A)$ is a group morphism.

We now turn to the study of a very particular class of generalized Rees rings where the condition of Remark 3.9. is automatically fulfilled, the normalizing Rees rings satisfying a polynomial identity. Again, let A be a $\sigma$-maximal order, let G be a torsionfree abelian (hence ordered) group, let $\varphi : G \to \mathrm{Aut}(A)$ be a group homomorphism and $c : G \times G \to U(Z(A))$ a 2-cocycle. Assume that for all $\gamma \in G$ there exists an $a_\gamma \in N_\sigma(A) = \{y \in Q_\sigma(A);\ Ay = yA \in F_\sigma(A)\}$ with $\varphi(\gamma)(x) = a_\gamma x a_\gamma^{-1}$ for all x. We let $\Phi : G \to D_\sigma(A)$ be a group homomorphism and call the ring $\check{A}(\Phi,\varphi,c)$ a normalizing Rees ring. By definition $\Phi(\lambda)$ is $\varphi(\gamma)$-invariant (globally) for all $\lambda,\gamma \in G$ (see Remark 3.9.) To simplify notation, we will write $\check{A}(\Phi)$ for $\check{A}(\Phi,\varphi,c)$ from now on. We want to investigate whether $\check{A}(\Phi)$ is a relative maximal order, resp. a Krull $\check{\sigma}$-order with respect to some well-determined $\check{\sigma}$, whenever A is a relative maximal order, resp. a Krull $\sigma$-order. *In the sequel we assume that* $\check{A}(\Phi)$ *is a* PI-*ring*. In a very general setting, one may derive necessary and sufficient conditions for generalized Rees rings to satisfy polynomial identities, however, we only hint at this problem in the special case of normalizing Rees rings here. Actually, in the next Section, it will be established that one can get away from PI-theory at the cost of introducing an interesting new theory of (non PI-) Krull orders.

If I is an ideal of $\check{A}(\Phi)$ and $\gamma \in G$, put $C_\gamma(I) = \{r \in Q_\sigma(A);\ i^\sim = ru_\gamma$ for some $i \in I\}$, where $\sim$ is as in the proof of (2.2.). Since we have assumed that $\varphi(\gamma) \in \mathrm{Aut}(A)$ for all $\gamma \in G$, it follows that $C_\gamma(I)$ is an A-bimodule in $Q_\sigma(A)$.

3.10. Lemma. If I is an ideal in $\check{A}(\Phi)$, then $C_e(I) \in \mathcal{L}(\sigma)$ if and only if $C_\gamma(I) \in \mathcal{F}_\sigma(A)$ for all $\gamma \in G$.

Proof. One implication is obvious. For each $\gamma \in G$ we have $C_e(I)u_eI_\gamma u_\gamma =$ $C_e(I)\varphi(e)(I_\gamma)c_{e,\gamma}u_\gamma$ and since $\varphi(e)$ leaves the elements of $\mathcal{D}_\sigma(A)$ invariant and since $c_{e,\gamma} \in U(Z(A))$, the foregoing yields that $C_e(I)I_\gamma \subset C_\gamma(I) \subset I_\gamma$. Since both extremes are in $\mathcal{F}_\sigma(A)$, so is $C_\gamma(I)$. □

3.11. Corollary. The filter $\mathcal{L}^2(\check{\sigma})$ consisting of all ideals I of $\check{A}(\Phi)$ such that $C_\gamma(I) \in \mathcal{F}_\sigma(A)$ for all $\gamma \in \mathcal{F}_\sigma(A)$ for all $\gamma \in G$ is multiplicatively closed. □

3.12. Lemma. For every ideal I of $\check{A}(\Phi)$ with $C_e(I) \in \mathcal{L}(\sigma)$, we have $Q_\tau(C_\gamma(I)) = Q_\tau(C_e(I)I_\gamma)$.

Proof. From $I_{\gamma^{-1}}I_\gamma C_e(I) \subset I_{\gamma^{-1}}C_\gamma(I) \subset C_e(I)$, it follows that $Q_\tau(I_{\gamma^{-1}}I_\gamma C_e(I)) =$ $Q_\tau(C_e(I))\,Q_\tau(I_{\gamma^{-1}}C_\gamma(I)) \subset Q_\tau(C_e(I))$ and therefore $Q_\tau(C_e(I)) = I_{\gamma^{-1}} \star Q_\tau(C_\gamma(I))$. Note that we have used the fact that the $I_\gamma$ are $\varphi(\lambda)$-invariant. Since $\mathcal{D}_\sigma(A)$ is commutative, we obtain that $Q_\tau(C_e(I))\star I_\gamma = Q_\tau(C_\gamma(I))$ i.e. $Q_\tau(C_e(I)I_\gamma) =$ $Q_\tau(C_\gamma(I))$, indeed. □

3.13. <u>Lemma</u>. If R is a graded ring of type G satisfying the identities of n x n-matrices such that the center Z(R) is a gr-field (i.e. homogeneous elements of Z(R) are invertible), then R is an Azumaya algebra.

<u>Proof</u>. The multilinear Razmyslov polynomial, f say, cannot vanish for every homogeneous substitution for the variables. The hypotheses on Z(R) imply that the Formanek center of R equals Z(R) (cf. [116] for details on PI - theory) and thus R is an Azumaya algebra.

3.14. <u>Theorem</u>. Let A be a $\sigma$-maximal order and suppose that $\check{A}(\Phi)$ satisfies polynomial identities, then $\check{A}(\Phi)$ is a $\check{\sigma}$-maximal order, where $\check{\sigma}$ is defined as in Corollary 3.11.

<u>Proof</u>. Since $\check{A}(\Phi)$ is a graded P.I.-ring, its graded ring of quotients $Q^g(\check{A}(\Phi))$ is obtained by inverting central homogeneous elements and it is a graded simple Artinian ring in the sense of [ 108 ]. The graded version of Wedderburn's Theorem (cf. [108]) yields that $Q^g(\check{A}(\Phi)) \cong M_n(\Delta)(\underline{\lambda})$ for some graded skewfield $\Delta$, and $\underline{\lambda} \in G^n$ defining the gradation by $((M_n(\Delta)(\underline{\lambda})_\gamma)_{i,j} = (\Delta_{\lambda_i \gamma \lambda_j^{-1}})_{i,j}$. From Lemma 4.2. of [ 107 ] it follows that $Z(Q^g(\check{A}(\Phi))) = Z(\Delta)$ is a gr-field graded by the group $H = \{\gamma \in G;\ \Delta_\gamma \cap Z(\Delta) \neq 0\}$. The lemma states that $Q^g = Q^g(\check{A}(\Phi))$ is an Azumaya algebra over Z(A). Because H is again a torsion free abelian group, it follows from [ 3 ] that $Z(\Delta)$ is a completely integrally closed domain.

First we show now that $Q^g$ is a maximal order over $Z(\Delta)$ in $Q = Q(\check{A}(\Phi))$. Consider an ideal I of $Q^g$ and let $q = c^{-1}r \in Q$ be such that $Iq \subset I$. Then $c^{-1}\, Q^g(I \cap Z(\Delta))(Q^g\, r\, Q^g \cap Z(\Delta)) \subset Q^g(I \cap Z(\Delta))$, and therefore $(I \cap Z(\Delta))c^{-1}\,(Q^g\, r\, Q^g \cap Z(\Delta)) \subset I \cap Z(\Delta)$. This yields that $Q^g r\, Q^g$ $Q^g(Q^g r\, Q^g \cap Z(\Delta)) \subset Q^g$ c and hence $c^{-1} r \in Q^g$. So $I \,\dot{.}\, I = Q^g$ and in a similar way $I \,\dot{.}\, I = Q^g$.

Next, consider $J \in L(\check{\sigma})$ and suppose that $Jq \subset J$ for some $q \in Q(\check{A}(\Phi))$, then $Q^g Jq \subset Q^g J$ and $Q^g J \subset Q^g$ is an ideal, hence $q \in Q^g$ follows. We may thus decompose q as $q = q_{\lambda_1} u_{\lambda_1} + \dots + q_{\lambda_n} u_{\lambda_n}$, with $\lambda_1 < \dots < \lambda_n$. The relation $Jq \subset J$ yields $(\star)$ : $C_e(J)\varphi(e)(q_{\lambda_n}) \subset C_{\lambda_n}(J)$. By definition it follows that $C_\gamma(J) \in F_\sigma(A)$ for all $\gamma \in G$, hence we may deduce from $(\star)$ that $Q_\tau(C_e(J))\varphi(e)(q_{\lambda_n}) \subset Q_\tau(C_{\lambda_n}(J))$ or $\varphi(e)(q_{\lambda_n}) \in Q_\tau(C_e(J))^{-1} \star Q_\tau(C_{\lambda_n}(J))$ By Lemma 3.12. this means that $\varphi(e)(q_{\lambda_n}) \in I_{\lambda_n}$ and thus $q_{\lambda_n} \in I_{\lambda_n}$. Replacing q by $q - q_{\lambda_n} u_{\lambda_n}$ and repeating the argument, one finally arrives at $q \in \check{A}(\Phi)$, and thus $J \mathbin{.^{\boldsymbol{\cdot}}} J = \check{A}(\Phi)$. In a similar way, one may establish $J \mathbin{^{\boldsymbol{\cdot}}.} J = \check{A}(\Phi)$. □

3.15. Corollary. If A is a maximal order and $\check{A}(\Phi)$ a normalizing Rees ring satisfying a polynomial identity, then $\check{A}(\Phi)$ is a maximal oder.

Proof. It is clear that A is a maximal order in Q(A) if and only if A is a $\sigma$-maximal order with respect to $L(\sigma) = L(A-\{o\})$ consisting of all nonzero ideals of A. Since $L(\check{\sigma}) = L(\check{A}(\Phi)-\{o\})$ in this case, the proof is obvious. □

3.16. Theorem. Let A be a Krull $\sigma$-order and let $\check{A}(\Phi)$ be a normalizing Rees ring satisfying a polynomial identity. If G satisfies the ascending chain condition on cyclic subgroups then $\check{A}(\Phi)$ is a Krull $\check{\sigma}$-order.

Proof. (adopting notation as in the proof of Theorem 3.14.) We know that $Q^g$ is an Azumaya algebra over its center $Z(\Delta)$, which is a gr-field graded by subgroup H of G. Since H also satisfies the ascending chain condition on cyclic subgroups, $Z(\Delta)$ is factorial (by Corollary 3.4. of [ 3 ]. Since $Q^g$ is an Azumaya algebra, divisorial ideals correspond bijectivity to divisorial ideals of the center and consequently every divisorial ideal of $Q^g$ is generated by a central element. If $\{A_n ; n \in \mathbb{N}\}$ is an ascending chain of divisorial $\check{A}(\Phi)$-ideals contained in $\check{A}(\Phi)$, then the ascending chain $\{Q_\tau(Q^g A_n);\ n \in \mathbb{N}\}$ is stationary in $Q^g$, i.e. there is a positive integer n' such that $Q_\tau(Q^g A_{n'}) = Q_\tau(Q^g A_m)$ for every $m \geqslant n'$. On the other hand, the fact that A is a Krull $\sigma$-order yields that there is a positive integer n" such that $Q_\tau(C_e(A_{n''})) = Q_\tau(C_e(A_m))$ for all $m \geqslant n''$. Let $N = \sup\{n',n''\}$ and take $k \geqslant N$. Since $\tau$ is a localization in degree e, $Q_\tau Q^g = Q^g$. Therefore, if $q A_N \subset A_k$, then $q A_N Q^g \subset A_k Q^g = A_N Q^g$ follows. But $A_N Q^g = Q^g c$ for some $c \in Z(\Delta)$, $Q^g$ is a maximal order in Q, hence $q \in Q^g$ follows.

Write $q = q_{\lambda_1} u_{\lambda_1} + \dots + q_{\lambda_n} u_{\lambda_n}$ with $\lambda_1 < \dots < \lambda_n$, then the relation $q A_N \subset A_k$ yields that $(\star)$ : $q_{\lambda_n} \varphi_{\lambda_n}(C_e(A_N)) \subset C_{\lambda_n}(A_k)$.

Note that A is a Krull $\sigma$-order satisfying a polynomial identity and that we have assumed that $\sigma$ is a central kernel functor; the $\varphi(\gamma)$ with $\gamma \in G$, are induced by normalizing elements of $Q_\sigma(A)$ hence $\varphi(\gamma)$ leaves the center of A invariant and thus $\sigma$ is $\varphi(\gamma)$ invariant. Note also that $A_n \in D_{\check{\sigma}}(\check{A}(\Phi))$ yields $A_n \in L(\check{\sigma})$ i.e. $C_e(A_n) \in L(\sigma)$. Because of this $(\star)$ yields :

$(\star\star)$ : $q_{\lambda_n} Q_\tau(C_e(A_N)) \subset Q_\tau(C_{\lambda_n}(A_k))$.

By assumption, $Q_\tau(C_e(A_N)) = Q_\tau(C_e(A_k))$, and applying Lemma 3.12. to this equality yields that $Q_\tau(C_{\lambda_n}(A_k)) = Q_\tau(C_{\lambda_n}(A_N))$.

From $(\star\star)$ and Lemma 3.12. again, it follows that $q_{\lambda_n} \in I_{\lambda_n}$. Replacing q by $q - q_{\lambda_n} u_{\lambda_n}$ and repeating the foregoing agument, we obtain $q \in \check{A}(\Phi)$.

Thus $A_k \,{}^{\bullet}\!.\, A_N \subset \check{A}(\Phi)$, whence $A_N^{-1} \star A_k \subset \check{A}(\Phi)$ and also, $A_k \subset A_N$ follows. But then $A_k = A_N$ and this finishes the proof. □

3.17. Corollary. If A is a Krull ring in the sense of M. Chamarie [29 ]) and $\check{A}(\Phi)$ is a normalizing Rees ring satisfying a polynomial identity, then $\check{A}(\Phi)$ is a Krull ring (in the P.I. case the different definitions of Krull rings given in [ 50 ], [ 75 ] or [ 92 ] coincide!)

Proof. If A is a Krull ring, then A is a Krull $\sigma$-order with $L(\sigma)$ consisting of all nonzero ideals of A. Then $\check{A}(\Phi)$ is a Krull $\check{\sigma}$-order with $L(\check{\sigma})$ consisting of all nonzero ideals of $\check{A}(\Phi)$. This entails that $Z(\check{A}(\Phi))$ is a Krull domain. Now, a maximal order (P.I. !) over a Krull domain is a Krull ring in the sense of [ 50 ] (or, of [ 75 ], [ 92 ],...) □

3.18. Remark. A normalizing Rees ring $\check{A}(\Phi) \subset Q_\sigma(A)\,[u_\gamma, \gamma \in G; \varphi, c\,]$ is a PI ring if c satisfies the following property : for every $\gamma \in G$, there is a positive integer n such that $c_{\lambda,\gamma^n} = c_{\gamma^n,\lambda}$ for all $\lambda \in G$. Upto changing the variables in $Q_\sigma(A)\,[u_\gamma, \gamma \in G; \varphi, c\,]$, replacing $u_\gamma$ by $u'_\gamma = a_\gamma^{-1} u_\gamma$, where $a_\gamma$ is the normalizing element of $Q_\sigma(A)$ inducing $\varphi(\gamma)$ in $Q_\sigma(A)$, we get an embedding of $\check{A}(\Phi)$ in $Q_\sigma(A)\,[u'_\gamma, \gamma \in G; 1, c'\,]$, where $c'_{\gamma,\lambda} = a_\gamma^{-1}\,\varphi(\gamma)(a_\gamma^{-1})a_{\gamma\lambda}\,c_{\gamma,\lambda}$ with $a_\gamma^{-1}\varphi(\gamma)\,(a_\lambda^{-1}) = a_\lambda^{-1}\,a_\gamma^{-1}$. The statement of the remark now becomes straightforward to verify.

In the non-normalizing case one has to add the condition on $\varphi(\gamma)$ that some power of $\varphi(\gamma)$ is inner in $Q_\sigma(A)$, for every $\gamma \in G$. This can be proved using the restriction technique as in Proposition 3.8. to get rid of the cocycle.

Let us now focus on HNP PI-rings and tame orders. Recall that a tame order over a Krull domain C in a central simple algebra $\Sigma$ over the field of fractions K of C is a C-order A such that $A = \bigcap_{p \in X^{(1)}(C)} A_p$ and for each $p \in X^{(1)}(C)$ we have that $A_p$ an HNP (P.I.)ring.

3.19. Lemma. Let A be an HNP P.I. ring then the following statements are equivalent :

1. $P \in \mathrm{Spec}(A)$ is a localizable prime ideal;
2. P is invertible or zero;
3. P extends to an ideal $Q_{A-P}(A)P$ of $Q_{A-P}(A)$;

4. $\kappa_{A-P}$ is a central kernel functor;

5. $P \in U$, where U is the open set of birationality in the sense of [152].

Proof. cf. [110]. □

We note also (cf. results of J. Robson, L. Small [128]) that the center of an HNP PI-ring is a Dedekind domain.

Let Inv(A), resp. Id(A) be the set of invertible, resp. idempotent prime ideals of A; the lemma states that Inv(A) is exactly the open set of birationality for the inclusion $Z(A) \to A$.

Recall also an HNP PI-ring A is a relative maximal order with respect to a central, perfect $\sigma$ such that $\tau$ is trivial, cf. (1.18.c). We now have :

3.20. Theorem. Let A be an HNP PI ring and let G be a torsionfree abelian group satisfying the ascending chain condition on cyclic subgroups. If $\check{A}(\Phi)$ is a normalizing Rees ring with polynomial identity, then $\check{A}(\Phi)$ is a tame order.

Proof. Rather technical, cf. [ 86 ]. We shall return to this result and related problems after having derived the main results of the next section, cf. (4.25.). □

The usefulness of Rees rings of different types will become apparent, when we investigate the class groups of orders in a subsequent chapter.

## D.4. Strongly Graded Rings of Type $\mathbb{Z}$ over Arithmetical Orders.

This Section follows logically after Section 2 and notations and terminology are the same as there.

4.1. Theorem. Let A be a strongly graded ring of type $\mathbb{Z}$ such that $A_0$ is a prime left and right Goldie ring. If $A_0$ is a maximal order in its ring of fractions $Q_o$, then A is a maximal order in its classical ring of fractions Q.

Proof. Consider a subring T in Q containing A and such that $Ta \subset A$ for some regular element a of Q lying in A. The proof proceeds in three steps.

1. $T \subset Q^g$. Indeed, since $0 \neq Ta \subset A$, we have that $Q^g Ta\, Q^g$ is a nonzero ideal of $Q^g$ and thus $Q^g\, Ta\, Q^g = Q^g u$ for some $u \in Q^g$, which may be chosen in $Ta\, Q^g$ - to see this, one has to repeat the argument used in Proposition 2.7., to prove a similar claim. Since $Q^g$ is a prime Goldie ring, u is regular in Q. For any $t \in T$ we obtain that $tu \in T(Ta)Q^g = Ta\, Q^g \subset Q^g u$, so $tu = qu$ for some $q \in Q^g$ and therefore $T \subset Q^g$ follows.

2. Because of 1. we can define $C_n(T)$ to be the set of homogeneous components of elements of T, of highest degree equal to n, with 0 added. It is clear that $A_n \subset C_n(T)$, hence $A_{-n}C_n(T) \subset Q_o$. One easily checks that $A_{-n}C_n(T)$ is a ring, containing $A_o$ as a subring. Since $C_n(T)$ is a right $A_o$-module, the set $I = \{a \in A_o; A_{-n}\; C_n(T)a \subset A_o\}$ is an ideal of $A_o$. Indeed, from $Ta \subset A$ it follows that $t_n a_k \in A_{k+n}$ for all $t_n \in C_n(T)$, where $a = a_k$ + lower degree terms. Consequently $A_{-n}C_n(T)a_k \subset A_k$ and thus $A_{-n}C_n(T)a_k A_{-k} \subset R_o$. Since $a_k A_{-k} \neq 0$ because $A_k A_{-k} = A_o$, it follows that I is a nonzero ideal of $A_o$. The hypotheses on $A_o$ entail that I contains a regular element of $A_o$. If $A_o$ is a maximal order, then $A_o = A_{-n}C_n(T)$ follows. Hence $A_n = C_n(T)$ and $A = T$ follows, because if $t = q_n + q_{n-1} + \dots + q_m \in T$ then $q_n \in A$ yields $t-q_n \in T$ and we repeat the argument.

3. If T is an overring of A such that $bT \subset A$ for some regular element b of Q such that $b \in A$ then, just as in 1, $T \subset Q^g$ follows and we repeat 2. This proves the theorem. □

4.2. Theorem. Let A be a strongly $\mathbb{Z}$-graded ring, such that $A_o$ is a prime left and right Goldie ring. If $A_o$ is a Krull order then A is a Krull order

(Krull order is used here in M. Chamarie's sense, cf. [29]).

<u>Proof</u>. It follows from Theorem 4.1. that A is a maximal order in $Q_{cl}(A)$ so we have to verify the A.C.C. on closed left (right) ideals of A. Consider an ascending chain of closed left ideals of A, say $L_1 \subset L_2 \subset \dots$. Denote $\bigoplus_{n \geq 0} A_n$ by $A_+$, then $L_j \cap A_+ \neq 0$ if $L_j \neq 0$ and $A_{-i}\, C_i(L_j \cap A_+)$ is a left ideal of $A_0$, which is nonzero if $C_i(L_j \cap A_+)$ is nonzero. For all $i \geq 0$, we have an ascending chain $A_{-i}\, C_i(L_1 \cap A_+) \subset A_{-i}\, C_i(L_2 \cap A_+) \subset \dots$. Moreover it is easy to see for all $0 \leq k \leq l$ that $A_{-k}\, C_k(L_j \cap A_+) \subset A_{-l}C_l(L_j \cap A_+)$. Now, consider the family $\{cl_{A_0}(A_{-i}\, C_i(L_j \cap A_+));\ i,j \in \mathbb{N}\} = \mathcal{B}$, where $cl_{A_0}$ is the closure operation in $A_0$ described as follows. Let $\mathcal{L}_0$ be the filter of left ideals H of $A_0$ such that $(r \cdot.\, H) \cdot.\, A_0 = A_0$; the closure of a left ideal L of $A_0$ with respect to this filter is denoted by $cl_{A_0}(L)$, i.e. $cl_{A_0}(L) = \{x \in A_0; Hx \subset L$ for some $H \in \mathcal{L}_0\}$. The A.C.C. on closed left ideals in $A_0$ entails that $\mathcal{B}$ has a maximal element $cl_{A_0}(A_{-p}C_p(L_q \cap A_+))$ for some p,q i.e. for all $i \geq p$ and $j \geq q$ we have $cl_{A_0}(A_{-i}\, C_i(L_j \cap A_+)) = cl_{A_0}(A_{-p}\, C_p(L_q \cap A_+))$. For $i < p$, there exists a positive integer m(i) such that for all $m \geq m(i)$ we have $cl_{A_0}(A_{-i}C_i(L_m \cap A_+)) = cl_{A_0}(A_{-i}C_i(L_{m(i)} \cap A_+))$, again by the A.C.C. on closed left ideals of $A_0$. Putting $M = \max\{q, m(0),\dots,m(p-1)\}$, we obtain for all $n \geq 0$ and $m \geq M$ that $cl_{A_0}(A_{-n}\, C_n(L_m \cap A_+)) = cl_{A_0}(A_{-n}\, C_n(L_M \cap A_+))$. On the other hand, $Q^g L_1 \subset Q^g L_2 \subset \dots$ is an ascending chain of left ideals of $Q^g$, hence there is a positive integer s such that $Q^g L_t = Q^g L_s$ for all $t \geq s$. Put $M' = \max\{s,M\}$. We claim that $L_m = L_{M'}$ for all $m \geq M'$, i.e. if $I \subset J$ are closed left ideals of A such that for all $n \geq 0$ we have $cl_{A_0}(A_{-n}\, C_n(I \cap A_+)) = cl_{A_0}(A_{-n}\, C_n(J \cap A_+))$, while $Q^g I = Q^g J$, then this implies that $I = J$. Assume $J \not\subset I$ and pick $j \in J - I$ and $j \in A_+$. This is possible, because

if $j = r_{n_1} + \dots + r_{n_m}$ with $n_1 > \dots > n_m$, then $A_{-n_m} j \subset J \cap A_+$, while it cannot be in $I$, since otherwise $j \in A_{n_m} A_{-n_m} j \subset I$. Take $j$ with minimal filtration degree $s$ say, such that $j \in (J \cap A_+) - I$, i.e. $j = r$ + lower degree terme, where $r \in A_s$. Since $A_s A_{-s} = A_o$, we may write $1 = \sum_i b_i a_i$, where $b_i \in A_s$ and $a_i \in A_{-s}$. For all $i$, we have that $a_i r \in A_{-s} C_s (J \cap A_+) \subset$ $\mathrm{cl}_{A_o}(A_{-s} C_s (J \cap A_+)) = \mathrm{cl}_{A_o}(A_{-s} C_s (I \cap A_+))$. Therefore, there is an $H \in L$, 1 the filter of the Krull order $A_o$, such that $H a_i r \subset A_{-s} C_s (I \cap A_+)$, so $A_s H a_i r \subset C_s (I \cap A_+)$. Take $x \in A_s H$, then there is $y \in I \cap A_+$, such that $y = x a_i r$ + lower degree terms. Since $I \subset J$ and $x a_i \in A_o$, we have $x a_i j - y \in J \cap A_+$. But, by the assumption on the filtration degree of $j$, it follows that $x a_i j - y \in I$ and thus, $x a_i j \in I$. Consequently, $A_s H a_i j \subset I$, i.e. $A A_s H a_i j \subset I$, hence $A H a_i j \subset I$.

On the other hand, $a_i j \in J$ and $Q^g J = Q^g I$ entail that $a_i j = c^{-1} z$, where $z \in I$ and $c$ is a regular element of $A_o$. Therefore, $(AH + Ac) a_i j \subset I$, so if we write $H' = H + A_o c$, then $H' \in L$ and $H'$ is an essential left ideal of $A_o$, such that $AH' a_i j \subset I$. Obviously, $\mathrm{cl}_A(AH') a_i j \subset \mathrm{cl}_A(I) = I$.

<u>Claim</u> : $\mathrm{cl}_A(AH') = A$.

If this holds, then $A a_i j \subset I$ yields $a_i j \in I$ and $b_i a_i j \in I$, so $\sum_i b_i a_i j \in I$ yields the contradiction $j \in I$. Hence $I = J$, indeed. So we have to show that $\mathrm{cl}_A(AH) = A$ for $H$ an essential left $A_o$-ideal in $L$, i.e. for each $r \in A$ we have to prove that $A \mathbin{.^{\bullet}} (AH \mathbin{^{\bullet}.} r) = A$, where we write $AH \mathbin{^{\bullet}.} r$ for $AH \mathbin{^{\bullet}.} \{r\}$. Note that $A \subset A \mathbin{.^{\bullet}} (AH \mathbin{^{\bullet}.} r)$.

Let us first deal with the situation where $r$ is an homogeneous element of $A$, say $r \in A_t$. Then $AH \mathbin{^{\bullet}.} r$ is a graded left ideal of $A$ and it is essential because $AH$ is an essential graded left ideal of $A$. Therefore $AH \mathbin{^{\bullet}.} r$ contains a regular element of degree zero, and $Q^g(AH \mathbin{^{\bullet}.} r) = Q^g$. If $q \in Q$ is such that $(AH \mathbin{^{\bullet}.} r) q \subset A$, then $Q^g(AH \mathbin{^{\bullet}.} r) q \subset Q^g$, i.e. $Q^g q \subset Q^g$ follows. We then have that $((AH \mathbin{^{\bullet}.} r) \cap A_n) q_{i_j} \subset A$ for all positive integers $n$ and

all $i_j$. We have reduced the problem to showing that for q homogeneous the fact that $((AH \cdot. r) \cap A_n)q \subset A$ for all n, implies that $q \in A$.
Since $A_tA_{-t} = A_o$, we may decompose $1 = \Sigma a_jb_j$ with $a_j \in A_t$ and $b_j \in A_{-t}$ where $t = \deg(r)$. Fix j for a moment and consider $b_jr \in A_o$. Put $\deg(q) = s$, then $(H \cdot. b_jr)b_jq$ is contained in $A_{s-t}$, because if $xb_jr \in H$ for $x \in A_o$, then $xb_j \in (AH \cdot. r) \cap A_{-t}$ and thus $xb_jq \in A$.
Decompose 1 as a sum $\Sigma c_i d_i$ with $c_i \in A_{t-s}$ and $d_i \in A_{s-t}$, then we obtain $(H \cdot. b_jr)b_jqc_i \subset A_o$ for all i and $b_jqc_i \in Q_o$. Since $H \in L$, we have $H \cdot. b_jr \in L$ too, hence $A_o \ .\cdot (H \cdot. b_jr) = A_o$. Consequently $b_jqc_i \in A_o$ for all i and thus $b_jqc_i \in A_{s-t}$ so $b_jq(\Sigma c_id_i) \in A_{s-t}$ and hence $b_jq \in A_{s-t}$. We thus find that $b_jq \in A_{s-t}$ for all j, hence $a_jb_jq \in A_s$ and $(\Sigma_j a_jb_j)q \in A_s$ i.e. $q \in A_s$. Conclusion : if r is homogeneous, then $A \ .\cdot(AH \cdot. r) = A$, indeed.

Now, let r be arbitrary. To prove that $A \ .\cdot(AH \cdot. r) = A$, consider $q \in Q$ such that $(AH \cdot. r)q \subset A$, then $(\bigcap_i (AH \cdot. r_i))q \subset A$, say $((AH \cdot. r_1) \cap \dots \cap (AH \cdot. r_m))q \subset A$, where the $r_i$ are the homogeneous components of r. Let us show that this implies $(AH \cdot. r_m)q \subset A$ and then that we have returned to the situation dealt with before, i.e. the homogeneous case. Consider $x \in (AH \cdot. r_2) \cap \dots \cap (AH \cdot. r_m) \cap A_o$, then $(AH \cdot. xr_1)xq \subset A$. Indeed, for $y \in AH \cdot. xr_1$ we have $yxq \in A$, since $yxr_1 \in AH$ yields that $yx \in AH \cdot. r_1$, while $yx \in (AH \cdot. r_2) \cap \dots \cap (AH \cdot. r_m)$ by the choice of x. However, $xr_1$ is homogeneous, therefore $A \ .\cdot(AH \cdot. xr_1) = A$ and $xq \in A$ follows. So, $((AH \cdot. r_2) \cap \dots \cap (AH \cdot. r_m) \cap A_o)q \subset A$. But, since A is strongly graded, every graded left ideal is generated by its part of degree zero, thus $((AH \cdot. r_2) \cap \dots \cap (AH \cdot. r_m))q \subset A$. Repeating this argument, we finally obtain $(AH \cdot. r_m)q \subset A$ indeed. □

4.3. Corollary (of the proof. If A is strongly $\mathbb{Z}$-graded such that $A_0$ is a prime left Goldie ring satisfying A.C.C. on ideals, then A satisfies A.C.C. on ideals. □

In the sequel we consider in particular the situation where $A_0$ is an Asano order. The definition of a graded left A-ideal in $Q^g$ entails that such a left A-ideal L contains a regular homogeneous element of $Q^g$ which may be chosen in A, while $Lb \subset A$ for some regular homogeneous element b of $Q^g$ which may also be chosen in A. However, since A has regular homogeneous elements of any degree, it is clear that both a and b as above may actually be chosen in $A_0$. If $A_0$ is only a left Goldie ring then the foregoing is still true for the (two-sided) graded A-ideals. In order to avoid any confusion we repeat the following definitions. A strongly $\mathbb{Z}$-graded ring A such that $A_0$ is (semi)prime left Goldie is said to be a graded Asano left order in $Q^g$ if the graded A-ideals form a group under multiplication; A is a gr-maximal left order in $Q^g$ if for each graded overring $A \subset T \subset Q^g$, such that $bTa \subset A$ for some regular homogeneous a,b in $Q^g$, it follows that A = T. Again, a,b may be taken in $A_0$ here !

4.4. Proposition. Let A be a strongly $\mathbb{Z}$-graded ring such that $A_0$ is a prime left Goldie ring. The following conditions on A are equivalent :

1. A is a graded Asano left order in $Q^g$;
2. A is a gr-maximal left order in $Q^g$ and every nonzero graded ideal of A is projective in A-mod;
3. for each nonzero graded ideal I of A there exists a graded A-ideal J in $Q^g$ such that IJ = JI = A;
4. The graded A-ideals form an abelian group under multiplication (of A-ideals in $Q^g$).

Proof. A straightforward grading of a result of J.C. Robson cf. [126]. □

Continuing in the same vein one easily deduces the following :

7.5. Corollary. If A is a graded Asano left order in $Q^g$, then :

1. every nonzero graded prime ideal of A is a gr-maximal ideal;
2. every nonzero graded ideal of A is in a unique way a product of graded prime ideals of A;
3. A satisfies A.C.C. for graded ideals;
4. A satisfies D.C.C. for graded ideals containing a fixed nonzero graded ideal;
5. every graded A-ideal in $Q^g$ is projective and finitely generated both as a left and right R-module. □

The example $\mathbb{Z}[X,X^{-1}]$ shows that R is not necessarily an Asano order, if $A_o$ is such. However we do have :

4.6. Proposition. If A is a strongly $\mathbb{Z}$-graded ring such that $A_o$ is a prime left Goldie ring then :

if $A_o$ is an Asano left order in $Q_o$, then A is a graded Asano left order in $Q^g$.

Proof. □

At this point is is necessary to introduce some terminology concerning v-ideals or reflexive ideals. Let A be any order in a simple artinian ring Q. An A-ideal I is invertible if $I(A \,.^{\boldsymbol\cdot}\, I) = A = (A \,^{\boldsymbol\cdot}.\, I)I$. Now, for any left ideal L, we define ${}_vL = A \,^{\boldsymbol\cdot}.(A \,.^{\boldsymbol\cdot}\, L)$ and if $L = {}_vL$, then we say that L is a left v-ideal (or a reflexive ideal). The right v-ideals may be defined similarly. An A-ideal I is a v-ideal if ${}_vI = I = I_v$. The A-ideals I which are v-ideals and are contained in A will be called v-ideals of A. We say that the v-ideal I is v-invertible if $(I(A \,.^{\boldsymbol\cdot} I))_v = A = {}_v((A \,^{\boldsymbol\cdot}.\, I)I)$. From Lemma 1.1. of [54 ] it follows that a v-ideal I is v-invertible if and only if $O_l(I) = A = O_r(I)$ and in this case we have again that $A \mathbin{\dot{.}} I = A \mathbin{.\dot{}} I = I^{-1}$. From 1.2. in [ 54 ] it follows that the v-invertible v-ideals form a group under the multiplication $I \star J = (IJ)_v = {}_v(IJ)$ and we denote it by $\mathcal{D}(A)$. The ring A is said to have enough v-invertible ideals if each v-ideal of A contains a v-invertible ideal of A.

Now we return back to the situation where A is a strongly graded ring of type $\mathbb{Z}$ such that $A_o$ is a prime noetherian asano order in $Q_o$, hence A is a prime noetherian maximal order in Q, its ring of fractions, and a graded Asano order in $Q^g$. Any nonzero graded ideal of A is invertible, hence a v-ideal of A. Furthermore, any nonzero ideal I of A with $I \cap A_o = 0$ and $I = Q^g I \cap A$ is also a v-ideal of A because of Lemma 2.2. in [ 29 ].

4.7. Proposition. Let A be a strongly $\mathbb{Z}$-graded ring such that $A_o$ is a prime Noetherian Asano order in $Q_o$, then :

1. If P is a nonzero prime ideal of A, then P is a v-ideal of A if and only if P is graded or $P \cap A_o = 0$. So the prime v-ideals of A are exactly the height one prime ideals of A.
2. If I is a v-ideal of A, then I is either graded or $I \cap A_o = 0$.
3. If I is a non-zero ideal of A such that $I \cap A_o = 0$ and $I = Q^g I \cap A$, then $I = ((P_1 P_2 \cdots P_k)^{-1})^{-1}$ where $P_i$ are prime ideals of P lying over zero in $A_o$ and $Q^g I = Q^g P_1 \cdots Q^g P_k$.

Proof.

1. If $P \cap A_o = 0$ then $P = Q^g P \cap A_o$ by Proposition 2.7. So if P is graded or $P \cap A_o = 0$, then P is a v-ideal of A as remarked above. Conversely, if P is not graded and $P \cap A_o \neq 0$, then $P \supsetneq P_g \neq 0$. But $P_g$ is a graded prime ideal of A, therefore P cannot be a v-ideal of A because in that case ht(P) = 1.

2. Suppose $I \cap A_o \neq 0$. If I is a v-ideal of A then we may write $I = {}_v(P_1 P_2 \cdots P_m)$ for some prime v-ideals $P_1, \ldots, P_m$ of A. Since $I \subset P_i$ for all i,

it follows that $P_i$ is a prime v-ideal of A such that $P_i \cap A_o \neq 0$. Because of 1., $P_i$ must be graded. However, a nonzero graded ideal of A is invertible thus $I = P_1 P_2 \dots P_m$ and I is graded.

3. We may write $Q^g I = Q^g P_1 Q^g P_2 \dots Q^g P_k$, where $P_1, \dots, P_k$ are nonzero prime ideals of A lying over zero in $A_o$. Obviously $I = Q^g I \cap A \supset P_1 \dots P_k$. Since I is a v-ideal of A and $Q^g I = Q^g P_1 \dots Q^g P_k$, one easily shows that $I = {}_v(P_1 P_2 \dots P_k)$. □

4.8. Remark. Proposition 4.7. is still true if we replace the noetherian condition by A.C.C. on one-sided integral v-ideals and if we assume that A/P is a Goldie ring for each prime ideal P of A such that $P \cap A_o = 0$.

4.9. Corollary. Let A be a strongly graded ring of type $\mathbb{Z}$ such that $A_o$ is a Dedekind prime ring, then A is a prime noetherian maximal order in Q such that graded left (and right) ideals of A are projective left (right) A-modules. In this case A is also a graded Asano order in $Q^g$. □

We now consider strongly graded rings A of type $\mathbb{Z}$ for which $A_o$ is an HNP ring. So far we have already established that such a ring A is a prime noetherian gr-hereditary ring. Consequently, since nonzero graded ideals of A are projective A-modules, it follows that every nonzero graded ideal I of A is a v-ideal of A, while $I(A \,{}^{\cdot}\!.\, I) = O_l(I)$ and $(A \,.\!{}^{\cdot}\, I)I = O_r(I)$, cf. [46,126] . The group of v-invertible graded A-ideals will be denoted by $D_g(A)$. Recall that an ideal I is eventually idempotent if there exists a strictly positive integer such that $I^n$ is idempotent. Let us summarize some properties of graded ideals of A.

4.10. Proposition. Let A be as before :

1. A non-trivial graded ideal of A is gr-maximal if and only if it is a prime ideal of A.
2. If M is gr-maximal, then M is either invertible or idempotent.

3. If P is a graded ideal of R, then P is a maximal invertible graded ideal of A if and only if P is the intersection of a graded cycle. A <u>graded cycle</u> of A is a finite set $\{M_1,\dots,M_n\}$ of distinct idempotent gr-maximal ideals of A, such that $O_r(M_1) = O_l(M_2),\dots,O_r(M_n) = O_l(M_1)$. We will also consider a single invertible gr-maximal ideal M (i.e. $O_r$ $O_r(M) = O_l(M) = A$) as a graded cycle.

4. Two graded cycles of A either coincide or are disjoint.

5. Let $D'_g(A)$ be the subgroup of $D_g(A)$ generated by all invertible graded ideals of A, then $D'_g(A)$ is a free abelian group generated by the maximal invertible graded ideals of A.

6. If A has <u>enough invertible</u> graded ideals, i.e. each nonzero graded ideal of A contains an invertible graded ideal, then $D'_g(A) = D_g(A)$.

7. Let I be a nonzero graded ideal of A not contained in any invertible graded ideal of A. If $M_1,\dots,M_n$ is the finite set of gr-maximal ideals containing I then $I^n = (M_1 \cap \dots \cap M_n)^n$ and it is idempotent.

8. Any graded ideal I of A is of the form I = JK for some eventually idempotent graded ideal J and an invertible ideal K of A.

<u>Proof</u>.

1. A satisfies the descending chain condition on graded left (right) ideals containing a fixed nonzero graded ideal., so if P is a nonzero graded prime ideal of A, then A/P is prime and gr-Artinian, hence gr-simple as in the ungraded case.

2. Proposition 2.2. of Eisenbud, Robson [ 46 ].

3. Proposition 2.4., 2.5. and Theorem 2.6. loc. cit.

4. Proposition 3.7. loc.cit.

5. Theorem 2.9. loc cit.

6. Corollary 2.11. loc. cit.

7. Proposition 4.3. and 4.5. loc. cit.

8. Theorem 4.2. loc. cit. □

One easily produces graded versions of these results. For instance let us prove 8. Let I be a non-trivial ideal of A, then by Theorem 4.2. in [46] we have $I_o = XJ$ where X is minimal among the invertible ideals of $A_o$ containing $I_o$ and J is an eventually idempotent ideal of $A_o$. Obviously $A_nXA_{-n}$, $n \in \mathbb{Z}$, is also minimal among the invertible ideals of $A_o$ containing $I_o = A_nI_oA_{-n}$. But $X \cap A_nXA_{-n}$ is again an invertible ideal of $A_o$ since these are of the form $P_1^{n_1} \cap \ldots \cap P_l^{n_l} \ldots P_k^{n_k}$, where the $P_i$ are distinct maximal invertible ideals of $A_o$. So $X\, A_nX\, A_{-n} = X = A_n\, XA_{-n}$ for all $n \in \mathbb{Z}$, hence $A\, X = X\, A$, and hence AX is an invertible ideal of A. Now, $I = AI_o = AXAJ$ and it follows that $AJ = JA$ and AJ is an eventually idempotent ideal of A.

*From now on, we assume that $A_o$ is an HNP-ring with enough invertible ideals.*

4.11. Lemma. If $M_1$ is a gr-maximal ideal of A which is idempotent, then there exist idempotent gr-maximal ideals $M_2,\ldots,M_n$ such that $\{M_1,\ldots,M_n\}$ is a graded cycle.

Proof. We have $M_1 = A(M_1)_o = (M_1)_oA$, where $(M_1)_o$ is idempotent and semiprime. Hence, by Corollary 5.5. of [127], $(M_1)_o = M_{11} \cap \ldots \cap M_{1k}$ for some idempotent maximal ideals $M_{11},\ldots,M_{1k}$ in $A_o$ such that $O_r(M_{1j}) \neq O_l(M_{1i})$ for all $i,j \in \{1,\ldots,k\}$. Since $A_o$ has enough invertible ideals, there exist idempotent maximal ideals of $A_o$, say $M_{21},\ldots,M_{2k}$ such that $O_r(M_{2j}) \neq O_l(M_{2i})$ for all $i,j \in \{1,\ldots,k\}$. Again by Corollary 5.5. of [127], it then follows that $M_{20} = M_{21} \cap \ldots \cap M_{2k}$ is idempotent. Since $(M_1)_o$ is idempotent, we have $(M_1)_o = M_{11} \cdot \ldots \cdot M_{1k}$ and similarly for

$M_{20}$, and thus Lemma 3.2. of [ 90 ] yields that $O_r((M_1)_o) = O_l(M_{20})$.
Let us now show that $M_{20}A$ is not contained in any proper graded ideal of A. So, assume that $M_{20}A \subset I$ for some graded A-ideal I. For the idempotent ideals $(M_1)_o$ and $M_{20}$, we have that $A_o \,.\!\!\cdot\, (M_1)_o = O_r((M_1)_o) = O_l(M_{20}) = A_o \,\cdot\!.\, M_{20}$, therefore $A \,.\!\!\cdot\, M_1 = A(A_o \,.\!\!\cdot\, (M_1)_o) = A(A_o \,\cdot\!.\, M_{20}) = A \,\cdot\!.\, M_{20}A$. So $M_{20}A \subset I \subset A$ yields $A \subset A \,\cdot\!.\, I \subset A \,.\!\!\cdot\, M_1$, hence $M_1 = A \,\cdot\!.(A \,.\!\!\cdot\, M_1) \subset A \,\cdot\!.(A \,\cdot\!.\, I) \subset A$. Note that $A \,\cdot\!.\, I$ is a graded A-*ideal* ! So, either $M_1 = A \,\cdot\!.(A \,\cdot\!.\, I)$ or $A \,\cdot\!.(A \,\cdot\!.\, I) = A$. In case $M_1 = A \,\cdot\!.(A \,\cdot\!.\, I)$, we obtain $A \,.\!\!\cdot\, M_1 = A \,.\!\!\cdot(A \,\cdot\!.(A \,\cdot\!.\, I)) = A \,\cdot\!.\, I$, since $A \,\cdot\!.\, I$ is a v-ideal, because it is projective. So $A(A_o \,\cdot\!.\, M_{20}) = A \,.\!\!\cdot\, M_1 = A \,\cdot\!.\, I = A(A_o \,\cdot\!.\, I_o)$. and thus $A_o \,\cdot\!.\, M_{20} = A_o \,\cdot\!.\, I_o$, entailing $M_{20} = I_o$, hence $M_{20}A = I_oA = I$. On the other hand $A \,\cdot\!.(A \,\cdot\!.\, I) = A$ implies $A \,\cdot\!.\, I = A$, hence $I = A$. As a consequence, we get that either $AM_{20}A = M_{20}A$ or $AM_{20}A = o$. But the latter yields that $A \supset A(A_o \,\cdot\!.\, M_{20})M_{20}A = (A_o \,\cdot\!.\, M_{20})AM_{20}A = A_o \,\cdot\!.\, M_{20})A$, entailing $A_o = A_o \,\cdot\!.\, M_{20}$, hence $A_o = M_{20}$, a contradiction.
Now, we put $M_2 = M_{20}A = RM_{20}A$. Since $(M_2)_o = M_{20}$, we have $M_{20}A = AM_{20}$. So we have obtained that $M_2$ is an idempotent gr-maximal ideal of A such that $O_r(M_1) = O_l(M_2)$. Repetition of this argument yields idempotent gr-maximal ideals $M_2, M_3, \ldots, M_m, \ldots$ of A together with idempotent maximal ideals $M_{2j}, M_{3j}, \ldots, M_{mj}$ of $A_o$ with $j = 1,\ldots,k$, such that $O_r(M_{m-1}) = O_l(M_m)$, $O_r(M_{m-1,j}) = O_l(M_{mj})$ and $(M_m)_o = M_{m1} \cap \ldots \cap M_{mk}$ for $m = 2,\ldots$ . Since $A_o$ has enough invertible ideals, there is a positive integer $n_j$ such that $\{M_{1j},\ldots,M_{n_j j}\}$ is a cycle. Put $p = \text{l.c.m}(n_1,\ldots,n_k)$. Then $M_{11} = M_{p+1,1},\ldots,M_{1k} = M_{p+1,k}$. But then $M_1 = M_{p+1}$, and $O_r(M_1) = O_l(M_2),\ldots,O_r(M_p) = O_l(M_{p+1}) = O_l(M_1)$. If q is the smallest integer m such that $M_1 = M_{m+1}$, then it is now obvious that $\{M_1,\ldots,M_q\}$ is a graded cycle. □

4.12. Proposition. A strongly $\mathbb{Z}$-graded ring A such that $A_o$ is an HNP ring with enough invertible ideals is graded hereditary with enough invertible graded ideals.

Proof. This follows from Lemma 4.11. as in the ungraded case. □

4.13. Lemma. Let P be a maximal invertible graded ideal of A, then :

1. $\mathcal{C}(P)$ is a regular Ore set of A;

2. $Q_P(A)$ is a semi-local HNP ring.

If $P = M_1 \cap \dots \cap M_n$, where $M_1,\dots,M_n$ is a graded cycle, then $J(Q_P(A)) = Q_P(A)P = \bigcap_{i=1}^{n} Q_P(A)M_i$, $Q_P(A)M_i = M_iQ_P(A)$ and $\{Q_P(A)M_1,\dots,Q_P(A)M_n\}$ is a cycle. So the Jacobson radical is the unique maximal invertible ideal of $Q_P(A)$.

Proof. Similar to the proof of Proposition 2.7. in [95 ]. □

4.14. Remark. If P is an invertible gr-maximal ideal of A, then $Q_P(A)$ is a bounded prime principal left and right ideal ring with a unique maximal ideal.

We now define $S^g = \cup\{B^{-1};$ B invertible graded ideal of A$\}$. Clearly $A \subset S^g \subset Q^g$ and $S^g$ is a strongly $\mathbb{Z}$-graded ring. Note also that $Q^g$ (resp. Q) is the ring of fractions of $S^g$ with respect to the set of regular homogeneous (resp. regular) elements of $S^g$.

$(S^g)_o = \cup\{B'^{-1}; B'$ an invertible ideal of $A_o$ such that $B'A = AB'\}$. Indeed the map $B \to B_o$ is a bijective correspondence between the set of all invertible graded ideals B of A and the set of all invertible ideals B' of $A_o$ such that $B'A = AB'$. Since $(S^g)_o$ is an overring of $A_o$ in $Q_o$, it follows by a result of S. Singh that $(S^g)_o$ is an HNP ring with enough invertible ideals.

4.15. Lemma.

1. $S^g$ does not have non-trivial graded ideals.

2. $S^g$ is a prime Noetherian Asano order in Q.

Proof.

1. Let I be a nonzero graded ideal of $S^g$, then $S^g(I \cap R) = I$. Indeed, if $q \in I$, then $Bq \subset I \cap A$ for some invertible graded ideal B of A, and

thus $q \in B^{-1}(I \cap A) \subset S^g(I \cap A)$. But then $I \cap A$ is a nonzero graded ideal of A and so it contains an invertible graded ideal X of A. Since $1 \in X^{-1}X \subset S^gX$ we find $S^gX = S^g$, hence $I \supset S^gX = S^g$, and thus $I = S^g$.

2. Any prime ideal of $S^g$ intersects $(S^g)_0$ trivially because of 1. Now $\mathcal{C}(P)$ is a regular Ore set of $S^g$ for any nonzero prime ideal P of $S^g$ and $Q_P(S^g)$ is a Dedekind prime ring. Then according to Theorem 2.6. of [64 ], $S^g$ is an Asano order. □

4.16. Lemma. $A = \bigcap_P Q_P(A) \cap S^g$, where P runs over the maximal invertible graded ideals of A.

Proof. Let $q \in \bigcap_P Q_P(A) \cap S^g$, then $qB \subset A$ for some invertible graded ideal B of A, and we choose B maximal with respect to this property. If $B \neq A$, then $B \subset P$ for some maximal invertible graded ideal P of A. Clearly $B \subsetneq BP^{-1} \subset A$ and $BP^{-1}$ is an invertible graded ideal of A. On the other hand, $cq \in A$ for some $c \in \mathcal{C}(P)$, so we have :
$qBP^{-1}P \subset qP \cap A \subset c^{-1}P \cap A \subset P$. So $qBP^{-1} \subset O_l(P) = A$, which contradicts the choice of B. Consequently $B = A$ and $q \in A$! □

4.17. Remarks.

1. The relation between ideals of A and ideals of $Q^g$ has been given before; in particular there is a bijective correspondence between prime ideals of A lying over zero in $A_0$ and proper prime ideals of $Q^g$.

2. The same holds for $S^g$ and $Q^g$. Moreover, $Q^gI \cap S^g = I$ for any ideal I of $S^g$. Indeed, put $I' = Q^gI \cap S^g$. Since $S^g$ is right Noetherian , $sI' \subset I$ for some regular element s of A of degree zero. But $S^g$ is gr-simple and thus $S^gsS^g = S^g$. So $I' = S^gI' = S^gsS^gI' \subset I$, hence $I' = I$. Therefore there is a bijective correspondence between ideals of $S^g$ and ideals of $Q^g$.

3. As in the proof of Lemma 4.15. 1, $S^g(I \cap A) = I$ for any ideal I of $S^g$.

Furthermore, $Q^gI \cap S^g = S^gI$ for any ideal I of A; in particular $S^gI$ is an ideal of $S^g$ and $S^gI = IS^g$. Indeed, let $q \in Q^gI \cap S^g$, then $Bq \subset A$ for some nonzero graded ideal B of A. Pick a regular element b of degree zero in B. Since A is right Noetherian, $bAqA = a_1A + \dots + a_nA$, for some $a_i \in A$; hence $AqA = b^{-1}a_1A + \dots + b^{-1}a_nA$. So $q \in Q^gI$ yields $sAqA \subset I$ for some regular element s of degree zero of A. But AsA is a nonzero ideal of A and thus contains an invertible graded ideal X of A. Consequently $Xq \subset I$ and $q \in S^gI$.

4.18. Lemma. Let J" be a non-trivial prime ideal of $Q^g$, and put $J' = J'' \cap S^g$ and $J = J' \cap A$, then :

1. $Q_J(A) = Q_{J'}(S^g) = Q_{J''}(Q^g)$ is a local Dedekind prime ring;
2. J is a prime v-ideal of A and is v-invertible;
3. if I" is any non-zero ideal of $Q^g$, then $I = I'' \cap A$ is a v-ideal of A and is v-invertible.

Proof.

1. Follows from Proposition 2.8. and Remark 4.17.

2. First we show that $Q_P(A)J = Q_P(A)$ for any maximal invertible graded ideal P of A. Assume $Q_P(A)J \neq Q_P(A)$. Since J is a prime ideal, $Q_P(A)J = JQ_P(A)$ is a non-trivial prime ideal of $Q_P(A)$. The structure of $Q_P(A)$ yields that $Q_P(A)J = Q_P(A)M$ for some gr-maximal ideal M of A. Then $J = Q_P(A)J \cap A = Q_P(A)M \cap A = M$, a contradiction. By Lemma 2.3. of [ 95], J is a v-ideal of A.

Let now $q \in O_r(J)$. For any maximal invertible graded ideal P of A, $Q_P(A)Jq \subset Q_P(A)J$, hence $q \in Q_P(A)$. Moreover, $S^gJq \subset S^gJ$ yields $q \in O_r(S^gJ) = S^g$. Therefore $q \in A$ by Lemma 4.16. Similarly $O_l(J) = A$, and it follows that J is v-invertible because of Lemma 1.1. in [ 54 ].

3. Again, I is a v-ideal of A by Lemma 2.3. of [95 ]. Since $Q^g$ is a prime

principal left and right ideal ring, $I'' = J_1''^{n_1} \dots J_k''^{n_k}$ for some maximal ideal $J_1'', \dots, J_k''$ of $Q^g$ and some $n_i > 0$. Obviously $I \supset J_1^{n_1} \dots J_k^{n_k}$, where $J_i = J_i'' \cap A$. But then $Q_P(A)I = Q_P(A)$ and, as in 2., $I$ is v-invertible. □

4.19. <u>Lemma</u>. Consider maximal invertible graded ideals $P$, $P_1$ of $A$, let $J''$, $J_1''$ be a non-trivial prime ideals of $Q^g$ and put $J = J'' \cap A$, resp. $J_1 = J_1'' \cap A$, then :

1. $Q_P(A)J = Q_P(A)$;
2. $Q_P(A)P_1 = Q_P(A)$;
3. $Q_J(A)J_1 = Q_J(A)$;
4. $Q_J(A)P = Q_J(A)$;

<u>Proof</u>.

1. See the proof of Lemma 4.18.2.

2. Write $P = M_1 \cap \dots \cap M_m$, $P_1 = N_1 \cap \dots \cap N_n$ as intersection of a graded cycle. Assume $Q_P(A)P_1 = P_1Q_P(A) \neq Q_P(A)$, then $Q_P(A)N_j \neq Q_P(A)$ for some j. Since $N_j$ is a prime ideal, $Q_P(A)N_j$ is a non-trivial prime ideal, and the structure of $Q_P(A)$ yields that $Q_P(A)N_j = Q_P(A)M_i$ for some i. Therefore $N_j = Q_P(A)N_j \cap A = Q_P(A)M_i \cap A = M_i$, entailing $P_1 = P$, a contradiction.

3. $Q_J(A)J_1 \neq Q_J(A)$ yields $Q_J(A)J_1 = Q_J(A)J$, hence $J_1 = J$, a contradiction.

4. Clear from $Q_J(A)P = Q_{J''}(Q^g)P \supset Q^gP = Q^g$. □

4.20. <u>Lemma</u>. $S(A) = S(S^g)$ and it is a simple noetherian ring.

<u>Proof</u>. Let $q \in S(A)$, then $Xq \subset A$ for some (v-invertible) ideal $X$ of $A$. According to Remark 4.17.3. $S^g$ is a nonzero ideal of $S^g$, and $S^gXq \subset S^g$. hence $q \in S(S^g)$. For the converse inclusion consider $q \in S(S^g)$ with $X'q \subset S^g$ for some nonzero ideal $X'$ of $S^g$. Because of Lemma 4.18.3. and Remark 4.17.2., $X = X' \cap A$ is a v-invertible v-ideal of $A$. Since $A$

is left noetherian, $Xq \subset S^g$ yields $XqB \subset A$ for some invertible graded ideal $B$ of $A$. Consequently $qB \subset X^{-1}$ and $q \in X^{-1}B^{-1} \subset S(A)$. Since $S^g$ is a prime noetherian Asano order, $S(S^g)$ is a simple noetherian ring, cf. [ 64 ]. □

We are finally ready to state the main theorem of this section.

4.21. Theorem Let $A$ be a strongly $\mathbb{Z}$-graded ring such that $A$ is an HNP ring with enough invertible ideals, then $A$ is a vHC-order with enough v-invertible ideals.

Proof. Lemma's 4.13, 4.15 and 4.16 yield $A = \bigcap_P Q_P(A) \cap S^g$, where $P$ runs over the maximal invertible graded ideals of $A$ and where $Q_P(A)$ is a semi-local HNP-ring with unique maximal invertible ideal and $S^g$ is a prime noetherian Asano order. Furthermore, $S^g = \bigcap_{J'} Q_{J'}(S^g) \cap S(S^g)$, where $J' = J'' \cap S^g$ and $J''$ runs through the maximal ideals of $Q^g$, cf. Remark 4.17. So we have :

$$A = (\bigcap_P Q_P(A)) \cap (\bigcap_J Q_J(A)) \cap S(A) \qquad (\star)$$

where $J = J'' \cap A$, where $Q_J(A) = Q_{J'}(S^g)$ is a local Dedekind prime ring and $S(A) = S(S^g)$ is a simple noetherian ring.

Now, consider an arbitrary regular element $c$ of $A$. Since $A$ is left noetherian, $\sum_B (B^{-1}c \cap A) = \sum_{i=1}^n (B_i^{-1} \cap A)$, where $B$ runs through the $P$ and $J$ in $(\star)$. Suppose $B \neq B_i$, $i = 1,\dots,n$, then $B^{-1}c \cap A \subset \sum_{i=1}^n (B_i^{-1}c \cap A) \subset (\sum_{i=1}^n B_i^{-1})c \subset (A \,{}^{\bullet}.\, B_1 \cdot \dots \cdot B_n)c$. Now assume $rc \in B$, $r \in A$, then $B^{-1}rc \subset B^{-1}B \subset A$, and $B^{-1}rc \subset B^{-1}c$. Thus $B^{-1}c \subset B^{-1}rc \cap A \subset (R \,{}^{\bullet}.\, B_1 \dots B_n)c$, hence $rB_1 \dots B_n \subset {}_vB = B$ . By Lemma 4.19. we obtain that $r \in r\, Q_B(A) = rB_1 \dots B_n Q_B(A) \subset BQ_B(A)$, and $r \in BQ_B(A) \cap A = B$. This means that $Q_B(A)c = Q_B(A) = cQ_B(A)$ for all $B$ different from $B_1,\dots,B_n$.

It follows that $A$ is a vHC-order with enough v-invertible ideals because of Corollary 4.6. of [ 54 ]. □

4.22. Corollary. Let A be as in Theorem 4.21 then $\mathcal{D}(A) \cong \mathcal{D}_g(A) \oplus \mathcal{D}(Q^g)$.

Proof. It is sufficient to prove that an ideal X of A is a maximal v-invertible v-ideal of A if and only if X = P of X = J with P and J as in Theorem 4.21. Lemma 1.2. of [ 96 ] states that in a vHC-order with enough v-invertible ideals, the maximal v-ideals are exactly the prime v-ideals. This entails that all P and J as in Theorem 4.12. are maximal v-invertible v-ideals of A. Let now $N = N_1 \cap \dots \cap N_{\bar{n}}$ be a maximal v-invertible v-ideal of A written as the intersection of a cycle. Since $N = {}_vN =$
$= (\bigcap_P Q_P(A)N) \cap (\bigcap_J Q_J(A)N) \cap S(A)$, we have that for some P (resp. some J) $Q_P(A)N \neq Q_P(A)$ (resp. $Q_J(A)N \neq Q_J(A)$).
Assume $Q_P(A)N \neq Q_P(A)$, then $Q_P(A)N_i \neq Q_P(A)$ holds for some i.
Write $P = M_1 \cap \dots \cap M_m$ as the intersection of a cycle. Then the structure of $Q_P(A)$ entails that $Q_P(A)N_i = Q_P(A)M_j$ for a certain index j, yielding $N_i = M_j$. Because of Lemma 1.8. of [ 95 ], N = P follows. □

4.23. Remarks.

1. One easily checks that if I is a v-invertible v-ideal of R, then I is either graded or $I \cap A_o = 0$.
Let I" be any nonzero ideal $Q^g$, i.e. $I" = J"_1^{n_1} \dots J"_k^{n_k}$ where $A"_i$ is a maximal ideal of $Q^g$. Then one can show that $I" \cap A = (J_1^{n_1} \dots J_k^{n_k})_v =$
$= {}_v(J_1^{n_1} \dots J_k^{n_k}) = {}_v(J_k^{n_1}) \cap \dots \cap {}_v(J_k^{n_k})$, where $J_i = J"_i \cap A$.

4.24. Note. Let A be a strongly $\mathbb{Z}$-graded ring such that $A_o$ is a prime noetherian Asano order. As in the case when $A_o$ is an HNP ring, one can show that $A = (\bigcap_P Q_P(A)) \cap (\bigcap_J Q_J(A)) \cap S(A)$, where P runs through the gr-maximal ideals of A and J through the prime ideals of A lying over zero in $A_o$; here $Q_P(A)$ and $Q_J(A)$ are bounded prime principal left and right ideal rings with a unique maximal ideal and S(A) is a simple noetherian ring (compare to [ 29 ]). Furthermore, for any nonzero ideal I of A we have $S(A)I = S(A) = IS(A)$ because $S(A) = S(S^g)$. So A is also a Krull ring in the sense of Marubayashi, see [ 29 ] and [ 92 ] .

4.25. Remark. In the situation of Theorem 4.21., if A is moreover a PI-ring (then $A_o$ is a HNP-ring with polynomial identity, so it automatically has enough invertible ideals!), then A is a vHC-order and a PI-ring, hence a tame order. Furthermore, Theorem 4.21. may be extended to the case of divisorially graded rings of type $\mathbb{Z}$ (cf. a recent paper by H. Marubayashi). An easy induction argument then allows to prove Theorem 3.21 in a much wider context, but for torsion-free abelian and finitely generated groups G (i.e. $G \cong \mathbb{Z}^n$ for some positive integer n) :

4.26. Theorem. Let A be a PI ring, which is divisorially graded of type $\mathbb{Z}^n$ such that $A_o$ is an HNP ring (hence a tame order), then A is a tame order. □

In combination with some results in the case of gradations by finite groups, the above theorem allows to treat the case of gradations of type G, where G is polycyclic-by-finite, but we do not go into the details here.

## D.5. Orders and Generalized Crossed Products for Finite Groups.

In the theory of (relative) Azumaya algebras, crossed products and generalized crossed products for finite groups appear very naturally and quite often. That the tools of graded ring theory may come in handy has been noted in Section C.1. However, it is also evident that different tools of graded theory were necessary to derive the structural results in the foregoing sections in this chapter, where the grading groups were mostly ordered or at least torsion-free. On the other hand, whereas relative Azumaya algebras may be the nicest orders from the relative point of view, it is certainly interesting to study the generalized crossed product structure of certain maximal orders (which need *not* be relative Azumaya algebras), if it exists at all. So, the aim of this section is to provide some important results in the vein of the main results of Section 3 and Section 4, but for gradations by finite groups and next, to show that generalized crossed product structures may occur in maximal orders which need not be (relative) Azumaya algebras, and to relate it to ramification properties. Finally, let us point out (and this is probably the best argument to motivate the consideration of gradations by finite groups in some detail) that when applying the construction of Rees rings in order to obtain new, well behaved rings, with nicer invariants, the finite groups evolve naturally !

Let A be a strongly graded ring of type G. For each $\sigma \in G$, we have that $A_\sigma A_{\sigma^{-1}} = A_e$ and hence we may fix for each $\sigma \in G$ a decomposition $1 = \sum_i u_\sigma^{(i)} v_{\sigma^{-1}}^{(i)}$, where $u_\sigma^{(i)} \in A_\sigma$ and $v_{\sigma^{-1}}^{(i)} \in A_{\sigma^{-1}}$. If $c \in Z(A_e)$, then the action of $\sigma \in G$ on c is given by $\sigma(c)t = tc$ for all $t \in A_\sigma$; alternatively, this action may be described by $\sigma(c) = \sum_i u_\sigma^{(i)} c\, v_{\sigma^{-1}}^{(i)}$. The ring fixed in $Z(A_e)$ under this action of G will be denoted by $A_o$ and it is clear that $A_o$ is contained in the center of A. Let $\Phi : G \to \mathrm{Pic}(A_e)$ be the group homomorphism defined by $\Phi(\sigma) = [A_\sigma]$ for any $\sigma \in G$. We define the $\Phi$-trace on $Z(A_e)$ by $\mathrm{tr}_\Phi(c) = \sum_{\sigma \in G} \sigma(c)$. The following is a modification of a result in [27] :

5.1. Proposition. Let A be strongly graded by a finite group G and suppose there exists an element $c \in Z(A_e)$ with $\mathrm{tr}_\Phi(c) = 1$. Let $\varphi : M \to N$ be an $A_e$-linear map between left A-modules M and N. There exists a canonical left A-linear map $\tilde{\varphi} : M \to N$, only depending upon $\varphi$ and c. If $\varphi$ is A-linear, then $\tilde{\varphi} = \varphi$. If $f : N \to T$ is a left A-linear map for some $T \in$ A-mod, then

$f_{\tilde{\varphi}} = (f_{\varphi})^{\sim}$.

Proof. For any $m \in M$, define $\tilde{\varphi}(m) = \sum_{\sigma \in G} \sum_i u_\sigma^{(i)} c\, \varphi(v_{\sigma^{-1}}^{(i)} m)$. The proof that $\tilde{\varphi}$ is left A-linear is identical to the proof of the similar statement in the generalized Maschke's Theorem as given in [158], but taking into account that $1 = \sum_{\sigma \in G} \sigma(c)$. Let us establish that the definition of $\tilde{\varphi}$ is independent of the choice of the decompositions $1 = \sum_i u_\sigma^{(i)} v_{\sigma^{-1}}^{(1)}$ for each $\sigma \in G$. Therefore, consider other decompositions $1 = \sum_j u'^{(j)}_\sigma\, \sigma\, v'^{(j)}_{\sigma^{-1}}$. A straightforward calculation yields for $m \in M$ that

$$\begin{aligned}
\tilde{\varphi}(m) &= \sum_{\sigma \in G} \sum_i u_\sigma^{(i)} c\, \varphi\, (v_{\sigma^{-1}}^{(i)} m) \\
&= \sum_{\sigma \in G} \sum_i u_\sigma^{(i)} c\, \varphi\, (v_{\sigma^{-1}}^{(i)} \sum_j u'^{(j)}_\sigma\, v'^{(j)}_{\sigma^{-1}} m) \\
&= \sum_{\sigma \in G} \sum_i \sum_j u_\sigma^{(i)} c\, v_{\sigma^{-1}}^{(i)}\, u'^{(j)}_\sigma\, \varphi(v'^{(j)}_{\sigma^{-1}} m) \\
&= \sum_{\sigma \in G} \sum_i \sum_j u_\sigma^{(i)} v_\sigma^{(i)}\, v_{\sigma^{-1}}^{(i)}\, u'^{(j)}_\sigma c\, \varphi(v'^{(j)}_{\sigma^{-1}} m) \\
&= \sum_{\sigma \in G} \sum_j u'^{(j)}_\sigma c\, \varphi(v'^{(j)}_{\sigma^{-1}} m) = \tilde{\varphi}'(m).
\end{aligned}$$

The other assertions in the proposition are easily verified. □

5.2. Corollary. Assume we are in the situation of the foregoing Proposition then :

1. Nauwelaerts) if $A_0$ is left hereditary, then A is left hereditary
2. if $A_e$ is left regular, then A is left regular;
3. if $A_e$ is separable over $A_0$, then A is separable over $A_0$.

Proof. 1. Basically, the lifting of a left $A_e$-linear splitting map in an exact sequence of left A-modules to an A-linear splitting map.

2. Left as an exercise;

3. Similar to the proof of the analogous statement in [ 158]. □

Note also that the condition $|G|^{-1} \in A_o$ implies that $tr_\Phi$ is surjective.

From now on we restrict attention to the case where $A_e$ is commutative, i.e. we have $A_e = Z(A_e)$. Moreover, we extend $tr_\Phi$ to A in a rather trivial way. by putting $tr_\Phi(r) = tr_\Phi(r_e)$, if $r = r_e + \dots + r_\sigma$ is the decomposition of r into homogeneous components. We also define a bilinear $A_o$-form $\tau_\Phi$ by putting $\tau_\Phi(s,t) = tr_\Phi(st)$. In order to prove that $\tau_\Phi$ is a symmetric associative bilinear form on A, we need the following key-lemma :

5.3. Lemma. Let A be strongly graded by a finite group G such that $A_e$ is commutative. For arbitrary $s \in A_\sigma$ and $t \in A_{\sigma^{-1}}$ we have $st = \sigma(ts)$.

Proof. Let $p_o$ be a maximal ideal of $A_o$. The localization $A_{p_o}$ of A at $p_o$ is strongly graded by G, with $(A_{p_o})_e = (A_e)_{p_o}$. Since taking the fixed ring for the action of G commutes with localization at prime ideals in the invariant ring, we may identify $(A_o)_{p_o}$ and $(A_{p_o})_o = A'_o$. Let $\overline{s}, \overline{t}$ be the images of s,t respectively in $A_{p_o}$. We first prove that $\overline{s}\,\overline{t} = \sigma(\overline{t}\,\overline{s})$ $(=\overline{\sigma(ts)})$. Write $A'_e$ for $(A_{p_o})_e$ and $p'_o$ for $A'_o p_o$ and consider the set of maximal ideals $\{P'_1,\dots,P'_m\}$ of $A'_e$ lying over $p'_o$. If we show that G acts transitively on this set, then m is finite. Suppose G does not act transitively and suppose that $Q'_1$ and $Q'_2$ are maximal ideals of $A'_e$ such that $Q'_2 \notin \{\sigma(Q'_1);\sigma \in G\}$. Put $q'_1 = \cap \{\sigma(Q'_1);\sigma \in G\}$. By the Chinese Remainder Theorem, there exists an element $a' \in A_e$ such that that $a' \equiv 0 \bmod Q'_2$ resp. $a' \equiv 1 \bmod q'_1$. Then $N(a') = \prod_{\sigma \in G} \sigma(a')$ also has the property that $N(a') \equiv 0 \bmod Q'_2$ and $N(a') \equiv 1 \bmod q'_1$. Since $N(a') \in A'_o$, the first relation implies that $N(a') \in p'_o$, whereas it follows from the second that $N(a') \notin p'_o$, a contradiction. Consequently, $A'_e$ is a semilocal ring and $Pic(A'_e) = 1$. It follows from this that $A' = \bigoplus_{\sigma \in G} A'_e u'_\sigma$, where $u'_\sigma$ is a unit of A', since $1 \in A'_e u'_\sigma A'_e u'_{\sigma^{-1}} = A'_e u'_\sigma u'_{\sigma^{-1}}$. Obviously, we may assume that $u'_e = 1$ and now $\overline{s} = x' u'_\sigma$ resp.

$\overline{t} = y'(u'_\sigma)^{-1}$, so $\overline{st} = x'\sigma(y')$, while $\overline{ts} = y'\sigma^{-1}(x')$. Consequently, $st-\sigma(ts)$ maps to zero in $A_{p_o}$ for every maximal ideal $p_o$ of $A_o$, hence $st-\sigma(ts) = 0$ follows.

5.4. Corollary. In the situation of the Lemma, if $1 = \sum_i u_\sigma^{(i)} v_{\sigma^{-1}}^{(i)}$, then also $1 = \sum_i v_{\sigma^{-1}}^{(i)} u_\sigma^{(i)}$.

Proof. We have that $\sigma(ts) = \sum_i u_\sigma^{(i)}\, ts\, v_{\sigma^{-1}}^{(i)}$ for $t \in A_\sigma$. But then we may commute $u_\sigma^{(i)} t \in A_e$ with s and obtain that $\sigma(ts) = \sum_i s\sigma^{-1}(u_\sigma^{(i)} t) v_{\sigma^{-1}}^{(i)} = \sum s\, v_{\sigma^{-1}}^{(i)}(u_\sigma^{(i)} t) = \sum_i s(v_{\sigma^{-1}}^{(i)} u_\sigma^{(i)})t = \sum_i st\, \sigma(v_{\sigma^{-1}}^{(i)} u_\sigma^{(i)}) = st\, \sigma(\sum v_{\sigma^{-1}}^{(i)} u_\sigma^{(i)})$. By the Lemma, and since the equalities hold for all s,t, we have $1 = \sigma(\sum_i v_{\sigma^{-1}}^{(i)} u_\sigma^{(i)})$, and the corollary follows. □

5.5. Remark. The corollary yields a generalization of the fact that a left inverse for $v \in A$ is also a right inverse (note that A is a PI-ring, since it is finite over a commutative ring).The authors were puzzled by the fact that it seems to be unavoidable to use the property the $A_e$ is semilocal in proving this. In the general strongly graded situation putting $a = \sigma(\sum_i v_{\sigma^{-1}}^{(i)} u_\sigma^{(i)})$ one may easily calculate that $a^2 = 1$ and that $a = \sigma(a) = \sum_i v_{\sigma^{-1}}^{(i)} u_\sigma^{(i)}$. To conclude from this that $a = 1$ seems impossible unless one uses the lemma.

5.6. Proposition. If A is strongly graded by a finite group G such that $A_e$ is commutative, then $\tau_\Phi$ defines a symmetric associative $A_o$-bilinear form on A.

Proof. Associativity and $A_o$-bilinearity are evident from the definition : Symmetry of $\tau_\Phi$ only has to be checked for homogeneous elements, say $s \in A_\sigma$ and $t \in A_\tau$. If $\tau \neq \sigma^{-1}$, then $\tau_\Phi(s,t) = tr_\Phi(st) = tr_\Phi(\sigma(ts)) = tr_\Phi(ts) = \tau_\Phi(t,s)$. □

Let us now consider first a very specific case. Let K be a field and let $A = \bigoplus_{\sigma \in G} Ku_\sigma$ be strongly graded by a finite group G such $A_e = K$. Denote $K^G$ by $K_o$ and let $H = \{\sigma \in G; \sigma|K = 1_K\}$. Consider an integrally closed domain $R_o$ in $K_o$. Write $u_\sigma u_\tau = f_{\sigma,\tau}\, u_{\sigma\tau}$ for all $\sigma, \tau \in G$, where $f_{\sigma,\tau}$ is representing a 2-cocycle and we may assume that $u_e = 1$. We also consider the strongly graded ring $A' = \sum_{\sigma \in H} Ku_\sigma$ of type H and we assume further that $|H|^{-1} \in K_o$ and that $tr_\Phi$ is non degenerate, $\Phi$ denoting the morphism which associates $[Ku_\sigma]$ to $\sigma$. Note that the argument used after Corollary 5.2. allows to view $tr_\Phi$ as a map defined on A. These assumptions together with the general version of Maschke's Theorem, which we have proved in Proposition 5.1. imply that A is $K_o$-separable. In this situation it makes more sense (even if the notation is somewhat more complicated) to write $tr^o_{A/K_o}$ instead of $tr_\Phi$ in order to make some of the dependencies more explicit.

5.7. Proposition. If $a \in A$ is integral over R, then $tr^o_{A/K_o}(a) \in R_o$.

Proof. First we deal with the case $H = G$ and $A' = A$. Let $\{a_1,\dots,a_n\}$ be a fixed $K_o$-basis for K, then in the left regular representation of A in terms of the $K_o$-basis $\{a_i u_\sigma; i = 1,\dots,n,\ \sigma \in G\}$, we easily calculate that the common trace $T_{A/K_o}$ in this representation coincides with $tr^o_{A/K_o}$, i.e. if $a \in Ku_\sigma$ and $\sigma \neq e$, then $T_{A/K_o}(a) = 0$ and for $a \in Ku_e$ we have $T_{A/K_o}(a) = na$ ! Therefore the statement is clear in this case.

In general write $B = K \otimes_{K_o} A = \sum_{\sigma \in G} (K \otimes_{K_o} K)u_\sigma$. Since K is a Galois extension of $K_o$ with group G/H, it follows that $K \otimes_{K_o} K \cong \sum_{\tau \in G/H} Ke_\tau$, where each $e_\tau$ is an idempotent, $\sigma(e_\tau) = e_{\sigma\tau}$ for any $\sigma \in G$ and $e_\tau e_\gamma = 0$ if $\gamma \neq \tau$. For $a \in K$, we have the identifications $a \otimes 1 = \sum_{\tau \in G/H} a\, e_\tau$ and $1 \otimes a = \sum_{\tau \in G/H} \tau(a)e_\tau$. Direct calculation yields $e_{\mu'}Be_\mu = \sum_{\sigma \in H\, \mu'\mu^{-1}} Ku_\sigma$

and we write $B_{\mu'\mu}$ for the latter. It is clear that $B_{\mu\mu} = A'$ and that each $B_{\mu'\mu}$ is an A'-bimodule. Let $i : A \to K \otimes_{K_o} A$ be the canonical inclusion. Calculating the image $i(\lambda u_\sigma) \in h(A)$ in the representation of B as a matrix algebra $(S_{\mu'\mu})$ with $\mu, \mu' \in G/H$ yields that $e_{\mu'}(1 \otimes \lambda u_\sigma)e_\mu = e_{\mu'}(\sum_{\tau \in G/H} \tau(\lambda)e_\tau u_\sigma e_{\mu'})$ and this is zero if $\sigma \notin H\mu'\mu^{-1}$ and equal to $\mu'(\lambda)u_\sigma$ otherwise. From the first part it follows that $tr^o_{A/K_o}(a) = T_{A'/K_o}(\sum_{\mu \in G/H} e_\mu i(a)e_\mu)$ for all $a \in A$. Since A' is $K_o$-separable, it is a semisimple artinian algebra, so we write $A' = \sum_{i=1}^m A_i'e_i$ for central idempotents $e_i \in L_i(A_i')$ and we put $[A_i' : L_i] = n_i^2$. Now we obtain

$$T_{A'/K_o}(\sum_{\mu \in G/H} i(a)_{\mu\mu}) = \sum_{i=1}^m n_i \, tr_{red}(Be_i/K_o)(i(a)e_i), \qquad (\star)$$

where $tr_{red}(Be_i/K_o)$ is the usual reduced trace. Consequently, if a is integral over $R_o$, then so are all $i(a)e_i$ and hence so is $(\star)$, proving our claim. □

5.8. <u>Note</u>. If A is a common crossed product, then $tr^o_{A/K_o}$ is the usual reduced trace.

Let us now deal with the case of maximal orders in separable algebras over a field. Consider a Dedekind ring R with field of fractions K and let $\Delta$ be strongly graded by the finite group G such that $\Delta_e = R$. We write A for $Q(\Delta)$ and hence we have that $A = \sum_{\sigma \in G} Ku_\sigma$. We assume that A satisfies the conditions imposed at the beginning of this section. Since R is a Dedekind ring, the fixed ring $R_o = R^G$ is again a Dedekind ring. We define the inverse different of $R/R_o$ with respect to $tr^o_{K/K_o}$ by $\mathcal{D}^{-1}_{R/R_o} = \{x \in K; tr^o_{K/K_o}(xR) \subset R_o\}$.

5.9. <u>Proposition</u>. If $\Gamma \supset \Delta$ is a maximal $R_o$-order in A, then $\Delta \subset \Gamma \subset \mathcal{D}^{-1}_{R/R_o}.\Delta$.

<u>Proof</u>. Since the theorem is "local", we may assume that $R_o$ is a discrete valuation ring. In this case R is semilocal and it follows that $\Delta$ has

the form $\sum_{\sigma \in G} Ru_\sigma$. From $\Delta \subset \Gamma$ it then follows that for all $r \in R$ and $\gamma \in G$ we have $\sum_r r\, a_\sigma\, u_\sigma\, u_\gamma^{-1} \in \Gamma$, hence $tr^o_{A/K_o}(\sum_r r\, a_\sigma u_\sigma u^{-1}) \in R_o$ or $tr_{K/K_o}(Ra_\gamma) \subset R_o$. So for all $\gamma \in D^{-1}_{R/R_o}$ we have $a_\gamma \in D^{-1}_{R/R}$ or $\sum_{\sigma \in G} a_\sigma u_\sigma \in D^{-1}_{R/R_o}.\Delta$. □

In order to improve this result we calculate the discriminant of $\Delta/R_o$ if terms of $tr^o_{A/K_o}$. Let $d^o(\Delta/R_o)$ be the ideal generated by the set of all $\det(tr^o(x_i x_j))$ with $x_i \in \Delta$ and $i,j = 1,\dots,n$ where $n = \dim_{K_o}(A) = rk_{R_o}(\Delta)$. □

5.10. <u>Theorem</u>. With notations and hypotheses as above, $d^o(\Delta/R_o) = d^o(R_e/R_e)^{|G|}$.

<u>Proof</u>. Again, we may reduce the theorem to the case where $R_o$ is local, i.e. a discrete valuation ring and as in the foregoing Proposition, $\Delta$ is of the form $\sum_{\sigma \in G} Ru_\sigma$. It will suffice to calculate $\det(tr(x_i x_j))$ for some $R_o$-basis $\{x_1,\dots,x_n\}$ of $\Delta$. So consider $\{a_i u_\sigma\}$ as such a basis. A straightforward calculation now yields that $\chi = \det(tr^o_{R/R_o}(a_i u_\sigma a_j u_\tau)) = \prod_{\sigma \in G} \det(tr^o_{R/R_o}(a_i \sigma(a_j) f_{\sigma,\tau}\, u_{\sigma\tau}))$. The latter equality holds because in the first member only entries with $\tau = \sigma^{-1}$ contribute. Moreover, $\{\sigma(a_i) f_{\sigma,\sigma^{-1}}; i = 1,\dots,r\}$ is an $R_o$-basis for R because $f_{\sigma,\sigma^{-1}} \in U(R)$, so we obtain further equalities $\chi = \det(tr^o_{R/R_o}(a_i \sigma(a_j) f_{\sigma,\sigma^{-1}}))$ and thus $d^o(\Delta/R_o) = d^o(R_e/R_o)^{|G|}$. □

5.11. <u>Corollary</u>. If $\Gamma \supset \Delta$ is a maximal $R_o$-order in A then $\Delta = \Gamma$ if and only if $d^o(\Gamma/R_o) = d^o(R/R_o)^{|G|}$.

<u>Proof</u>. In view of the foregoing result, the Corollary may be proved in the classical way, cf. I. Reiner [122]. □

5.12. Remark. The obvious advantage of the foregoing Theorem and the Corollary is that we obtain from it a criterion for maximality of $\Delta$ in terms of R, $R_o$ and A alone. In the sequel of the section, we are considering a kind of converse to this problem, i.e. we look for a criterion to decide whether certain maximal orders may be considered as strongly graded rings over certain commutative subrings.

An extension of Dedekind rings $R \supset R_o$ is said to be a *pseudo-Galois extension* if the following properties hold :

1. If the field of fractions of $R_o$ resp. R is $K_o$ resp. K then $K/K_o$ is a Galois extension;

2. The extension $R/R_o$ is tamely ramified, or equivalently : $tr_{R/R_o}$ is surjective.

In the following proposition we use the techniques of Gr-Dedekind rings, for which we refer to [145,155].

5.13. Proposition. If the extension of Dedekind domains $R/R_o$ ramifies in the primes $P_1,\dots,P_n$ with ramification indices $e_1,\dots,e_n$, then we may form $\check{R} = \sum_{i\in\mathbb{Z}} Q_1^i \cdot \dots \cdot Q_n^i X^i$ and $\check{R}_o = \sum_{i\in\mathbb{Z}} P^{[i/e_1]} \cdot \dots \cdot P^{[i/e_n]} X^i$, where $[\alpha]$ denotes the smallest integer larger than $\alpha$ and $Q_i = \mathrm{rad}(RP_i)$. Then $\check{R}/\check{R}_o$ is a graded Galois extension in the sense of Section I.3.

Proof. From [145] it follows that both $\check{R}_o$ and $\check{R}$ are Gr-Dedekind rings. We may reduce the Proposition to the gr-local case by localizing homogeneously at some $P_i$ in $R_o$; so we just write P and Q in the sequel of the proof. It is easy to verify that $\mathrm{rad}(\check{R})$ equals R $\mathrm{rad}(\check{R})$ and furthermore that $\check{S} = \check{R}/\mathrm{rad}(\check{R}) = \sum_{i\in\mathbb{Z}} (R/Q)X^i$ resp. $\check{S}_o = \check{R}_o/\mathrm{rad}(\check{R}_o) = \sum_{i\in\mathbb{Z}} (R_o/P)X^{ie}$, where e is the ramification index of P in Q. The fact that $R/R_o$ is tamely ramified implies that R/Q is a separable extension of $R_o/P$ and that $e \neq 0$ in $R_o/P$. Consequently, it follows that $\check{S}$ is separable over $\check{S}_o$ and thus also that $\check{R}$ is separable over $\check{R}_o$. Since Q is obviously G-invariant in R, the action of G on R can be extended in a natural way to an action of G on $\check{R}$, such that $(\check{R})^G = \check{R}_o$. That $\check{R}$ is a ($\mathbb{Z}$-graded) Galois extension of $\check{R}_o$ is thus obvious. □

If $R_o \subset R$ is an extension of Dedekind rings and $\Delta$ is a maximal $R_o$-order in A, then $\Delta$ is said to radicalize R is $\Delta \supset R$ and for every maximal ideal p of $R_o$ we have $\mathrm{rad}(\Delta_p) = \mathrm{rad}(R_p)\Delta = \Delta\ \mathrm{rad}(R_p)$. This condition is clearly satisfied if $\Delta$ is an Azumaya algebra and R is a Galois commutative subring, but also in the case where $R_o$ is local and A has a division ring for its residue algebra.

5.14. Proposition. Let $R_o \subset R$ be an extension of Dedekind rings such that the field extension $K_o \subset K$ obtained by taking fields of fractions is a Galois extension with group G. If $\Delta$ is a maximal $R_o$-order which radicalizes R, then there is a bijective correspondence between ideals of $\Delta$ and G-invariant ideals of R.

Proof. We may assume that $R_o$ is a discrete valuation ring and in this case the G-invariant ideals of R are just the powers of rad(R), whereas on the other hand the ideals of $\Delta$ are the powers of $\mathrm{rad}(\Delta)$. Now, first note that $\mathrm{rad}(R)^n\Delta = \mathrm{rad}(\Delta)^n = \Delta\ \mathrm{rad}(R)^n$. We claim that R is a direct summand of $\Delta$ in R-mod. To see this, it suffices to establish that $\Delta/R$ is a torsionfree R-module. Consider $a \in \Delta$ and $s \in R$ such that $sa \in R$ i.e. $a \in \Delta \cap s^{-1}R$ and a is then certainly integral over $R_o$. However, R is the integral closure of $R_o$ in K, so $a \in R$ and the claim follows. Write $\Delta = R \oplus \Delta'$ for some $\Delta' \in$ R-mod, then $\mathrm{rad}(\Delta)^n \cap R = \mathrm{rad}(R)^n\Delta \cap S = (\mathrm{rad}(R)^n \oplus \mathrm{rad}(R)^n\Delta') \cap R = \mathrm{rad}(R)^n$. □

The next result uses a fact on central class groups, which will be proved in Chapter E, however, we include it here, where it belongs.

5.15. Proposition. In the situation of the foregoing Proposition, if we assume moreover that R is a maximal commutative subring of Δ and that R is a pseudo-Galois extension of $R_o$ such that for all $p \in \Omega(R_o)$, we have that $\Delta_p/\mathrm{rad}(\Delta_p)$ is a separable extension of $R_o/p$, then Δ is a generalized crossed product over R graded by G.

Proof. Again, we may assume that $R_o$ is local. Define $\check{R}_o$ and $\check{R}$ as in (5.13.) and define $\check{\Delta} = \sum_{i \in \mathbb{Z}} \mathrm{rad}(\Delta)^i X^i$. Since the central class group of $\check{\Delta}$ vanishes, [88] yields that $\check{\Delta}$ is a graded Azumaya algebra over $\check{R}_o$. Proposition 5.13. entails that $\check{R}$ is a graded Galois extension of A. It is also clear that $\check{R}$ is a maximal commutative subring of $\check{\Delta}$ and that $\check{\Delta}$ is a generalized crossed product of $\check{R}$ and G with respect to some group morphism $\Phi : G \to \mathrm{Pic}(\check{R})$ and some $c \in H^2(G,U(R))$. Since the $\mathbb{Z}$-gradation and the G-gradation on $\check{\Delta}$ are compatible, it follows that $\Delta = \check{\Delta}_o$ is a generalized crossed product with respect to $R = (\check{R})_o$ and G and the group morphism $\Phi : G \to \mathrm{Pic}(\check{R}) = \mathrm{Pic}^g(\check{R}) = \mathrm{Pic}((\check{R})_o) = \mathrm{Pic}(R)$. Note that $\mathrm{Pic}(\check{R}) = \mathrm{Pic}^g(\check{R})$ holds for any $\mathbb{Z}$-graded Gr-Dedekind ring, whereas $\mathrm{Pic}^g(\check{R}) = \mathrm{Pic}((\check{R})_o)$ follows from the fact that $\check{R}$ is strongly graded. Note also that we used the fact that G is defined as a group of automorphisms of $\check{R}$ defined in $\mathbb{Z}$-degree zero, i.e. on R over $R_o$. □

In the foregoing, we have been dealing with the problem of deciding whether a given maximal order may be strongly graded over some commutative subring and it turned out that the conditions for this to happen are given in terms of ramification of certain primes in the commutative subring. We now aim to look at a converse problem. Given an order of a certain type, e.g. a maximal order, under what conditions is a strongly graded ring over that order again an order over the same type. Again the conditions for this to happen will be given in terms of non-ramification of primes; note that we now do not restrict to strongly graded rings over commutative subrings!

In the sequel of this section A is a strongly graded ring for a finite group G and we suppose that $A_e$ is a prime PI- ring, where e is the unit of G. Let R be the center of $A_e$ and let L be its field of fractions. The classical ring of quotients of $A_e$, $Q_{cl}(A_e) = LA_e$, is a central simple L-algebra. The regular elements in $A_e$ form a regular Ore set S of A and $Q = A_S$ is strongly graded by G such that $Q_e = Q_{cl}(A_e)$. It is clear that Q is an Artinian ring and therefore Q coincides with the classical ring of quotients of A. The action of G on R defined by the canonical morphisms $G \to \mathrm{Pic}(A) \to \mathrm{Aut}(R)$ extends to an action of G on L. Obviously L is a finite Galois extension of $L^G$ and R is integral over $R^G$; moreover, L may be obtained by localizing R at $R^G-\{0\}$ and it follows that $L^G$ is the field of fractions of $R^G$. Note that $Q = L^G A$. Furthermore Q is a finite dimensional $L^G$-algebra since it is a finitely generated $Q_e$-module; consequently Q, and hence also A, is a PI- ring. If A is assumed to be a prime ring, with center C, say (let K be the field of fractions of C), then Q is a central simple K-algebra such that KA = Q. Note that $R^G = C \cap A_e$ and $L^G = K \cap Q_e$.

5.12. Theorem. (E. Nauwelaerts, F. Van Oystaeyen). Let A be strongly graded by the finite group G such that $tr_\Phi(a_o) = 1$ for some $a_o \in R$ (where $tr_\Phi$ is as defined in the remarks preceding Proposition 5.1.). If A is a prime ring and $A_e$ is a tame order in $Q_e$, then C is a Krull domain and A is a tame C-order in Q.

Proof. 1° Claim : A is integral over $R^G$, and hence also over C. Since $A_e$ is a PI-ring which is an integral extension of R, while R is an integral extension of $R^G$ it follows from R. Paré, W. Schelter [ 115 ] that $A_e$ is integral over $R^G$.

If $a_\sigma \in A_\sigma$, $\sigma \in G$, then $a_\sigma^n \in A_e$ for some n dividing $|G|$ and thus $a_\sigma$ is integral over $R^G$. If $a \neq 0$ in A, let $a = a_{\sigma_1} + \dots + a_{\sigma_m}$ be the homogeneous decomposition of a and let $T = R^G\{a_{\sigma_1},\dots,a_{\sigma_m}\}$ be the $R^G$-subalgebra of A generated by $\{a_{\sigma_1},\dots,a_{\sigma_m}\}$. Clearly, T is a PI ring and the monomials in $a_{\sigma_1},\dots,a_{\sigma_m}$ are integral over $R^G$ since they are homogeneous. From C. Procesi [116] p. 152 it follows that T is finitely generated as an $R^G$-module i.e. T is integral over $R^G$. This entails that a is integral over $R^G$ and so the claim has been established.

2° From $R^G = R \cap L^G$ we retain that $R^G$ is a Krull domain. Since R is an integral extension we have the lying-over and incomparability properties for prime ideals. Furthermore since R is a domain and $R^G$ is integrally closed we also have the going-down property and the extension $R^G \to R$ has P.D.E. . If we consider $q \in X^{(1)}(R^G)$ then the localization $R_q$ is the integral closure of $(R^G)_q$ in L and as such, $R_q$ is a Dedekind domain. The ring $(A_e)_q$ is an $R_q$-module. Now $(A_e)_q$ is a noetherian $R_q$-algebra and if m is a maximal ideal of $R_q$, say $m = pR_q$ for some $p \in X^{(1)}(R)$ lying over q, then $(A_e)_q$ localized at $R_q$-m is isomorphic to $(A_e)_p$ and hence it is hereditary. Since $(A_e)_q$ is a finitely generated torsionfree $R_q$-module it follows that $(A_e)_q = \cap \{((A_e)_q)_m;\ m \text{ a maximal ideal of } R_q\}$. Therefore, utilizing the fact that $A_e$ is a divisorial R-order by assumption, we obtain $A_e = \cap \{(A_e)_q;\ q \in X^{(1)}(R^G)\}$. For each $q \in X^{(1)}(R^G)$, $A_q$ is left and right hereditary because of Corollary 5.2.1. However, $\cap \{A_q; q \in X^{(1)}(A^G)\}$ is a strongly graded ring containing A and such that its initial subring equals $A_e$ , hence we obtain : $A = \cap \{A_q;\ q \in X^{(1)}(R^G)\}$.

3° From the above equality we derive that : $C = \cap \{C_q;\ q \in X^{(1)}(R^G)\}$, and since each $C_q$ is a Dedekind domain it is clear that C is integrally closed in L. Hence C is the integral closure of $R^G$ in L and C is a Krull domain because $R^G$ is. It remains to show that A is a tame C-order in Q. By 1°, C is an integral extension of $R^G$ so we have (as in 2°) that

$P \in X^{(1)}(C)$ yields $P \cap R^G \in X^{(1)}(R^G)$. On the other hand, if $q \in X^{(1)}(R^G)$ then we may find a $P \in X^{(1)}(C)$ lying over q. For such P the ring of quotients of A at C-P, $A_P$ say, is hereditary because $A_q$ is hereditary and $A_P$ is an intermediate ring between $A_q$ and Q, cf. Kuzmanovich [ 84 ] . Finally let us verify that A is a hereditary prime PI-ring and $Z(A_q) = C_q$. According to J. C. Robson, L.W. Small [ 128 ] (cf. Theorem 3), $C_q$ is Dedekind and $A_q$ is a finitely generated $C_q$-module. Argumentation as in 2° yields :

$A_q = \cap \{(A_q)_M;$ M a maximal ideal of $C_q\}$, and

$A_q = \cap \{A_P;\ P \in X^{(1)}(C),$ P lies over q$\}$

From $A = \cap \{A_q;\ q \in X^{(1)}(R^G)\}$ the equality :

$A = \cap \{A_P; P \in X^{(1)}(C)\}$ follows. □

We stick to notations introduced above but now we do not assume that A is a prime ring but allow it to be a semiprime ring. The following lemma may be proved in a way similar to the proof of a similar property in D. Eisenbud, J. Robson [ 46 ] .

5.17. Lemma. Let A be a semiprime Goldie hereditary ring (hence A is also Noetherian). If the left ideal I of A contains a regular element of A then A/I is an Artinian left A-module. □

As a consequence of this lemma we have that a nonzero prime ideal of a semiprime Noetherian hereditary ring is maximal if it contains a regular element.

5.18. Theorem. (E. Nauwelaerts, F. Van Oystaeyen). Let A be a ring strongly graded by a finite group G such that the order of G is a unit in A. Assume that $A_e$ is a prime PI - ring which is a maximal R-order in $Q_e$. If for every prime ideal q of $R^G$ we have : $qA_e = P_1 \cdot \dots \cdot P_t$, for distinct prime ideals of $A_e$, then A is a semiprime Dedekind order in Q (in the sense of Robson [126 ] .)

Proof. Since A has no $|G|$-torsion it is easily seen that the fact that $A_e$ is prime makes A into a semiprime ring. Since A is finitely generated as a left (or right) $A_e$-module it follows that A is left and right Noetherian. Corollary 5.2. entails that A is left and right hereditary. Now consider a maximal integral A-ideal M. From $Q = L^G A$ it follows that $M \cap R^G \neq 0$ and since the latter is a prime ideal pf $R^G$ our assumptions entail $(M \cap R^G)A_e = P_1 \cdot \ldots \cdot P_t = P_1 \cap \ldots \cap P_t$, where the $P_i$ are distinct maximal ideals of $A_e$. The ring $A/(M \cap R^G)A$ is strongly graded by G and Corollary 5.5. of Cohen, Montgomery [37 ] entails that $(M \cap R^G)A$ is a semiprime ideal. So we have that $(M \cap R^G)A = M \cap J$ where J is an ideal of A not contained in M and such that $(M \cap R^G)A = M\,J = J\,M$. There exists an inverse for $(M \cap R^G)A_e$ in $Q_e$ and this extends to an inverse for $(M \cap R^G)A$ in Q. Thus, $JM = MJ$ has an inverse in Q and so does M. For an arbitrary integral A-ideal I (i.e. containing a regular element of A). Observe that $I \subset IM_1^{-1} \subset A$ for any maximal integral A-ideal $M_1$ containing I. If $I = IM_1^{-1}$ then $I = IM_1 = IM_1^2 = \ldots$ and $I \subset M_1^m$ for all $m \in \mathbb{N}$. By the lemma $A/I$ is an Artinian ring, hence $M_1^k = M_1^{k+1}$ for some $k \in \mathbb{N}$ and one easily derives that $M_1 = A$, a contradiction. So $IM_1^{-1} \subsetneq A$ and we may repeat the procedure until after a finite number of steps (because A is Noetherian) we obtain an integer $t \geq 1$ such that $IM_t^{-1} \cdot \ldots \cdot M_2^{-1} M_1^{-1} = A$. Hence $I = M_t \cdot \ldots \cdot M_2 M_1$ and by the first part of this proof it follows that I is invertible in Q. □

5.19. Corollary. If we assume that A is prime in the foregoing theorem then A is a maximal C-order in Q. As special cases we recover two well-known results for twisted group rings and crossed products (cf. I. Reiner, [122]).

5.20. Theorem. (Williamson, Harada) let R be a Dedekind domain with field of fractions K and let S be the integral closure of R in a Galois extension L of K with finite Galois group G. The crossed product $S \star G$ with respect to some factor set $f : G \times G \to U(S)$ and the Galois action of G on S is a hereditary order if and only if S/R is tamely ramified (i.e. if and only if there exists an $s \in S$ with $tr(s) = \sum_{\sigma \in G} \sigma(s) = 1$).

<u>Proof</u>. Direct from Corollary 5.2. applied to this speccial case. □

5.21. <u>Theorem</u>. (Auslander, Goldman, Rim). The twisted group ring S o G with respect to the Galois action of G on S is a maximal R-order if and only if S/R is an unramified extension, i.e. if and only if the discriminant ideal d(S/R) = R.

<u>Proof.</u> Apply Theorem 5.18. or Corollary 5.19. The other implication is easily checked. □

# E CLASS GROUPS OF KRULL ORDERS

## E.1. Some General Facts on Krull Orders.

In Section A.2. we reviewed some results on orders over Krull domains and since these orders were only assumed to be integral over their centers, we have stressed the fact that this theory generalizes some earlier results of L. Silver. However, the results of Section A.2. could also be viewed as a special case of results on general Krull orders or even vHC-orders. There are three more or less independent theories for noncommutative Krull rings. The noncommutative Krull rings introduced by M. Chamarie generalize the commutative theory from the "global" point of view, i.e. one studies maximal orders satisfying the ascending chain condition on certain "divisorial" ideals. The Krull rings in H. Marubayashi's sense are obtained through "local" considerations, i.e. as intersections of some suitable localizations. Whereas both theories are based on the theory of (maximal) orders in simple artinian rings, the third theory, developed mainly by E. Jespers and P. Wauters, starts from the relatively new concept of a symmetrical order in a simple ring and then uses these symmetrical orders in the study of noncommutative Krull rings in a way rather similar to Marubayashi's approach. It turns out that for PI -rings the three definitions coincide, so in the case we will study, we may draw from each one of the above theories. Whereas the properties mentioned in Section A.2. stand like stalagmites on the older theory, the properties in this section hang like stalactites from the recent general theories; so, although they describe the phenomena, their differences stem from the conceptual differences in the theory they are rooted in. The reader who has a special interest in the general theory of Krull orders may consult [ 29,30,50,54,74,75,76,77,78,92,93,94,95 ].

Let A be a maximal order in a simple artinian ring $\Sigma$. An ideal I of A is reflexive as a left ideal if and only if it is reflexive as a right ideal and so we may simply say I is reflexive (c-ideal in Chamarie's terminology). The set of reflexive ideals with product given by $I \star J = (IJ)^{\star\star}$ is an abelian group $D(A)$. If A satisfies the A.C.C. on (integral) reflexive ideals, then a reflexive ideal is prime if and only if it is a maximal reflexive ideal, moreover, $D(A)$ is the free abelian group generated by the reflexive prime ideals of A, each of these being a prime ideal of height 1 (all this is already in Asano's paper [ 10 ]). If A and B are maximal orders in $\Sigma$ which are context-equivalent (in particular, if they are Morita equivalent), then $D(A)$ and $D(B)$ are canonically isomorphic. Note also that every nonzero ideal of A is reflexive if and only if every ideal of A is invertible, i.e. if and only if A is an Asano order in $\Sigma$.

1.1. Lemma. If A is a maximal order in a simple Artinian ring $\Sigma$, then Z(A) is completely integrally closed. If A satisfies the A.C.C. on reflexive ideals of A, then Z(A) is a Kru l domain.

Proof. Easy, cf. [ 30 ]. □

1.2. Definition. (M. Chamarie [ 29 ]) A maximal order A in a simple Artinian ring $\Sigma$ is said to be a left Krull order if for every essential left ideal I of A there exists an essential left ideal J of finite type such that $I^{\star\star} = J^{\star\star}$. A left and right Krull order is said to be a Krull order.

From the above definition it follows easily that if one of two context-equivalent maximal orders is a Krull order, then so is the other. Consequently, every Krull order will be context-equivalent to a Krull order which is also a domain.

Now we restrict to orders in a central simple algebra $\Sigma$, i.e. we may assume that A is a prime PI ring. If A is a maximal order and a PI ring, then the fact that Z(A) is completely integrally closed entails that A is integral over its center (see W. Schelter [137] for example), i.e. A is a Z(A)-order in the sense of A.2. Let us now show that Krull PI orders in Chamarie's sense are just Fossum's Krull orders, as introduced in A.2.

1.3. Lemma. If A is a maximal order in the central simple algebra $\Sigma$, then A is a reflexive Z(A)-module of finite rank.

Proof. Since A is a torsion-free Z(A)-module of finite rank, we have $A^{**} = \{x \in \Sigma; \alpha(x) \in Z(A) \text{ for all } \alpha \in A^{*}\}$. It is also clear that A contains an equivalent order $A_1$ which is a free Z(A)-module of finite rank, hence $c\,A \subset A_1$ for some $0 \neq c \in Z(A)$. Consequently, $c\,A^{**} \subset A_1^{**} = A_1 \subset A$. Maximality of A and the equivalence of A and $A^{**}$ leads to a contradiction. □

1.4. Theorem. For any prime PI ring A, the following assertions are equivalent :

1. A is a Krull ring;
2. Z(A) is a Krull domain and A is a maximal Z(A)-order.

Proof. 1. ⇒ 2. Follows from the above remarks.
2. ⇒ 1. That A is a maximal order in $\Sigma$ follows from the fact that Z(A) is completely integrally closed. □

From the Lemma one easily deduces that A satisfies the A.C.C. on reflexive right (and left) ideals.
In general the noetherian hypothesis on A does not imply Z(A) to be noetherian (on the other hand, if Z(A) is noetherian, then A is a finitely generated Z(A)-module and hence A is certainly noetherian).
The following result provides a useful sufficient condition.

1.5. Proposition. Let A be a noetherian maximal order in a central simple algebra $\Sigma$ of PI degree n. If n is invertible in A, then Z(A) is noetherian and A has finite type as a Z(A)-module.

Proof. Since A is integral over Z(A) and Z(A) is completely integrally closed, $Tr(A) \subset Z(A)$ follows (where Tr is the reduced trace of $\Sigma$). If I is an ideal of Z(A) and $z \in AI \cap Z(A)$, then $Tr(z) \in I$ and $z = n^{-1}Tr(z) \in I$ follows; thus, $I = AI \cap Z(A)$. It is now readily verified that Z(A) is noetherian. A result of G. Cauchon's then entails that A is a finitely generated Z(A)-module. □

The $\Omega$-Krull rings considered by E. Jespers, P. Wauters are "divisorially Zariski central" so in the PI case we find the related properties for maximal orders over Krull domains by considering PI $\Omega$-Krull rings and showing that these are exactly the maximal orders over Krull domains. In order to avoid introducing a lot of new definitions we derive these results in the context of Chamarie Krull orders, following M. Chamarie's approach [30]. □

1.6. Lemma. (Going down). Let A be a Krull order in a central simple algebra $\Sigma$. Let $p \subset q$ be prime ideals of Z(A) and assume that there exists a prime ideal Q of A such that $Q \cap Z(A) = q$, then there is a prime ideal P of A such that $P \subset Q$ and $P \cap Z(A) = p$.

Proof. Let S be the multiplicatively closed subset of A generated by Z(A)-p and $\mathcal{C}(Q)$. As in the commutative case one shows that $A_p \cap S = \emptyset$ (e.g. Theorem 6, p. 262 in O. Zariski - P. Samuel [179]). Let P be an ideal of A containing Ap and maximal with respect to the property of not intersecting S. One easily verifies that P is a prime ideal of A such that $P \cap Z(A) = p$. □

1.7. Lemma. Let A be a Krull order in a central simple algebra $\Sigma$ and let P be a prime ideal of A, then P is reflexive if and only if P has height one or equivalently, if $p = P \cap Z(A)$ has height one. If this holds, then P is the unique prime ideal of A lying over p.

Proof. Since A is a maximal Z(A)-order as in Section A.2., the terms "reflexive" and "divisorial" (or "$\sigma_1$-closed") have the same meaning on Z(A)-lattices, and therefore, the first assertion is obvious. Lemma 1.6. implies the P.D.E. condition, i.e. ht(p) = 1. Now, $Z(A)_p$ is a discrete valuation ring and $A_p = Z(A)_p \otimes_{Z(A)} A$ has a unique maximal ideal $Z(A)_p \otimes_{Z(A)} P$ (since $A_p$ is a classical maximal order of $\Sigma$ over the discrete valuation ring $Z(A)_p$). The final assertion is an immediate consequence of this. □

In the situatiin of the foregoing lemmas A is integral over Z(A), hence this extension has the well-known properties : going up, incomparability and lying over. If one has going down and unique lying over, then the ring A would be Zariski central or globally birational in the sense of [150,152]. Recall that a ring extension $A \subset B$ is said to be birational if intersecting prime ideals of B with A defines a homeomorphism (for the Zariski topologies on Spec(B) and Spec(A) of some open $U \subset$ Spec(B) onto some open $V \subset$ Spec(A). If U = Spec(B) then the extension is said to be globally birational. In [119] E. Nauwelaerts and F. Van Oystaeyen showed that a prime PI ring is a birational extension of its center and related the open set of birationality to localizability properties of certain prime ideals. The definition of birationality may be modified by replacing Spec by $X^{(1)}$ and thus one obtains the notion of divisorial birationality. Since we have going down on $X^{(1)}(A)$ and unique lying over for $X^{(1)}(Z(A))$, one easily deduces that a Krull PI order A is globally divisorially birational extension of Z(A). From this, it is evident that localization of A at $P \in X^{(1)}(R)$ is a central localization (on all modules !), cf. [150]

for the proof of the birational case, and that A satisfies the left and right Ore conditions with respect to the multiplicative set $\mathcal{C}(P)$. We include the proof of the following Proposition, because it is independent of birationality properties (and also because it gives a complete characterization of the open set of birationality).

1.8. Proposition. Let A be a Krull order in the central simple algebra $\Sigma$ and let P be a prime ideal of A resp. $p = P \cap Z(A)$, then

1. $A_P = Z(A)_p \otimes_{Z(A)} A$, but this need not hold for arbitrary A-modules ;

2. The left (and right) Ore conditions with respect to $\mathcal{C}(P)$ hold if and only if P is the unique prime ideal lying over p,

Proof. 1. That $A_p$ is contained in $A_P$ is clear. Conversely, let $x \in A_P$, then AxA is a fractional A-ideal and therefore, there is a left A-ideal I of finite type such that $I \subset AxA$ and $I^{-1} = (AxA)^{-1}$. Therefore, $Iz \subset A$ for some $z \in \mathcal{C}(P)$ and so there is an ideal J of A such that $J \not\subset P$ and $x J \subset A$. In $\mathcal{D}(A)$ we write $J^{**} = \prod_{i=1}^{n} P_i^{(n_i)}$, where the $P_i$ are reflexive prime ideals and $(n_i)$ denotes the $n_i$-fold product in $\mathcal{D}(A)$. If $J^{**} \cap Z(A) \subset p$, then $P_i \cap Z(A) \subset p$, for some $i \in \{1,\dots,n\}$. By "going down" and the foregoing lemma, $P_i \subset P$ follows; therefore $J \subset P$, a contradiction. But $J^{**} \cap Z(A) \not\subset p$ entails that $x \in A_p$, hence $A_P = A_p = Z(A)_p \otimes_{Z(A)} A$.

2. If A satisfies the left Ore condition with respect to $\mathcal{C}(P)$, then $\mathcal{C}(P)$ consists of regular elements of A. It then follows that $P_P$ is the unique maximal ideal of $A_P$ and the first part yields that P is the unique prime ideal of A lying over p. Conversely, if P is the unique prime ideal of A lying over p, then $P_p$ is the unique maximal ideal of $A_p$ and thus $P_P$ is the unique maximal ideal of $A_P$. Since $A_p$ is a PI ring, it follows that $A_P$ has Jacobson radical $P_P$ and by a result of A. Heinicke [ 67 ] , it follows that A satisfies the (left) Ore condition with respect to $\mathcal{C}(P)$. □

1.9. Corollaries. 1. The kernel functor $\kappa_P$ is a central kernel functor (i.e. localization at P is the same as localization at p) if and only if A satisfies the Ore conditions to $\mathcal{C}(P)$.

2. If M is a divisorial Z(A)-lattice and an A-module, then $M_P \cong M_p$ for any prime ideal P of R.

3. If I is an essential right ideal of A, then $I^{**} = A$ if and only if $(I \cap Z(A))^{**} = Z(A)$. Indeed, $I^{**} = A$ if and only if $I A_P = A_P$ for any reflexive prime ideal P of A, hence $I \cap Z(A) \not\subset p$ for all $p \in X^{(1)}(Z(A))$ or $(I \cap Z(A))^{**} = Z(A)$.

4. If M is torsionfree of finite rank, then it is reflexive as a right A-module if and only if M is reflexive as a Z(A)-module. Indeed, from the assumption it follows that M is torsion free of finite rank as a Z(A)-module, so reflexivity of $M_A$ is equivalent to $M = \bigcap_{P \in X^{(1)}(A)} \underset{A}{\otimes} A_P = \bigcap_{p \in X^{(1)}(Z(A))} M_p$, i.e. to the reflexivity of $M_{Z(A)}$. □

We include the following result in order to present a very easy class of Krull orders e.g. reflexive Azumaya algebras over Krull domains.

1.10. Proposition. If A is a reflexive Azumaya algebra over a Krull domain C, then A is a maximal order. A reflexive Azumaya algebra over a Krull domain is an Azumaya algebra if and only if A is flat as a C-module.

Proof. The first statement appears also in [ 89 ] . A short proof may be obtained as follows. Suppose B is a maximal C-order containing A, then $A_p = B_p$ for every $p \in X^{(1)}(C)$ because $A_p$ is an Azumaya algebra over $C_p$. Since $A = \cap\{A_p ; p \in X^{(1)}(C)\}$, we have $A = B$. The second statement has essentially been proved in M. Orzech's paper [ 112] . We include the proof for completeness' sake. Assume first that A is flat in C-mod, then $A \underset{C}{\otimes} A^{opp} \to (A \underset{C}{\otimes} A^{opp})^{**} = Q_{\sigma_1}(A \underset{C}{\otimes} A^{opp}) = S$ is monomorphic. Consider

$\alpha = c^{-1}(\Sigma\lambda_i \otimes \mu_i) \in S$, where $c \in C$, $\lambda_i \in A$ and $\mu_i \in A^{opp}$. By the finite character property for C, the set $Y = \{p \in X^{(1)}(C);\ c \notin U(C_p)\}$ is finite. Write $\Gamma = \cap\{A_p^{opp}; p \in Y\}$ and $\Gamma' = \cap\{A_p^{opp}; p \in X^{(1)}(C)-Y\}$. Clearly, $\alpha \in A \otimes_C \Gamma'$. On the other hand flatness of A as a C-module entails that tensoring with A commutes with "finite intersections", $\alpha \in \cap\{A \otimes A^{opp})_p;$ $p \in X^{(1)}(C)-Y\} = A \otimes_C (\cap\{A_p^{opp}; p \in X^{(1)}(C)-Y\})$.

Therefore $\alpha \in (A \otimes_C \Gamma) \cap (A \otimes_C \Gamma')$ or $\alpha \in A \otimes_C A^{opp}$. The converse is obvious, since an Azumaya algebra over C is already a projective C-module. □

1.11. Proposition. Let A be a reflexive Azumaya algebra over a Krull domain C, then

1. $\mathcal{D}(A) \cong \mathcal{D}(C)$;
2. $\mathrm{Pic}_C(A,\sigma_1) \cong \mathrm{Cl}(C)$.

Proof. Taking into account the fact that $\mathrm{Cl}(C) = \mathrm{Pic}_C(C,\sigma_1)$, this is an easy consequence of (B.5.6.) □

1.12. Note. If A is a maximal order over a Krull domain C, such that the map $\rho^\star : \mathcal{D}(A) \to \mathcal{D}(C)$ which sends $I \in \mathcal{D}(A)$ to $I^A \in \mathcal{D}(C)$ is an isomorphism, then for every integral divisorial ideal J of A we have that $J = Q_{\sigma_1}(A(J \cap C))$, hence for every $p \in X^{(1)}(C)$ we have that all ideal of $A_p$ are generated by their central parts.

The above indicates that $\mathrm{Pic}_C(A,\sigma_1)$ is a first candidate to play the role of a class group in the noncommutative case. In the case of a reflexive Azumaya algebra this group reduces to $\mathcal{D}(A)/\mathcal{P}(A)$ where $\mathcal{P}_C(A)$ is the subgroup of $\mathcal{D}(A)$ of principal ideals generated by a central element. For more general maximal orders this equality breaks down and it is exactly the latter which we will study further as the "central class group" in the next section.

## E.2. The Central Class Group.

In this section we introduce the central class group of a maximal order over a Krull domain and derive some of its elementary properties. The results mentioned are given in the PI case, but they extend to general central $\Omega$-Krull rings in the sense of [76,77 ].

Let A be a maximal order over a Krull domain C in some central simple algebra $\Sigma$ with center K. In the free abelian group $D(A)$, we may consider $P_C(A)$ the subgroup generated by the principal ideals generated by a single central element and $I(A)$, the subgroup of invertible ideals (with respect to the common "product" of fractional ideal. We now define the central class group of A as :

$$\mathrm{CCl}(A) = D(A)/P_C(A).$$

By Corollary (A.3.6.) it follows that $I(A)/P_C(A) = \mathrm{Picent}(A) = \mathrm{Pic}_C(A)$, so we have a canonical embedding of Picent(A) into the central class group CCl(A) and a commutative diagram.

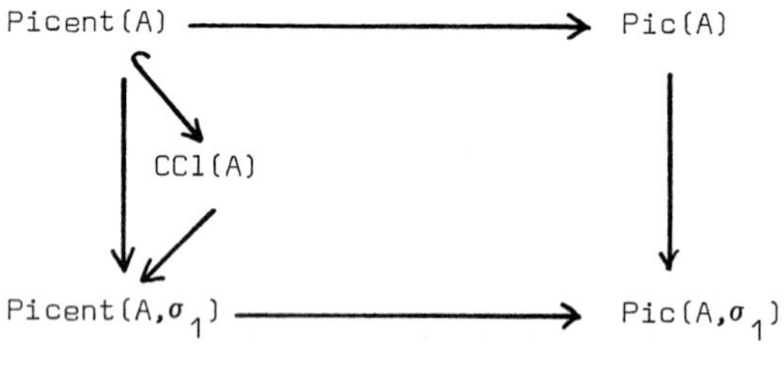


Consider a maximal order B over a Krull domain D and assume that $A \subset B$ is an extension in the sense of C. Procesi [116], i.e. $B = A.B^A$. We may define a set map $\psi : D(A) \to D(B)$ which sends I to $(BI)^{**}$. This $\psi$ need not be a group morphism, even in the case where A and B are commutative Krull domains ! However, $\psi|I(A)$ defines a morphism and it induces a group morphism $\tilde{\psi} : \mathrm{Picent}(A) \to \mathrm{Picent}(B)$. Another quite natural map

$\Phi: \mathcal{D}(A) \to \mathcal{D}(B)$ is obtained from the "divisor" point of view in the following way. If $P \in X^{(1)}(A)$ and $Q \in X^{(1)}(B)$, then we write $\overline{P}$ resp. $\overline{Q}$ for the unique maximal ideal of $A_P$ resp. $B_Q$. If Q lies over P, i.e. if $Q \cap A = P$, then $B_Q P = (\overline{Q})^{e_{P,Q}}$ for some positive integer $e_{P,Q}$, called the ramification index of Q over P. If $Q_1,\dots,Q_n$ are the primes in $X^{(1)}(B)$ lying over P, then we define $\Phi(P) = \prod_{i=1}^{n} Q_i^{e_i}$ where $e_i = e_{P,Q_i}$. Clearly, $\Phi$ extends to a map $\mathcal{D}(A) \to \mathcal{D}(B)$, which is a group morphism by definition. The difficulty here, is that it is not clear at all whether $\Phi(Ax) = Bx$ for any $0 \neq x \in Z(A)$. As in the commutative case, we say that the extension $A \subset B$ satisfies condition PDE if for all $P \in X^{(1)}(B)$ we have $ht(A \cap P) \leqslant 1$.

2.1. Proposition. The following statements are equivalent :

1. $\Phi = \psi$;
2. $\Phi(Ax) = Bx$ for all $0 \neq x \in Z(A)$;
3. The extension $A \subset B$ satisfies condition PDE.

Proof. 1. ⇒ 2. Indeed, $\Phi(Ax) = \psi(Ax) = (Bx)^{**} = Bx$.

2. ⇒ 3. Consider $Q \in X^{(1)}(B)$ and assume that $ht(Q \cap A) > 1$. Pick $x \neq 0$ in $Z(A) \cap Q$, then $Bx = \Phi(Ax) = \prod_{i=1}^{n} Q_1^{t_1} \cdot \dots \cdot Q_n^{t_n}$ in $\mathcal{D}(B)$ and since $ht(Q \cap A) > 1$, some of the appearing $Q_i$ equals Q. On the other hand, $Bx = \psi(Ax)$ has a decomposition in which Q must appear since $Bx \subset Q$ and $ht(Q) = 1$; this leads to a contradiction and therefore $ht(Q \cap A) \leqslant 1$. follows.

3. ⇒ 1. Consider $I = P_1^{s_1} \cdot \dots \cdot P_m^{s_m}$ in $\mathcal{D}(A)$. First we calculate $\psi(I) = \cap \{B_Q I ; Q \in X^{(1)}(B)\} (= Q_{\kappa_B}(BI))$. If $P_i = Q \cap A \in X^{(1)}(A)$, then $B_Q I = B_Q(A_{P_i} I) = B_Q(\overline{P}_i)^{s_i}$. If $Q \cap A = 0$ or if $I \not\subset Q \cap A$, then $B_Q I = B_Q$.

Write $Q_{j,i}$ for the primes in $X^{(1)}(B)$ lying over $P_i$ and $e_{j,i}$ for the ramification index of $Q_{j,i}$ over $P_i$. From the foregoing, it follows that

$\psi(I)$ and the ideal $J = \prod_j Q_{j,1}^{s_1 e_{j,1}} \cdot \ldots \cdot \prod_j Q_{j,m}^{s_m e_{j,m}}$ have the same localization at each $Q \in X^{(1)}(B)$, and as they are both divisorial, $\psi(I) = J$ follows. Secondly, $\Phi(P_1)^{s_1} \star \ldots \star \Phi(P_m)^{s_m}$ and by the definition of $\Phi$, it is clear that $\Phi(I) = J$, so $\Phi(I) = \psi(I)$. as desired. □

2.2. Corollary. 1. If the extension $A \subset B$ satisfies one of the equivalent conditions of Proposition 2.1., then $\psi$ induces a morphism $\tilde{\psi} : Cl(A) \to Cl(B)$ which restricts to $\tilde{\psi} :$ Picent(A) $\to$ Picent(B) introduced before.
2. If A is a maximal order over a Krull domain C, then $Cl(C) \subset CCl(A)$.

Proof. 1. Obvious.
2. In proposition 2.1., put B = A and A = C. Since for $P \in X^{(1)}(A)$ we have $p = P \cap C \in X^{(1)}(C)$ and moreover for any $p \in X^{(1)}(C)$ there is exactly one $P \in X^{(1)}(A)$ lying over p, condition PDE holds for the extension $C \subset A$. If $I \in D(C)$ is such that $\psi(I) = Ac$ for some $0 \neq c \in K$, then $\psi(Ic^{-1}) = A$. From $(AIc^{-1})^{\star\star} = A$ and the properties of the height one prime ideals mentioned above, it follows that $Ic^{-1} = C$, i.e. $I = Cc$. □

It is now very easy to describe the cokernel of the group morphism $Cl(C) \to CCl(A)$.

2.3. Proposition. For any maximal order A over a Krull domain C, the following sequence of groups is exact :

$$1 \to Cl(C) \to CCl(A) \to G = \bigoplus_P \mathbb{Z}/e_P\mathbb{Z} \to 1,$$

where the $e_P$ are the ramification indices of the essential calculations of C which ramify in $\Sigma = Q(A)$. The exponent of the group G is bounded by the PI degree of $\Sigma$.

Proof. If $p \in X^{(1)}(C)$, then $P = \psi(p)$ has the property that $(Ap)^{\star\star} = (P^{e_P})^{\star\star}$ for some positive integer $e_P$ and it is obvious that $e_P$ is the ramification index of the discrete valuation associated to $C_p$ in the central simple algebra $\Sigma$. Since $\psi(P(C)) = P_C(A)$, it follows easily from $1 \to D(C) \to D(A) \to \bigoplus_P \mathbb{Z}/e_P\mathbb{Z}$ that the sequence given in the statement of the proposition is exact. Note that the number of ramifying primes is finite i.e. G is finite. To verify this, it suffices to consider an arbitrary nonzero c in the Formanek center of A, cf. [116], and to note that $S = \{p \in X^{(1)}(C); c \in p\}$ is finite, while the localization at any $p \in X^{(1)}(C)-S$ is an Azumaya algebra (because its Formanek center will equal the center) and thus, these correspond to $P \in X^{(1)}(A)$ such that $e_P = 1$. The final claim is an easy consequence of PI theory e.g. it follows from a result of G. Bergman and L. Small's, but we sketch an alternative proof, using the reduced norm. For $P \in X^{(1)}(A)$, we have that $(nr(P))^{**} = (p^f)^{**}$ for some positive integer f, so from $(Ap^f)^{**} = (P^{e_P f})^{**}$ it follows that $(p^{fn})^{**} = (p^{(f^2 e_P)})^{**}$ and so, since $D(C)$ is the free group on the $p \in X^{(1)}(C)$, it follows that $n = fe_P$. □

2.4. <u>Corollary</u>. If A and B are maximal C-orders in the central simple algebra $\Sigma$, then $CCl(A) = CCl(B)$.

<u>Proof.</u> The group G in Proposition 2.3. depends only on certain ramification indices of valuations of K in $\Sigma$. □

2.5. <u>Notes</u> 1. It may be helpful to point out that an extension of Krull orders $A \subset B$ as in Proposition 2.1. satisfies condition PDE if and only if the extension $Z(A) \subset Z(B)$ does.

2. The foregoing corollary points out a weakness in the theory of the central class group, because, even if maximal orders need not be conjugated or even when they are not necessarily Morita equivalent, still they have the same class group.

3. If $\Sigma = M_n(K)$, then any maximal order A over a Krull domain C of K is a reflexive Azumaya algebra Indeed, $A_p$ being a maximal $C_p$-order for each

$p \in X^{(1)}(C)$, it is also an Azumaya algebra, so it follows easily that A is a reflexive Azumaya algebra.

4. As a consequence of Proposition 2.3. one easily derives that maximal orders over a Krull domain, which are Morita equivalent have the same class-group.

If A is a maximal order over a Krull domain C and $\kappa$ is an arbitrary idempotent kernel functor in A-mod, then $\kappa$ is symmetric because A is a PI ring. Since $Q_{\sigma_1}(A) = A$, we may assume that $\sigma_1 \leqslant \kappa$ and then, from the fact that A is $\sigma_1$-noetherian, we obtain $\kappa = \inf\{\kappa_{A-P} = \kappa_P; P \in \mathcal{P}\}$ for some $\mathcal{P} \subset X^{(1)}(A)$. From $Q_\kappa(A) = \bigcap_P Q_P(A)$ and Proposition 1.8.1., it follows that $Q_\kappa(A)$ is a subintersection of A and that $A \to Q_\kappa(A)$ is a (central) extension which satisfies condition PDE. The former statement may also be derived from M. Chamarie's result Theorem 4.2.4., p. 87 in [ 30 ]; the latter is a general fact about $\Omega$-Krull rings, cf. E. Jespers [ 74 ] . Also M. Chamarie's Theorem 4.2.4. in loc. cit. applied to the PI case, establishes that all localizations of A are subintersections and vice versa. For the reader's convenience, let us prove this special case here.

2.6. Lemma. Let A be a maximal order over a Krull domain C in the central simple algebra $\Sigma$ and consider an intermediate ring $A \subset B \subset \Sigma$. The following statements are equivalent :

1. B is a localization of A (on the left, say);
2. There is a subset $Y \subset X^{(1)}(A)$ such that $B = \cap \{A_P; P \in Y\}$.

Proof. We only have to prove that 1. implies 2. Let Y be the (possibly empty) subset of $X^{(1)}(A)$ consisting of all P such that $B \subset A_P$. Suppose $x \in \cap \{A_P; P \in Y\}$ and consider the largest ideal I of A such that $Ix \subset A$. Clearly I is divisorial. Since for each $P \in Y$ there is an ideal $J_P \not\subset P$ such that $J_P x \subset A$, it follows that $I \supset \sum_{P \in Y} J_P$ cannot be contained in any of the $P \in Y$. Write $I = P_1^{\nu_1} \cdot \ldots \cdot P_n^{\nu_n}$ for certain $P_i \notin Y$, i.e. $B \not\subset A_{P_i}$ for $i = 1,\ldots,n$. If $x_i \in B - A_{P_i}$ then $(A+Ax_iA)^{-1} \subset P_i$ and then $P_i^{-1} \subset (A+Ax_iA)^{**} \subset B$. Finally, this yields that $I^{-1} = (P_1^{-\nu_1} \cdot \ldots \cdot P_n^{-\nu_n})^{**} \subset B$ and as $x \in I^{-1}$ it follows that $x \in B$. □

2.7. Proposition. Assume that $A \subset B \subset \Sigma$ satisfies the conditions of the Lemma, then we have an exact sequence of the form

$$1 \to H \to CCl(A) \to CCl(B) \to 1$$

Moreover, H is generated by those $P \in X^{(1)}(A)$ such that $P \notin Y$.

Proof. From the remarks above, it is clear that $A \to B$ is an extension satisfying PDE, so we may consider the group morphism $\tilde{\psi} : Cl(A) \to Cl(B)$ defined in Corollary 2.2.. That $\psi : \mathcal{D}(A) \to \mathcal{D}(B)$ and $\tilde{\psi}$ are surjective is clear. If $[I] \in Ker(\tilde{\psi})$, then $\psi(I) = Bc$ for some $0 \neq c \in K$, i.e. $\psi(Ic^{-1}) = B$. Write $Ic^{-1} = P_1^{n_1} \star \ldots \star P_k^{n_k}$ in $\mathcal{D}(A)$. Now $\psi(Ic^{-1}) = B$ yields that $\psi(P_i) = B$ for all $i = 1,\ldots,k$, hence $P_i \in \mathcal{L}(\kappa)$. □

2.8. Some easy examples.

1. Let A be the ring of twisted polynomials $\mathbb{C}[X,-]$, i.e. defined by $zX = X\bar{z}$ for all $z \in \mathbb{C}$, then $C = \mathbb{R}[X^2]$. Here, $Cl(C) = \{1\}$ and $CCl(A) \cong \mathbb{Z}/2\mathbb{Z}$, because the only central valuation which ramifies is the one determined by $(X^2)$ and its ramification index is 2.

2. If $\Delta$ is a skewfield, then $CCl(\Delta[X,Y]) = 1$.

3. $C = \mathbb{Z}[\sqrt{-5}]$ and consider

$$A = \begin{pmatrix} \mathbb{Z}[\sqrt{-5}] & (2,1+\sqrt{-5}) \\ (2,1+\sqrt{-5})^{-1} & \mathbb{Z}[\sqrt{-5}] \end{pmatrix}$$

within $M_2(\mathbb{Q}[\sqrt{-5}])$, then $CCl(A) = Cl(C) = \mathbb{Z}/2\mathbb{Z}$.

4. The ring $\check{A}(\Phi)$ constructed in Section D.3., e.g. see Corollary 3.17., will have nicely behaved central class groups. To calculate these, we need some more machinery from the theory of graded rings and this is the topic of the next Section.

## E.3. Central Class Groups of Graded Krull Orders.

Whereas in the commutative theory, the fact that the Brauer group of a projective variety is given as a graded relative invariant (cf. Section VII.1. of [ 162 ]) provides a geometrical motivation to study graded relative invariants, the motivation to repeat this for noncommutative rings is of a more arithmetical nature. The aim of chapter D, in particular Section D.3., was to reduce problems about certain orders to problems about graded orders and then to solve the graded problems. For the latter we will only have information available on graded objects and this is the main reason why we have to consider graded invariants and compare them to their ungraded equivalents. We certainly do not strive for utmost generality, on the contrary, we will restrict attention to graded orders over Krull domains, i.e. all rings considered will satisfy a polynomial identity. In this case the graded invariants relate in the best possible way to their ungraded equivalents. For more general results, one may consult [ 97 ] .

The key lemma is the following generalization of Lemma D.3.13.

3.1. Lemma. Let A be a graded ring of type G satisfying the identities of n x n-matrices. If every central homogeneous element of A is invertible, then A is an Azumaya algebra over Z(A). (Note that we did *not* assume that Z(A) is graded too !).

Proof. Since A satisfies the identities of n x n matrices, it has a multilinear central polynomial f (e.g. the Razmylov-polynomial) which is <u>not</u> an identity. Hence f cannot vanish for every homogeneous substitution of the variables, say $c = f(\lambda_1,\dots,\lambda_r) \neq 0$. The assumptions imply that c is invertible and therefore the Formanek center of A equals Z(A), i.e. A is an Azumaya algebra. □

3.2. <u>Corollary</u>. If a PI ring A which is graded of type G is gr-simple, then it is an Azumaya algebra.

<u>Proof</u>. If $c \in Z(A)$ is homogenous in A, then $Ac = A$ and the lemma may be applied. □

3.3. <u>Lemma</u>. If A is a prime PI ring, which is graded of type G, then for every graded ideal I of A, there exists nonzero homogeneous elements in $I \cap Z(A)$.

<u>Proof</u>. Let $f(X_1,\dots,X_m)$ be the multilinear central Razmyzlov polynomial for A. If for every substitution by homogeneous $i_{\sigma_1},\dots,i_{\sigma_m}$ in I we have $f(i_{\sigma_1},\dots,i_{\sigma_m}) = 0$, then $f(i_1,\dots,i_m) = 0$ for every $i_1,\dots,i_m \in I$. Let $0 \neq c \in I \cap Z(A)$, then $c^N f(r_1,\dots,r_m) = 0$ for all $r_1,\dots,r_m \in A$, where N is the (total) degree of f. Since f is not an identity for A, it follows that $c^N = 0$, a contradiction. Therefore, $0 \neq f(i_{\sigma_1},\dots,i_{\sigma_m})$ for some homogeneous $i_{\sigma_1},\dots,i_{\sigma_m} \in I$ and it is clear that $f(i_{\sigma_1},\dots,i_{\sigma_m}) \in I \cap Z(A)$ is homogeneous.

3.4. <u>Corollary</u>. Let A be a maximal order over a Krull domain C. Let $Q^h$ be the graded ring of fractions obtained by inverting the elements of $Z(A) \cap h(A)$ and let $Q^g$ be the graded ring obtained by graded localization of A at the graded kernel functor $\tau$ with graded filter $\mathcal{L}(\tau)$ generated by the graded essential left ideals of A, then $Q^g = Q^h$ is an Azumaya algebra.

<u>Proof</u>. From Definition 1.2. it follows that $\tau$ has finite type, therefore $Q^g = Q^g_\tau(A) = Q_{\underline{\tau}}(A)$, where $\underline{\tau}$ is the (ungraded) kernel functor corresponding

to the filter of left ideals generated by the graded essential left ideals of A (for some detail on graded localization, see [ 109 ], II.9.,II.10, e.g. Proposition II.10.2.). By Lemma 2.6. we have $Q^g = \cap\{A_P; P \in X^{(1)}(A)$ and P contains no proper graded A-ideal$\}$ and $Q^h = \cap\{A_P; P \in X^{(1)}(A)$ and P contains no homogeneous central element of A$\}$. Lemma 3.1. yields that $Q^g$ is an Azumaya algebra. Bu Lemma 3.3. a prime $P \in X^{(1)}$ contains a graded ideal of A if and only if it contains an element $c \in h(A) \cap Z(A)$, hence $Q^g = Q^h$. □

3.5. Theorem. Let A be a maximal order over a Krull domain C and suppose that A is graded of type G, then we have an exact sequence

$$1 \to H \to CCl(A) \to Cl(Z(Q^h)) \to 1,$$

where H is the subgroup generated by the classes of the $P \in X^{(1)}(A)$ such that P contains an element of $h(A) \cap Z(A)$ i.e. p contains a nonzero element of $h(A)$, where $p = P \cap Z(A)$.

Proof. Apply Proposition 2.7. and note that $Q^h$ is an Azumaya algebra Note that $Z(Q^h) = S_h^{-1}C$, where $S_h$ is the multiplicative set of elements of C which are homogeneous in A, hence $Cl(Z(Q^h)) = Cl(C)/H_C$; where $H_C$ is the subgroup of Cl(C) generated by the classes of $p \in X^{(1)}(C)$ containing an element of $S_h$. □

3.6. Corollary. Let A be as in the theorem and assume that G is a torsion free abelian group, then Cl(A) is generated by the classes of $P \in X^{(1)}(A)$, which are graded.

Proof. The assumptions imply that C is graded too, hence $Z(Q^h)$ is a gr-field. Since $Z(Q^h)$ is also a Krull domain, it is clear that $Cl(Z(Q^h)) = 1$, cf. D. Anderson [ 3 ]. Moreover, if $P \in X^{(1)}(A)$ contains a graded ideal then $P_g \neq 0$ and since it is a(graded) prime ideal, as G is ordered, $P = P_g$ follows. □

For a maximal order A over a Krull domain C, which is graded of type G, we define the graded central class group $CCl_g(A)$ to be the subgroup of $CCl(A)$ generated by the classes of $P \in X^{(1)}(A)$ such that P contains a nonzero homogeneous element (or a graded ideal or *central* homogeneous element). Theorem 3.5. then yields the exact sequence

$$1 \to CCl_g(A) \to CCl(A) \to Cl(Z(Q^h)) \to 1$$

3.7. Example. Assume that A is as above and let G be torsion free abelian, then we may introduce the subgroup $\mathcal{D}_g(A)$ of $\mathcal{D}(A)$ consisting of the graded divisorial ideals of A. Formally in a way similar to the ungraded case, one may establish that $\mathcal{D}_g(A)$ is the free abelian group generated by $X_g^{(1)}(A)$, which consists of all $P \in X^{(1)}(A)$ with $P = P_g$. We may define $CCl_g(A)$ as $\mathcal{D}_g(A)/\mathcal{P}_c(A)_g$ and $Picent^g(A)$ as the subgroup of $Picent(A)$ generated by the graded invertible ideals of A. One easily checks that $CCl_g(A)$ defined this way is identical to the one defined before. In addition to the exact sequence deriving from Theorem 3.5., stating that $CCl_g(A) \cong CCl(A)$, we also derive that $Picent^g(A) \cong Picent(A)$. Moreover, if A is an extension of $A_e$, than $A_e$ is also a maximal order over a Krull domain.

Let us now turn to the study of some specific graded rings, in particular generalized Rees rings constructed over certain arithmetical rings like HNP rings, tame orders, Krull orders, ... . So, let us return to the terminology of relative maximal orders introduced in Section D.1. and we continue in the vein of Section D.3. Although the use of Lemma 3.1. (instead of the weaker version stated in Lemma D.3.13.) and of Lemma 3.3. enables us to get rid of a priori conditions on the grading group G, we do restrict attention to *torsion free abelian groups* in the rest of this section. This allows us to use the main results of Section D.3., in particular Theorem D.3.14, Corollary D.3.15, Theorem D.3.16 and Corollary D.3.17. Otherwise

of course, the gain in generality concerning G has to be paid for by imposing conditions on the generalized Rees ring constructed (e.g. like being a maximal order , a tame order, a hereditary order, ...).
Notations are as in Section D.1. and D.3; throughout A denotes a $\sigma$-Krull order and $\sigma$ and $\tau$ are supposed to be central G-invariant kernel functors.
The set of $\check{\sigma}$-divisorial ideals of $\check{A}(\Phi)$ which are graded submodules of $Q^g(\check{A}(\Phi))$ will be denoted by $D^g_{\check{\sigma}}(\check{A}(\Phi))$.

3.8. <u>Proposition</u>. The map $\chi : D_\sigma(A) \to D_{\check{\sigma}}(\check{A}(\Phi))$ defined by $L \to \check{L} = Q_\tau(\check{A}(\Phi)L)$ is a group isomorphism.

<u>Proof</u>. First check that $\check{L}$ is an ideal. Since $D_\sigma(A)$ is abelian, L commutes with each $I_\gamma$ and therefore, we only have to check that $LX_\gamma$ is in $Q_\tau(\check{A}(\Phi)L)$. Now, $LX_\gamma = X_\gamma(X_\gamma^{-1}LX_\gamma) = X_\gamma a_\gamma^{-1}La_\gamma$, where $a_\gamma$ induces $\varphi_\gamma$. Since $Aa_\gamma$ are $Aa_\gamma^{-1}$ are in $D_\sigma(A)$ we have $Q_\tau(Aa_\gamma^{-1}L) = (Aa_\gamma^{-1})\star L = L\star(Aa_\gamma^{-1}) = Q_\tau(La_\gamma^{-1})$, hence $LX_\gamma \subset Q_\tau(\check{A}(\Phi)L)$ follows. From $L \in F_\sigma(A)$ it follows that there is an $I \in L(\sigma)$ such that IL and LI are in $L(\sigma)$. By Lemma D.3.10 and Corollary D.3.11. we have $Q_\tau(\check{A}(\Phi)I) \in L(\check{\sigma})$ and $Q_\tau(\check{A}(\Phi)I)Q_\tau(\check{A}(\Phi)L) = (\bigoplus_{\gamma\in G} Q_\tau(I_\gamma I)X_\gamma)(\bigoplus_{\gamma\in G} Q_\tau(I_\gamma L)X_\gamma)$ $\subset \bigoplus_{\delta\in G} Q_\tau(I_\delta IL)X_\delta \subset \check{A}(\Phi)$, while $\check{A}(\Phi)I\check{A}(\Phi)L\check{A}(\Phi) = \bigoplus_{\lambda,\gamma,\delta\in G} I_\lambda II_\gamma LI_\delta X_{\lambda+\gamma+\delta} \in L(\check{\sigma})$. As it is contained in $\check{I}\check{L}$ we find $\check{I}\check{L} \in L(\check{\sigma})$ and similarly $\check{L}\check{I} \in L(\check{\sigma})$. Now, there exists $J \in L(\sigma)$ such that $J \subset L$, hence $\check{J} \subset \check{L}$ and $\check{J} \in F_{\check{\sigma}}(\check{A}(\Phi))$. If $\check{L}f \subset \check{A}(\Phi)$, then $Q^g(\check{A}(\Phi))f \subset Q^g(\check{A}(\Phi))$. But, $Q^g(\check{A}(\Phi))$ is a maximal order, hence $f \in Q^g(\check{A}(\Phi))$, say $f = \sum_{i=o}^{n} f_{\gamma_i}X_{\gamma_i}$ with $f_{\gamma_i} \in Q(A)$ and $\gamma_o < \dots < \gamma_n$. From $\check{L}f \subset \check{A}(\Phi)$, it follows that $Lf_{\gamma_n} \subset I_{\gamma_n}$, i.e. $f_{\gamma_n} \in L^{-1}\star I_{\gamma_n}$. This means that $\check{L}f_{\gamma_n} \subset \check{A}(\Phi)$ and $f_{\gamma_n} \in \check{A}(\Phi) .^{\cdot} \check{L}$, proving that $\check{A}(\Phi) .^{\cdot} \check{L}$ is a graded $\check{A}(\Phi)$-submodule of $Q^g(\check{A}(\Phi))$. Furthermore, an easy verification shows that $(\check{A}(\Phi) .^{\cdot} \check{L})_e = (A .^{\cdot} L)X_e$. It is now evident that $(\check{L})^{\star\star}$ is a graded $\check{A}(\Phi)$-submodule of $Q^g(\check{A}(\Phi))$, which is a divisorial $\check{A}(\Phi)$-ideal with part of degree e exactly $LX_e$, hence $\check{L} = (\check{L})^{\star\star}$ and $\check{L} \in D^g_{\check{\sigma}}(\check{A}(\Phi))$. One

One verifies similarly that for any B in $D^g_{\overset{\vee}{\sigma}}(\overset{\vee}{A}(\Phi))$ that we have $B_e = JX_e$ for some $J \in D_\sigma(A)$ and $Q_\tau(B) = \overset{\vee}{J}$. Since $\chi$ is clearly a group morphism, we only have to show that $Q_\tau(B) = B$ for any $B \in D^g_{\overset{\vee}{\sigma}}(\overset{\vee}{A}(\Phi))$. Actually, it will suffice to establish that $\overset{\vee}{A}(\Phi) \;.\!\!^{\cdot}\; H = B'$ is $\tau$-closed for any ideal H of $\overset{\vee}{A}(\Phi)$. We have $H_\gamma X_\gamma B'_\delta X_\delta \subset I_{\gamma\delta} X_{\gamma\delta}$ with obvious notation. Since $Q_\tau(I_{\gamma\delta} X_{\gamma\delta}) = I_{\gamma\delta} X_{\gamma\delta}$ and since $\tau$ is central, it follows that $H_\gamma X_\gamma Q_\tau(B'_\delta X_\delta) \subset I_{\gamma\delta} X_{\gamma\delta}$, hence $Q_\tau(B'_\delta X_\delta) \subset (\overset{\vee}{A}(\Phi) \;.\!\!^{\cdot}\; H)_\gamma$, or $Q_\sigma(B') = B'$ indeed. □

Let $P_{\overset{\vee}{\sigma}}(\overset{\vee}{A}(\Phi))$ be the subgroup of $D_{\overset{\vee}{\sigma}}(\overset{\vee}{A}(\Phi))$ consisting of principal ideals generated by a normalizing element of $N_{\overset{\vee}{\sigma}}(\overset{\vee}{A}(\Phi))$ and let $P^g_{\overset{\vee}{\sigma}}(\overset{\vee}{A}(\Phi))$ consist of all such principal ideals generated by an homogeneous normalizing element.

3.9. <u>Proposition</u>. For any $L \in D_\sigma(A)$, the following statements are equivalent :

1. $\overset{\vee}{L} \in P^g_{\overset{\vee}{\sigma}}(\overset{\vee}{A}(\Phi))$;
2. There exists a divisorial A-ideal $B \in \mathrm{Im}(\Phi)$, such that $L = Bx$ for some $x \in N_\sigma(A)$.

<u>Proof</u>. 1. ⇒ 2. Suppose $\overset{\vee}{L} = \overset{\vee}{A}(\Phi) x_\gamma X_\gamma$, where $x_\gamma X_\gamma \in N_{\overset{\vee}{\sigma}}(\overset{\vee}{A}(\Phi))$, then $AX_e x_\gamma X_\gamma = x_\gamma X_\gamma AX_e$, i.e. $A\varphi_e A\varphi_e(x_\gamma)c_{e,\gamma} = x_\gamma Ac_{\gamma,e}$. Thus, $Ax_\gamma = x_\gamma A$, since $\overset{\vee}{A}(\Phi)x_\gamma \in D_{\overset{\vee}{\sigma}}(\overset{\vee}{A}(\Phi))$, i.e. $x_\gamma \in N_\sigma(A)$. Furthermore, $(\overset{\vee}{L})_e = LX_e = I_{\gamma^{-1}} \varphi_{\gamma^{-1}}(x_\gamma)c_{\gamma^{-1},\gamma} X_e = I_{\gamma^{-1}} x_\gamma X_e$, hence $L = I_{\gamma^{-1}} x_\gamma$.

2. ⇒ 1. If $B \in \mathrm{Im}(\Phi)$, then $B = I_\gamma$ for some $\gamma \in G$. We claim that $\overset{\vee}{B} = \overset{\vee}{A}(\Phi)X_{\gamma^{-1}}$. Indeed. $\overset{\vee}{A}(\Phi)X_{\gamma^{-1}} = Q_\tau((\overset{\vee}{A}(\Phi)X_{\gamma^{-1}})_e \overset{\vee}{A}(\Phi)) = Q_\tau(I_\gamma \overset{\vee}{A}(\Phi)) = \overset{\vee}{B}$. Finally, if $L = I_\gamma x$ for some $x \in N_\sigma(A)$, then $\overset{\vee}{L} = Q_\tau(\overset{\vee}{A}(\Phi)I_\gamma)x = \overset{\vee}{A}(\Phi)X_{\gamma^{-1}}x$, hence $\overset{\vee}{L} \in P^g_{\overset{\vee}{\sigma}}(\overset{\vee}{A}(\Phi))$. □

3.10. <u>Theorem</u>. If we define $NCl_\sigma(A) = D_\sigma(A)$ and $NCl^g_{\overset{\vee}{\sigma}}(A) = D_{\overset{\vee}{\sigma}}(\overset{\vee}{A}(\Phi))/P^g_{\overset{\vee}{\sigma}}(\overset{\vee}{A}(\Phi))$, then we have the following exact sequence :

$$0 \to \pi(\mathrm{Im}(\Phi)) \to \mathrm{NCl}_\sigma(A) \to \mathrm{NCl}^g_{\check\sigma}(\check A(\Phi)) \to 0,$$

where $\pi : D_\sigma(A) \to \mathrm{NCl}_\sigma(A)$ is the canonical morphism.

Proof. This follows immediately from the foregoing. □

Recall that we assumed $\sigma$ to be central and inducing a perfect localization. From Section D.3. we retain that $Q_\sigma(A)[X_\gamma,\varphi,c]$ is a Krull $\tilde\sigma$-order, where $L(\tilde\sigma)$ consists of all ideals I of $Q_\sigma(A)[X_\gamma,\varphi,c]$ with $C_e(I) = Q_\sigma(A)$.

3.11. Proposition. With assumptions as before, the following sequence is split exact :

$$0 \to D^g_{\check\sigma}(\check A(\Phi)) \to D_{\check\sigma}(\check A(\Phi)) \to D_{\tilde\sigma}(Q_\sigma(A)[X_\gamma,\varphi,c]) \to 0$$

Proof. Let $\tilde\tau$ be associated to $\tilde\sigma$ and consider $P \in P(\tilde\sigma)$. We claim that $P \cap \check A(\Phi) \in P(\tilde\sigma)$. That $P \cap \check A(\Phi) \in L(\check\sigma)$ follows from the fact that $Q_\sigma(C_e(P \cap \check A(\Phi))) = C_e(P) = Q_\sigma(A)$ and thus $C_e(P \cap \check A(\Phi)) \in L(\sigma)$. It is also obvious that $P \cap \check A(\Phi)$ is a prime ideal. Now, first we check that any $\tilde\sigma$-divisorial ideal I of $Q_\sigma(A)[X_\gamma,\varphi,c]$ has the property that $I \cap \check A(\Phi)$ is a $\tilde\sigma$-divisorial ideal of $\check A(\Phi)$. Since $\check A(\Phi)$ is a Krull $\check\sigma$-order, the divisorial ideal of $\check A(\Phi)$. Since $\check A(\Phi)$ is a Krull $\check\sigma$-order, the divisorial closure of any ideal is given by letting $Q_{\check\tau}(-)$ act. Now, $Q_{\check\tau}(I \cap \check A(\Phi)) = I \cap \check A(\Phi)$ because, if $Jx \subset \check A(\Phi) \cap I$ for some $J \in L(\check\tau)$ and $x \in Q_\sigma(A)[X_\gamma,\varphi,c]$, then $\check A(\Phi)x = Q_{\check\tau}(J)x \subset Q_{\check\tau}(I \cap \check A(\Phi))$and we have to distinguish the following cases :

a. $J_g \in L(\check\sigma)$. Here $Q_\sigma(A)[X_\gamma,\varphi,c]\, Jx \subset I$ and from $Q_\sigma(A)[X_\gamma,\varphi,c]\, J = Q_\sigma(A)[X_\gamma,\varphi,c]$, it then follows that $x \in I \cap \check A(\Phi)$.

b. $J_g \notin L(\check\sigma)$.In this case, $J' = Q_\sigma(A)[X_\gamma,\varphi,c]\, J \in L(\tilde\tau)$, because $J' \in L(\tilde\sigma)$ and because $J' \subset P$ for some $P \in P(\tilde\sigma)$ implies that $J \subset P \cap \check A(\Phi) \in P(\check\sigma)$, a contradiction, hence $J'x \subset I$ and also $Q_{\tilde\tau}(J')x \subset I$, i.e. $x \in I \cap \check A(\Phi)$ !

Now, knowing that $P \cap \check{A}(\Phi)$ is divisorial, we derive that $P \cap \check{A}(\Phi) \in P(\check{\sigma})$, for if $P \cap \check{A}(\Phi)$ is not a maximal divisorial ideal contained in $\check{A}(\Phi)$, then $P \cap \check{A}(\Phi) \subset P_1$ for some $P_1 \in P(\check{\sigma})$ and so $P = Q_\sigma(A)[X_\gamma,\varphi,c]\ (P \cap \check{A}(\Phi))$ is contained in a divisorial ideal $P_1 Q_\sigma(A)\ [X_\gamma,\varphi,c] = P'$, leading to $Q_\tau(P_1' \cap \check{A}(\Phi)) = \check{A}(\Phi)$ and thus $P_1 = \check{A}(\Phi)$, a contradiction.

The exactness of the above sequence is easily verified and the splitting map $D_{\check{\sigma}}\ (Q_\sigma(A)[X_\gamma,\varphi,c]) \to D_{\check{\sigma}}(\check{A}(\Phi))$ is obtained by just intersecting down divisorial ideals. □

Let $CCl^g_{\check{\sigma}}(\check{A}(\Phi)) = D^g_{\check{\sigma}}(\check{A}(\Phi))/P^g_c(\check{A}(\Phi))$ resp. $CCl_{\check{\sigma}}(\check{A}(\Phi)) = D_{\check{\sigma}}(\check{A}(\Phi))/P_c(\check{A}(\Phi))$, where $P^g_c(\check{A}(\Phi))$ is the group of principal ideals in $\check{A}(\Phi)$ generated by a homogeneous central element and $P_c(\check{A}(\Phi))$ the group of principal ideals of $\check{A}(\Phi)$ generated by an arbitrary central element.

3.12. Lemma. There is a natural inclusion $CCl^g_{\check{\sigma}}(\check{A}(\Phi)) \to CCl_{\check{\sigma}}(\check{A}(\Phi))$.

Proof. Easy ! □

3.13. Theorem. As before, let A be a Krull $\sigma$-order, then

$$CCl_{\check{\sigma}}(\check{A}(\Phi)) = CCl^g_{\check{\sigma}}(\check{A}(\Phi)) \oplus CCl_{\tilde{\sigma}}(Q_\sigma(A)[X_\gamma,\varphi,c]).$$

Proof. From the split exact sequence

$$0 \to D^g_{\check{\sigma}}(\check{A}(\Phi)) \to D_{\check{\sigma}}(\check{A}(\Phi)) \to D_{\check{\sigma}}(Q_\sigma(A)[X_\gamma,\varphi,c]) \to 0$$

and the foregoing Lemma, we immediately deduce the following exact diagram :

$$\begin{array}{ccccccccc}
 & & 0 & & 0 & & 0 & & \\
 & & \downarrow & & \downarrow & & \downarrow & & \\
0 & \longrightarrow & P^g_c(\check{A}(\Phi)) & \longrightarrow & P_c(\check{A})(\Phi) & \longrightarrow & P_c(Q_\sigma(A)[X_\gamma,\varphi,c]) & \longrightarrow & 0 \\
 & & \downarrow & & \downarrow & & \downarrow & & \\
0 & \longrightarrow & D^g_{\check{\sigma}}(\check{A}(\Phi)) & \longrightarrow & D_{\check{\sigma}}(\check{A}(\Phi)) & \longrightarrow & D_{\tilde{\sigma}}(Q_\sigma(A)[X_\gamma,\varphi,c]) & \longrightarrow & 0 \\
 & & \downarrow & & \downarrow & & \downarrow & & \\
0 & \longrightarrow & \mathrm{CCl}^g_{\check{\sigma}}(\check{A}(\Phi)) & \longrightarrow & \mathrm{CCl}_{\check{\sigma}}(\check{A}(\Phi)) & \longrightarrow & \mathrm{CCl}_{\tilde{\sigma}}(Q_\sigma(A)[X_\gamma,\varphi,c]) & \longrightarrow & 0 \\
 & & \downarrow & & \downarrow & & \downarrow & & \\
 & & 0 & & 0 & & 0 & &
\end{array}$$

which proves the assertion. □

3.14. Remarks.

1. If $L(\sigma) = L(A-\{0\})$ then $Q_\sigma(A)[X_\gamma,\varphi,c]$ is just $Q(A)[X_\gamma,\varphi,c]$ is just $Q(A)[X_\gamma,\varphi,c]$ and every divisorial ideal is generated by a central element, because $Q(A)[X_\gamma,\varphi,c]$ is an Azumaya algebra over a factorial domain. Then $\mathrm{CCl}(Q_\sigma(A)[X_\gamma,\varphi,c]) = 1$ and $\mathrm{CCl}^g_{\check{\sigma}}(\check{A}(\Phi)) = \mathrm{CCl}_{\check{\sigma}}(\check{A}(\Phi))$. We conjecture that $\mathrm{CCl}_{\tilde{\sigma}}(Q_\sigma(A)[X_\gamma,\varphi,c])$ is always trivial in the situation we consider here, i.e. $\sigma$ is a central kernel functor inducing a perfect localization.

2. We have an exact sequence

$$0 \to \mathrm{Ker}(\pi) \to \mathrm{CCl}_{\check{\sigma}}(\check{A}(\Phi)) \to \mathrm{NCl}_{\check{\sigma}}(\check{A}(\Phi)) \to 0$$

where $\mathrm{Ker}(\pi)$ is a torsion group.

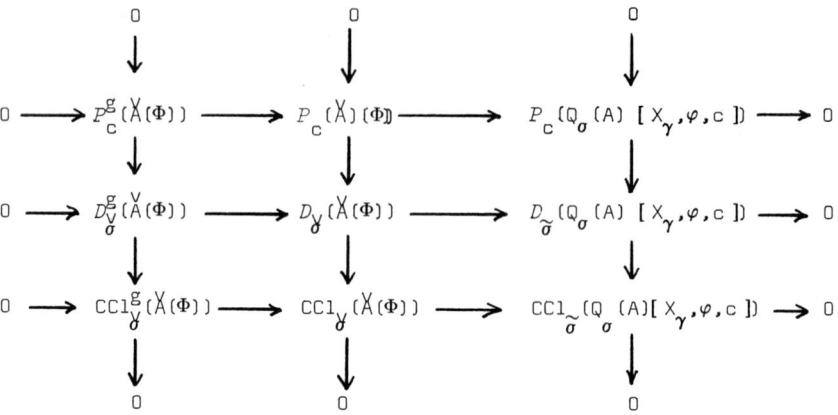

If we apply the relative results obtained above to the case where A is a maximal order over a Krull domain, i.e. $L(\sigma)$ is the filter generated by all nonzero ideals and $L(\tau)$ is the filter corresponding to $X^{(1)}(Z(A))$, then we recover particular cases of some results mentioned earlier, e.g. Theorem 3.5. The gain in the abstract relative presentation we have chosen is that we may also consider tame orders or HNP rings for A and still obtain the results for the relative class groups $CCl_\sigma$ and $NCl_\sigma$. In view of the results of Section D.4., in particular Remark 4.25., it is clear that one may generalize the results above to arbitrarily divisorially graded rings, if one restricts the group G to be of the form $\mathbb{Z}^n$; at least, this is true for all results dealing with $CCl_\sigma$, whereas one may expect some problems for the description of $NCl_\sigma$ because the latter depends by definition on the normalizing elements. Another possible generalization is obtained by considering divisorially graded rings A of type G, for arbitrary G, with the additional condition that A is a $\check{\sigma}$-Krull order, $A_e$ a $\sigma$-Krull order (e.g. in case A is of the form $\check{B}(\Phi)$); with the appropriate definition of $CCl_\sigma^g$ and $CCl_{\check{\sigma}}^g$ in the vein of the definition of $CCl^g$ preceding Example 3.7. one may then deduce modified versions of Lemma 3.12. and Theorem 3.13., but we see no point in going into these details here.

If $L(\sigma)$ is the filter generated by all nonzero ideals, in particular in case A is a maximal order over a Krull domain, we write NCl(A) for $NCl_\sigma(A)$ and call it the <u>normalizing class group</u> of A. We will come back to this new concept in a separate Section.

## E.4. Central Class Group versus Class Group of the Center.

The usefulness of the central class group depends on the fact whether one is able to describe those maximal orders A over Krull domains for which CCl(A) = Cl(Z(A)) holds. In view of Proposition 2.3. this condition holds if none of the essential valuations of Z(A) ramify in $\Sigma$ = Q(A); therefore one could at first sight be lead to believe that A has to be a reflexive Azumaya algebra in this case. On the other hand, it is almost obvious that this cannot be true if one knows about a class of rings introduced by G. Azumaya ; these rings need not be Azumaya algebras, but they are Zariski central, while they also satisfy an unramifiedness condition, as the one we encounter here, but even at *all* prime ideals.

In this section, we reprove some of the results stated in [88 ] (the title of this paper is left entirely to the responsability of the author), but we introduce some simplifications and corrections, where necessary.

The first case to consider is the local case.

4.1. Lemma. Let A be a maximal order over a discrete valuation ring C with maximal ideal m. If CCl(A) = 1, in other words : if A is unramified, then either A is an Azumaya algebra over C, or else the center of A/mA is a purely inseparable extension of C/mC.

Proof. If Z(A/mA) = C/m, then the PI degree of mA equals the PI degree of A and so by the Artin-Procesi theorem [ 5,116 ], it follows that A is an Azumaya algebra. The converse is obvious. Now, the existence of separable elements in Z(A/mA) over C/m entails the existence of nonzero polynomials p', q' in Z(A/mA)[ t ] lying over the same separable p in (C/mC)[ t ], hence it leads to prime ideals P,Q of A[ t ] such that $P \cap A = mA = Q \cap A$, while P and Q lie over the same prime ideal of C[t]. If A[t] satisfies the Ore conditions with respect to $\mathcal{C}(P)$, then we may apply Proposition 1.8.2. to the Krull order A [ t ] and derive a contradiction. From Proposition 1.8.1. it follows that $Q_P(P) = A[t]_P$, and that it is the unique maximal ideal of $A[t]_P$ is equally easy to verify. So $Q_P(P)$ is the Jacobson radical of A [ t ] and moreover $A[t]_P/Q_P(P) = A[t]/P = (A/mA)[t]/P'$ for some P' generated by an irreducible polynomial $p(t) \in Z(A/mA)[t]$. Since the elements of $\mathcal{C}(P)$ are invertible modulo the Jacobson radical, they are invertible in $A[t]_P$ and so it is a classical fact that this implies the Ore conditions with respect to $\mathcal{C}(P)$, cf. [67 ] for example, which we may apply, since A[ t ] is noetherian. □

4.2. Remark. There is another straightforward proof for the foregoing lemma, completely in terms of the theory of maximal orders. Indeed, first reduce to the complete case and then, using Hensel's lemma, one may lift a proper separable subextension $\overline{S}$ of C/mC in Z(A/mA) to an extension S of C. The tensorproduct of an unramified maximal C-order with a separable unramified extension of C yields again a maximal order. Therefore the assumption $\overline{S} \neq C/mC$ will lead to a contradiction if one observes that the residual algebra of the tensorproduct must split up, because of the appearance of nontrivial idempotents stemming from $\overline{S} \otimes_{C/mC} \overline{S}$.
In the absence of the condition CCl(A) = Cl(C) = 1, this proof may be modified, so as to yield equality between the degree of the separable closure of C/mC in Z(A/mA) and the ramification index of C in Q(A), i.e. the order of CCl(A).

4.3. Lemma. Let A be a maximal order over a discrete valuation ring C, the the following properties are equivalent :

1. A is an Azumaya algebra;

2. CCl(A) = 1 and there exists a separable splitting field L of $\Sigma$ = Q(A) such that $A \otimes_C D$ is hereditary, where D is the integral closure of C in L.

Proof. 1. ⇒ 2. If A is an Azumaya algebra, then so is $A \otimes_{\bar{C}} D$ and therefore $A \otimes_C D$ is a maximal order, hence a hereditary D-order.

2. ⇒ 1. Suppose that Z(A/mA) is a purely inseparable field extension of C/mC. From the fact that the canonical map Br(C/mC) → Br(Z(A/mA)) is epimorphic in this case, it follows that upto considering suitable matrix rings over A or Q(A), we may assume that A/mA contains contains a C/mC-central simple algebra B such that $A/mA = B \otimes_{C/mC} Z(A/mA)$. The inverse image $A_1$ of B in A is a C-order contained in A and mA is a common ideal of A and $A_1$. Actually, it is clear that mA is the unique nonzero prime ideal of $A_1$ and that $Z(A_1/mA)$ = C/mC. By definition D is a semilocal Dedekind

domain and its nonzero prime ideals $m_1,\dots,m_q$ are the ones lying over mD. Let $O_j$ be the discrete valuation ring $D_{m_j}$ for $j = 1,\dots,q$ and write $m_j'$ for the maximal ideal of $O_j$. Localizing the inclusions $(A \otimes_C D)m \subset A_1 \otimes_C D \subset A \otimes_C D$, at $m_j$ say, we obtain $(A \otimes_C O)m_j' \subset A_1 \otimes_C O_j \subset A \otimes_C O_j$. Since $(A_1 \otimes_C O_j)/(A \otimes_C O_j)m_j' = B \otimes_{C/mC} (D/m_jD)$ is central simple, it follows that $(A \otimes_C O_j)m_j'$ is the unique nonzero prime ideal of $A_1 \otimes O_j$. Therefore, ideals of $A \otimes_C O_j$ intersect $A_1 \otimes_C O_j$ in subsets of $(A \otimes_C O_j)m_j'$.
Now, the unique prime ideal of $Z(A/mA) \otimes_{C/mC} (D/m_jD)$ is the nilradical because $Z(A/mA)/(C/mC)$ is purely inseparable. Consequently, there is only one prime ideal of $A \otimes_C O_j$ lying over $(A \otimes_C O_j)m_j'$.
However, Harada's determination of the structure of hereditary orders in matrix rings (see also M. Artin [8,9]) learns that the number of prime ideals of $A \otimes_C O_j$ lying over $m_j'$ is at least as big as the number of diagonal blocks of the type $M_{n_j}(O_j)$ appearing in the structure of $A \otimes_C O_j$ (which is hereditary as $A \otimes_C D$ was). This proves that $A \otimes_C O_j = M_n(O_j)$, where n is the PI degree of A, but then an easy descent argument proves that A is an Azumaya algebra. Alternatively one deduces that $A \otimes_C D$ is a reflexive Azumaya algebra over the Dedekind domain D, hence it is an Azumaya algebra over D and so is an Azumaya algebra as well. Therefore, $Z(A/mA) = C/mC$ follows. Since we have checked both possibilities occuring in Lemma 4.1., this finishes the proof. □

4.4. <u>Corollary</u>. (of the proof). If in Lemma 4.3.2. one just requires $A \otimes_C D$ to be a tame D-order, the result remaines valid. □

4.5. <u>Remark</u>. It is possible to derive the equivalent of Lemma 4.3. in case 2. is replaced by 2. CCl(A) = 1 and $\Sigma$ may be split by an étale extension S of C. The proof remains unaltered at the crucial points. It is fundamental in all cases considered, that one obtaines a grip on the structure of

$(S/mS) \otimes_{C/mC} Z(A/mA)$.

After the foregoing, the treatment of the global case does not cause any new problems.

If A is a maximal order over a Krull domain C in the central simple algebra $\Sigma = Q(A)$ and if L is a separebla splitting field of $\Sigma$, then we say that A = L-tame if $A \otimes_C D$ is a tame D-order, where D is the integral closure of C in L. We say that A is Zariski-tamifiable (terminology of [87], also left to the responsibility of the author), if for every $p \in X^{(1)}(C)$, there exists a separable splitting field L(p) of $\Sigma$ such that $A \otimes_{C_p} S(p)$ is a tame S(p)-order, where S(p) is the integral closure of $C_p$ in L. We have :

4.5. Theorem. The following properties of A are equivalent :

1. A is a reflexive Azumaya algebra over C;
2. CCl(A) = Cl(C) and A is Zariski tamifiable;
3. CCl(A) = Cl(C) and A is L-tame for some separable splitting field L of $\Sigma$.

Proof. The implications 1. ⇒ 3. ⇒ 2. are obvious, so let us prove that 2. ⇒ 1. For any $p \in X^{(1)}(C)$, we have that $A_p$ is a maximal $C_p$-order such that $CCl(A_p) = 1$. The assumptions entail that $A_p$ satisfies the conditions of Corollary 4.4., hence $A_p$ is an Azumaya algebra over $C_p$. Being a maximal order over a Krull domain, A does satisfy the necessary relative finiteness conditions and so we may deduce from the local conditions that A is a reflexive Azumaya algebra. □

For any $p \in X^{(1)}(C)$ one may go to the complete local case at p and we use the fact that $\Sigma$ then has an unramified splitting field, $L_u$ say. If $S_u$ is the integral closure of $\hat{C}_p$ in $L_u$, then unramifiedness of both A and

$S_u$ entails that $A \otimes_{\hat{C}_p} S_u$ is a maximal order in $M_n(L_u)$, hence an Azumaya algebra.

If $S_u$ were separable or even projective over $\hat{C}_p$, then we would deduce from this that A is an Azumaya algebra too. Note that separability of $S_u$ contradicts the inseparability of Z(A/mA) over $\hat{C}_p/p\hat{C}_p = \overline{C}$, because otherwise, there would be a commutative subfield $Z(A/mA) \otimes_{\overline{C}} \overline{S}_u$ (linear disjointness) of rank n over the center of A/mA. This remark makes it clear that in theorem 4.6.2. there are obvious candidates for the fields L(p).

Let us now include the examples mentioned by D. Saltman in [ 135] , which show that pure inseparability phenomena can indeed prevent orders A with CCl(A) = Cl(C) from being reflexive Azumaya algebras.

4.7. Examples. 1. Equicharacteristic case. Let F be any field of characteristic $p \neq 0$ and put $K = F((t))$ resp. $C = F[[t]]$. Let {a,b} be contained in a p-basis for F over the prime field and let $\Delta$ be the cyclic algebra $[at^{-p},b]$. Pick $\alpha \in \Delta$ such that $\alpha^p - \alpha = at^{-p}$, then $K(\alpha)/K$ is a field extension with residual extension $F(a^{1/p})/F$. Since $b \notin F(a^{1/p}))^p$, clearly, be cannot be a norm of $K(\alpha)$ and therefore $\Delta$ is a skewfield. There exists a valuation ring A (in O.Schilling's sense [ 139] ) in $\Delta$ over C and with CCl(A) = 1, but one easily verifies that $A/At = F(a^{1/p}, b^{1/p})$.

2. Characteristic 0 case. Suppose char(K) = 0 and that K has a residue field F with char(F) = p. Up to adding one, we may assume that K contains a p-th root of unity, $\omega$ say. Let {a,b} be contained in a p-basis for F over the prime field and choose $a', b' \in K$ mapping to a,b resp. Let $\Delta$ be the cyclic algebra $[a',b']$ defined over K. Again the discrete valuation of K extends to an unramified valuation ring A of $\Delta$ such that A/Am equals $F(a^{1/p}, b^{1/p})$.

E.5. The Normalizing Class Group.

At the end of Section 3 we introduced the normalizing class group and a relative version of it. In [ 89 ] L. Le Bruyn started to study this invariant in some detail, but since we do not aim to connect these invariants to Weil and Cartier divisors on noncommutative structure sheaves as in [ 89 ], we only present the ring theoretical features here. The obvious ingredients of the theory of normalizing class groups are normalizing elements and conjugation of maximal orders. Therefore we start by giving an old result, probably due to C. Chevalley. We do not claim that our proof is original, but it is certainly short.

5.1. Lemma. Let R be a Dedekind domain with field of fractions K, then the following statements are equivalent:

1. all maximal R-orders in $M_n(K)$ are conjugated;
2. for every $[I] \in Cl(R)$, there is a $[J] \in Cl(R)$ such that $[J^n] = [I]$.

Proof. Since $M_n(R)$ is an Azumaya algebra, all maximal R-orders in $M_n(K)$ are Azumaya algebras too, since they are Morita equivalent to $M_n(R)$. The injectivity of $Br(R) \to Br(K)$ entails that every maximal order of $M_n(K)$ is of the form $End_R(P)$ form some projective R-module P of rank n. If $End_R(P) \cong M_n(R) \cong End_R(R^n)$, then $P \cong R^n \underset{R}{\otimes} J \cong J \oplus \dots \oplus J$ for some $[J] \in Cl(R)$, by Morita theory. On the other hand, if $I_1, \dots, I_n \in Cl(R)$, then $Q = I_1 \oplus \dots \oplus I_n$ yields a maximal order $\Lambda = End_R(Q)$ in $M_n(K)$. The latter order will be isomorphic to $M_n(R)$ if and only if $I_1 \oplus \dots \oplus I_n \cong J \oplus \dots \oplus J$ for some $[J] \in Cl(R)$. By Steinitz' theorem $I_1 \cdot \dots \cdot I_n = J^n$ will follow from the above isomorphism of R-modules. Since every projective R-module P of rank n is of the form $I_1 \oplus \dots \oplus I_n$ as above, the Lemma is evident now. □

5.2. Remark. If R is any ring, let $Aut_n(R)$ be the group of automorphisms of the R-algebra $M_n(R)$ and let $Inn_n(R)$ be the subgroup of inner automorphisms of $M_n(R)$, then $Aut_n(R)/Inn_n(R) \cong Pic_n(R)$. The proof of this uses the same techniques as the ones used above.

5.3. Corollary. If R is a Dedekind domain with field of fractions K then we obtain an exact sequence of abelian groups

$$1 \longrightarrow Cl_n(R) \longrightarrow Cl(R) \xrightarrow{\varphi} Cl(R) \longrightarrow t_R(M_n(K)) \longrightarrow 1,$$

where $\varphi([I]) = [I^n]$, and where for a central simple algebra $\Sigma$ we denote the set of conjugacy classes of maximal R-orders in $\Sigma$ by $t_R(\Sigma)$. The advantage of the proof given for Lemma 5.1. is that it extends easily to a more general situation of particular interest to us. In the sequel we suppose that R is a Krull domain. Let $Ref_n(R)$ be the set of isomorphism classes (of R-modules) of reflexive R-modules of rank n. The group Cl(R) acts on $Ref_n(R)$ as follows : $[I] \cdot [P] = [(P \otimes_R I)^{**}]$, where $[I] \in Cl(R)$ and $[P] \in Ref_n(R)$.

5.4. Proposition. There is a bijective correspondence between $t_R(M_n(K))$ and the orbits of the action of Cl(R) on $Ref_n(R)$.

Proof. If A is a maximal R-order in $M_n(K)$, then A is reflexive Azumaya since $A_p$ is an Azumaya algebra for every $p \in X^{(1)}(R)$. Since $\beta(R) \hookrightarrow Br(K)$, it follows that $A = End_R(P)$ for some reflexive R-module P of rank n. By the relative Morita theory developed in Chapter B and C, cf. (C.2.13.) for example, we know that $End_R(P) \cong End_R(Q)$ if and only if there exists an $[I] \in Cl(R)$ such that $P \cong (Q \otimes_R I)^{**}$. □

5.5. Corollary. 1. All maximal R-orders in $M_n(K)$ are conjugated if and only if every $[P] \in Ref_n(R)$ is of the form $[I \oplus \dots \oplus I]$ for some $I \in Cl(R)$.

2. If R is a regular local ring of global dimension at most 2, then all maximal R-orders in $M_n(K)$ are conjugated (M. Ramras [118].

Proof. 1. If $\mathrm{End}_R(P) \cong \mathrm{End}_R(R^n)$, then $P \cong (R^n \otimes_R I)^{**} = I \oplus \dots \oplus I$ and the converse is obvious.

2. The conditions on R imply that reflexive modules are projective. □

Conjugation defines an equivalence relation on the set of maximal orders and the orders in the conjugacy class of some maximal R-orders A in $\Sigma = Q(A)$ are parametrized by $\Sigma^*/N(A)$, whare $N(A)$ is the group of A-normalizing elements in $\Sigma$. It is possible to extend the terminology of K. Roggenkamp [132] to our situation. If M,N are torsion free left A-modules, which are divisorial R-lattices, then we say that M and N are in the same genus if for every $p \in X^{(1)}(R)$ we have that $M_p$ and $N_p$ are isomorphic as $A_p$-modules. The genus-class of M will be denoted by $G(M)$. It is possible to relate elements of $C(A)$ to idèles in $\Sigma$, where an idèle in $\Sigma$ is a family $\{q_p; p \in X^{(1)}(R),\ q_p \in \Sigma^*$ and almost all $q_p = 1\}$.

5.6. Remark. The elements of $G(A)$ correspond bijectively to the idèles in $\Sigma$. Indeed, if $M \in G(A)$ then, upto isomorphism, we may assume that $KM = \Sigma$, hence $M_p = \Lambda_p$ for almost all $p \in X^{(1)}(R)$, say for $p \in Y \subset X^{(1)}(R)$. For $p \notin Y$, we see that $M_p \cong A_p$, i.e. $M_p = A_p a_p$ for some $a_p \in \Sigma^*$. So, to $M \in G(A)$ we associate the idèle $\{a_p; p \in X^{(1)}(R)\}$, where we put $a_p = 1$ if $p \in Y$. If both M and N have the same idèle associated to them, then $M = \cap\{M_p; p \in X^{(1)}(R)\} = \cap\{N_p; p \in X^{(1)}(R)\} = N$. On the other hand, for a given idèle $\{a_p; p \in X^{(1)}(R)\}$ define $M_p = A_p a_p$ and let $Y \subset X^{(1)}(R)$ be the set of all p for which $a_p = 1$. If $B = \cap\{A_p; p \in Y\}$, then $N = B \cap M_{p_1} \cap \dots \cap M_{p_k}$, where $\{p_1,\dots,p_k\} = X^{(1)}(R) - Y$, is a left A-module, which is a divisorial R-lattice such that $N_p = M_p$ for all $p \in X^{(1)}(R)$ as is easily verified.

If h(A) denotes the pointed set of isomorphism classes of left ideals of A, then we have :

5.7. Proposition. If A is a maximal order over a Krull domain R, then there is an exact sequence of pointed sets

$$1 \longrightarrow \mathrm{NCl}(A) \xrightarrow{i} h(A) \xrightarrow{\rho} t_R(\Sigma) \longrightarrow 1$$

(it is obvious how these sets are to be considered as pointed sets).

Proof. Define a map of pointed sets $i : \mathrm{NCl}(A) \to h(A)$ by mapping $[I]$ to the (left) isomorphism class $< I >$ of I. If $[I]$ maps to $< A >$, then $I = An$ for some $n \in \Sigma$ and it is clear that $n \in N(A)$ since I is twosided and A is a maximal order. Consequently, i is a monomorphism of pointed sets. Define the map of pointed sets $\rho$ by mapping the isomorphism class $< L >$ of a left divisorial A-ideal to the conjugacy class of $O_r(L) = \{x \in \Sigma;\ Lx \subset L\}$. It is easily seen that $\rho$ is well defined. Moreover, if $\Gamma$ represents a class in $t_R(\Sigma)$, then $A \mathbin{.^{\boldsymbol\cdot}} \Gamma$ is a divisorial R-lattice, a left A-ideal and a right $\Gamma$-ideal; therefore $O_r(A \mathbin{.^{\boldsymbol\cdot}} \Gamma) = \Gamma$ and we may conclude that $\rho$ is surjective. It remains to verify exactness at h(A). Suppose $O_r(L) = x^{-1}Ax$ for some $x \in \Sigma^{\star}$, then $A = xO_r(L)x^{-1} \subset O_r(Lx^{-1})$ yields $A = O_r(Lx^{-1})$, or $Lx^{-1}$ is a (twosided) divisorial A-ideal. Therefore $< L > = < Lx^{-1} >$ and $\mathrm{Ker}(\rho) \subset \mathrm{Im}(i)$ follows. Since $\rho$ does not depend upon the choice of representative, the converse inclusion is obvious. □

In [ 89 ] L. Le Bruyn goes on to provide a (non abelian cohomological interpretation of the terms h(A) and $t_R(\Sigma)$ in the above sequence, but the value of this method remains to be established.

The natural morphisms $M_n(A) \to M_m(M_n(A))$ define a partial ordering on $\{M_n(A); n \in \mathbb{N}\}$. All of these ring morphisms are extensions in the sense of C. Procesi [ 116] and moreover, they satisfy the PDE condition. We define the asymptotical class group of A by $\mathrm{NCl}_\infty(A) = \varinjlim \mathrm{NCl}(M_n(A))$. If R is a Krull domain, then the kernel of the canonical morphism

$\varphi_n : Cl(R) \to NCl(M_n(R))$ consists of n-torsion elements as is easily seen by taking reduced norms in $M_n(I) = M_n(R)a$, where a is a normalizing element in $M_n(K)$. Note also that such an n-torsion element has to lie in Pic(R).

5.8. Proposition. (M. Van den Bergh). If R is a commutative Krull domain of finite Krull dimension, then $NCl_\infty(R) = Cl(R)/Tors(Pic(R))$.

Proof. It suffices to note that if $[I] \in Cl(R)$ has the property that $M_n(I) \cong M_n(R)$, then by Morita theory $I \oplus \dots \oplus I \cong R \oplus \dots \oplus R$, so $I \otimes P \cong P$ for $P = R \oplus I \oplus I^{\otimes 2} \oplus \dots \oplus I^{\otimes (n-1)}$, so, as P is faithfully projective, there is a Q with $P \otimes Q \cong R^p$, as R has finite Krull dimension. It follows that $I \oplus \dots \oplus I \cong R \oplus \dots \oplus R$, whence the assertion. □

In particular, for a Dedekind domain R we have $NCl_\infty(R) \cong Cl(R) \otimes_{\mathbb{Z}} \mathbb{Q}$.

5.9. Proposition. Let A be a maximal order over a Krull domain R of finite Krull dimension, then

1. $Cl_\infty(R) \hookrightarrow NCl_\infty(A)$;
2. the kernel of $\psi : NCl(M_n(R)) \to NCl(M_n(A))$ consists of m-torsion elements, where m is in the PI degree of A.

Proof. 1. If $[I] \in Cl(R)$ is such that $M_n(A)I = M_n(A)q_n$ for some $M_n(A)$-normalizing element $q_n$, then $[I]$ is nm-torsion, where m is the PI-degree of A. Thus it suffices to invoke Proposition 5.8. to finish the proof.

2. Since $D(M_n(R)) \cong D(R)$, we may choose an $I \in D(R)$ representing an element of $Ker(\psi)$. Then $[I]$ is mn-torsion in Cl(R) and since $Ker(\varphi_n) = Cl_n(R)$, it follows that $I^m$ represents the identity of $NCl(M_n(R))$. □

We conclude this section by proving an equivalent of Theorem 3.5. for the normalizing class group. After Corollary 3.6. we have defined a graded central class group $CCl_g(A)$; now, we define $NCl_g(A)$ for any maximal order A (over a Krull domain R) graded by an arbitrary group G, to be the quotient of $CCl_g(A)$ by the subgroup of principal ideals generated by a homogeneous normalizing element, and we call it the <u>graded normalizing class group</u> of A.

5.10. <u>Theorem.</u> If the K-central simple algebra $\Sigma$ is of the form $M_r(\Delta)$, where $\Delta$ is a skewfield, then we consider $A = M_r(B)$ for a maximal order B of $\Delta$ over the Krull domain R. If B (hence A) is graded by a t.u.p. group G (e.g. an ordered group), then we obtain the following exact sequences of abelian groups

a. $1 \to NCl_g(B) \to NCl(B) \xrightarrow{\pi_B} NCl(Q^g(B)) \to 1$

b. $1 — NCl_g(A) \to NCl(A) \xrightarrow{\tilde{\pi}_A} NCl(M_n(Q^g(B))) \to 1$

<u>Proof</u>. a. We only have to check that $Ker(\pi_B) \subset NCl_g(B)$. Since $1 \to CCl_g(B) \to CCl(B) \to CCl(Q^g(B)) \to 1$ is exact, we only have to verify that if $Bb = bB$, where b is a B-normalizing element, contains a homogeneous element $b_\tau$, then b is homogeneous. If $b_\tau = xb = (x_{\tau_1} + \ldots + x_{\tau_n})(b_{\sigma_1} + \ldots + b_{\sigma_m})$, then the t.u.p. property of G and the fact that B has no zero divisors entail that B is homogeneous.

b. Since for any maximal order A over a Krull domain R the morphism $NCl(A) \to NCl(M_n(A))$ is surjective, making use of the exact sequence in a, we obtain an exact commutative diagram

$$\begin{array}{ccccccccc}
1 & \to & NCl_g(B) & \to & NCl(B) & \xrightarrow{\pi} & NCl(Q^g(B)) & \to & 1 \\
 & & \downarrow & & \downarrow & & \downarrow & & \\
1 & \to & NCl_g(A) & \to & NCl(A) & \longrightarrow & NCl(Q^g(A)) & & \\
 & & & & & & \downarrow & & \\
 & & & & & & 1 & &
\end{array}$$

where $Q^g(A) = M_n(Q^g(B)) = Q^g(M_n(B))$. Indeed, it suffices to check exactness at NCl(A). If $J \in \mathcal{D}(B)$ represents an element mapping to $[I] \in Ker(\pi_A)$, then $\pi_B([J]$ is n-torsion, i.e. $[J^n] \in NCl_g(B)$, but then $[J] \in NCl_g(B)$ follows, hence $[I] \in NCl_g(A)$.

Of course, it follows that $\pi_A$ is surjective too, proving the exactness in b. □

5.11. Remark. Since $Q^g(B)$ is an Azumaya algebra, $NCl(Q^g(B))$ is very easy to calculate. Moreover, the calculation of $NCl_g(B)$ is also much easier than that of NCl(B), in particular, when G is torsion free abelian. Of course, the draw-back of this method is that not every maximal R-order in $M_n(\Delta)$ is of the form $M_n(B)$ as above. Although any other R-order is context-equivalent to one of the form $M_n(B)$, the behaviour of NCl(-) and $NCl_g(-)$ under context equivalence or even Morita equivalence is not clear. From here one should therefore strive towards more intrinsic characterizations of properties of NCl(-), which may lead towards the study of *one-sided* invariants. This seems a very promising topic for research, but it does not fall under the twosided scope of the present book.

# REFERENCES


[1] Amitsur, S.A., Homology Groups and Double Complexes for Arbitrary Fields, J. Math. Soc. Japan 14 (1962) 1-25.

[2] Anderson, D.D., Anderson, D.F., Divisibility properties of graded domains, Canad. J. Math. 24 (1982) 196-215.

[3] Anderson, D.F., Graded Krull Domains, Comm. in Algebra 7 (1979) 79-106.

[4] Anderson, F., Fuller K., Rings and Categories of Modules, Springer Verlag, New York, 1974.

[5] Artin M., On Azumaya algebras and finite dimensional representations of rings, J. of Algebra, 11 (1969) 532-536.

[6] Artin M., On the Joins of Hensel Rings, Adv. in Math. 7 (1971) 282- 296.

[7] Artin M., Integral homomorphisms of pi-rings, Proceedings Durham, 1979.

[8] Artin, M., Local structure of maximal orders on surfaces, LNM 917 146-181, Springer Verlag, 1982.

[9] Artin, M., Left ideals in maximal orders, LNM 917, 182-193, Springer Verlag, Berlin, 1982.

[10] Asano K., Arithmetische Idealtheorie in nichtkommutativen Ringen, Jap. J. Math. 15 (1939) 1-36

[11] Auslander B., Finitely generated reflexive modules over integrally closed noetherian domains, ph. d. thesis, Univ. of Michigan, 1963

[12] Auslander B., The Brauer group of a ringed space, J. of Algebra 4 (1966) 220-273.

[13] Auslander B., Central separable algebras which are locally endomorphism rings of free modules, Proc. A.M.S. 30 (1971) 395-404.

[14] Auslander M., Buchsbaum D., Unique factorization in regular local rings, Proc. Nat. Acad. of Sci. USA 45 (1959) 733-734.

[15] Auslander M., Goldman O., Maximal orders, Trans. A.M.S. 97 (1960) 1-24.

[16] Auslander M., Goldman O., The Brauer Group of a Commutative Ring, Trans. A.M.S. 97 (1960) 367-409.

[17] Azumaya G., On Maximally Central Algebras, Nagoya Math. J. 2 (1951) 119-150.

[18] Bass H. Algebraic K-Theory, Benjamin, New York, 1968.

[ 19 ] Bass H., Lectures on topics in algebraic K-theory, Tata Institute of Fundamental Research, Bombay, 1967.

[ 20 ] Bass H., Murthy P., Grothendieck groups and Picard groups of Abelian group rings, Ann. of Math. 86 (1967) 16-73.

[ 21 ] Bourbaki N., Algèbre, Eléments de Mathématique, 4,6,7,11,14,23, 24, Hermann, Paris, 1947-59.

[ 22 ] Bourbaki N., Algèbre Commutative, Eléments de Mathématique 27, 28, 30, 31, Hermann, Paris, 1961-66.

[ 23 ] Braun A., Affine P.I. rings and their generalizations, J. Algebra 33 (1975) 435-462.

[ 24 ] Caenepeel S., A Cohomological Interpretation of the Graded Brauer Group I, Comm. in Algebra 11(1983) 2129-2149.

[ 25 ] Caenepeel S., A Cohomological Interpretation of the Graded Brauer Group II, to appear.

[ 26 ] Caenepeel S., Van den Bergh M., Van Oystaeyen F., Generalized Crossed Products Applied to Maximal Orders, Brauer Groups and Related Exact Sequences, J. Pure Appl. Algebra, to appear.

[ 27 ] Caenepeel S., Verschoren A., A Relative Version of the Chase-Harrison-Rosenberg Sequence, to appear.

[ 28 ] Cartan E., Eilenberg S., Homological Algebra, Princeton University Press, Princeton, 1956.

[ 29 ] Chamarie M., Anneaux de Krull non commutatifs, J. of Algebra 72 (1981), 210-222.

[ 30 ] Chamarie M., Anneaux de Krull non commutatifs, Thèse, Université Claude Bernard, Lyon I, 1981.

[ 31 ] Chase S., Harrison P., Rosenberg A., Galois Theory and Cohomology of Commutative Rings, Mem. AMS 52 (1965) 1-19;

[ 32 ] Chase S., Rosenberg A, Amitsur Cohomology and the Brauer Group, Mem. AMS 52 (1965) 20-45.

[ 33 ] Chatters A.W., Heinicke A.G., Localization at a Torsion Theory in Hereditary Noetherian Rings, Proc. Cosidon Math. Soc. 27 (1973) 193-204.

[ 34 ] Chouinard L.G., Krull Semi groups and Divisor Class Groups, Canad. J. Math. 23 (1981) 1459-1468.

[ 35 ] Claborn L., Fossum R., Generalizations of the Notion of Class Groups, Illinois J. Math. 12 (1968) 228-253.

[ 36 ] Claborn L., Fossum R., Class Groups of n-Noetherian Rings, J. of Algebra 10 (1968) 263-285.

[ 37 ] Cohen M., Montgomery S., Group-graded Rings, Smash Products and Group Actions, Preprint.

[38] Cohn P., Algebra II, J. Wiley, London 1977.

[39] Dade E., Compounding Clifford's Theory, Ann. of Math., 91 (1970) 236-270.

[40] Dade E., Group Graded Rings and Modules, Math. Zeitschrift, 174 (1980) 241-262.

[41] De Meyer F., Projective Modules over Central Separable Algebras, Canad J. Math. 21 (1969) 39-43.

[42] De Meyer F., Ingraham E., Separable Algebras over Commutative Rings, LNM 181; Springer Verlag, Berlin 1971.

[43] Deuring M. Algebren, Springer Verlag, Berlin, 1935.

[44] Dieudonné J, Grothendieck A., Eléments de Géométrie Algébrique I, II, III, Publ. Math. IHES, 1960-63.

[45] Eisenbud D., Robson J.C., Modules over Dedekind Prime Rings, J. Algebra 16 (1970) 67-85.

[46] Eisenbud D., Robson J.C., Hereditary Noetherian Prime Rings, J. Algebra 16 (1970) 86-104.

[47] Faith C., Algebra, Rings, Modules and Categories I, Springer Verlag Berlin, 1973.

[48] Formanek E., Central Polynomials for Matrix Rings, J. Algebra, 23(1972) 129-132.

[49] Formanek E., Noetherian P.I. Rings, Comm. Algebra 1,(1974) 79-86.

[50] Fossum R., Maximal Orders over Krull Domains, J. Algebra 10 (1968) 321-332.

[51] Fossum R., The Divisor Class Group of a Krull Domain, Springer Verlag, Berlin 1973.

[52] Fröhlich A, The Picard Group of Noncommutative Rings, in particular of Orders, Trans. AMS, 180(1973) 1-45.

[53] Fröhlich A., Reiner I, Ullom S., Class Groups and Picard Groups of Orders, Proc. London Math. Soc. 180 (1973) 405-434.

[54] Fujita H., A Generalization of Krull Orders, to appear.

[55] Gabriel P., Des Catégories Abeliennes, Bull. Soc. Math. France, 90(1962) 323-448.

[56] Gabriel P.,Popescu, N., Caractérisation des catégories abéliennes avec générateurs et limites inductives exactes, C.R. Ac. Sc. Paris, 258 (1964) 4188-4190.

[57] Giraud J., Cohomologie non abélienne, Springer Verlag, Berlin, 1971.

[ 58 ] Godement R., Théorie des Faisceaux, Hermann, Paris, 1958.

[ 59 ] Golan J., Localization in Noncommutative Rings, M. Dekker, New York, 1975.

[ 60 ] Goldman O., Rings and Modules of Quotients, J. of Algebra 13 (1969) 10-47.

[ 61 ] Goldman O., Elements of Noncommutative Arithmetic I, J. Algebra 35 (1975) 308-341.

[ 62 ] Grothendieck A., Sur quelques points d'algèbe homologique, Tohoku Math. J. 9 (1957) 119-221.

[ 63 ] Grothendieck A, Local Cohomology, LNM 1, Springer Verlag, Berlin, 1967.

[ 64 ] Hajarnavis C.R., Lenagan T.H., Localization in Asano Orders, J. Algebra 21 (1972) 441-449

[ 65 ] Harada M. Hereditary Orders, Trans AMS, 107 (1963) 273-290.

[ 66 ] Hartshorne R. Algebraic Geometry, Springer Verlag, Berlin, 1977

[ 67 ] Heinicke A.G., On the Ring of Quotients at a Prime Ideal of a Right Noetherian Ring, Canadian J. Math. 24 (1972) 703-712.

[ 68 ] Herstein, I. Noncommutative Rings, Carus Math., Monographs 15 Math. Ass. Amer. 1968.

[ 69 ] Hutchinson J., Turnidge D. Morita Equivalent Quotient Rings, Comm in Algebra 4, (1976)669-675.

[ 70 ] Iversen, B., Noetherian Graded Modules, Aarhus Preprint 29, 1972.

[ 71 ] Jacobson N., Lectures in Abstract Algebra III, Van Nostrand, Princeton, 1964.

[ 72 ] Jacobson N., p.i. Algebras ,an Introduction, LNM 441, Springer Verlag, Berlin, 1975.

[ 73 ] Janusz, Separable Algebras over Commutative Rings, Trans AMS, 122 (1966) 461-479.

[ 74 ] Jespers E., On Geometrical $\Omega$-Krull Rings, Comm. in Algebra, to appear.

[ 75 ] Jespers E., Le Bruyn L., Wauters P, $\Omega$-Krull rings I, Comm. in Algebra 10(1982)1801-1818.

[ 76 ] Jespers E., Le Bruyn L., Wauters P., A characterization of Central $\Omega$-Krull Rings, J. Algebra 81 (1983) 165-179.

[ 77 ] Jespers E., Wauters P., On Central $\Omega$-Krull rings and their class group, to appear.

[78] Jespers E., Wauters P., Marubayashi-Krull Orders and Strongly Graded Rings, J. Algebra, 86 (1984) 511-521.

[79] Kanzaki T., On Generalized Crossed Products and Brauer Groups, Osaka J. Math. 5 (1968) 175-188.

[80] Kaplansky I., Commutative Rings, Allyn and Bacon, 1970, Boston, Mass.

[81] Knus M., A Teichmüller Cocycle for finite Extensions, preprint.

[82] Knus M., Ojanguren M., Théorie de la Descente et Algèbres d' Azumaya, LNM 389, Springer Verlag, Berlin, 1974.

[83] Knus M., Ojanguren M., Cohomologie étale et groupe de Brauer in LNM 844, 210-228, Springer Verlag, Berlin, 1981.

[84] Kuzmanovich J., Localization of Dedekind prime rings, J. of Algebra 21(1972) 378-393.

[85] Lambek J., Michler G., The Torsion Theory at a Prime Ideal of a Right Noetherian Ring, J. Algebra 25 (1973) 364-389.

[86] Le Bruyn L., Van Oystaeyen F., Generalized Rees Rings and Relative Maximal Orders Satisfying Polynomial Identities, J. Algebra 83 (1983) 409-436.

[87] Le Bruyn L., A note on maximal Orders over Krull Domains, Proc. E. Noether Days, North Holland, to appear.

[88] Le Bruyn L., On the Jespers-Van Oystaeyen Conjecture, J. Algebra to appear.

[89] Le Bruyn L., Class Groups of Maximal Orders over Krull Domains, Ph. D. Thesis, University of Antwerp, UIA, 1983.

[90] Lenegan T.H. Bounded Hereditary Noetherian Prime Rings, J. London Math. Soc. 6 (1973) 241-246.

[91] MacLane S., Categories for the Working Mathematician, Springer Verlag, Berlin, 1971.

[92] Marubayashi H., Noncommutative Krull Rings, Osaka J. Math. 13 (1976) 491-500.

[93] Marubayashi H., Polynomial Rings over Krull Orders in Simple Artinian Rings, Hokkaido Math. J., 9 (1980) 63-78.

[94] Marubayashi H., A characterization of Bounded Krull Prime Rings, Osake J. Math. 15 (1978) 13-20.

[95] Marubayashi H., A Krull Type Generalization of HNP Rings with enough Invertible Ideals, Comm. in Algebra, 11 (1983) 469-500.

[96] Marubayashi H., Remarks on v HC-Orders in a Simple Artinian Ring, Preprint.

[97] Marubayashi H., Nauwelaerts E., Van Oystaeyen, F., Graded Rings over Maximal Orders, Comm. in Algebra, to appear.

[98] Maury G., Raynaud J., Ordres Maximaux au sens de K. Asano, LNM 808, Springer Verlag, Berlin, 1980.

[99] Maury G., La condition "intégralement clos" III : sur les ordres maximaux au sens d'Asano non nécessairement réguliers, J. Algebra 25 (1973) 522-533.

[100] Michler G., Asano Orders, Proc. London Math. Soc. 19 (1969) 421-443.

[101] Milnor J., Introduction to Algebraic K-Theory, Princeton University Press, Princeton, New Jersey, 1971.

[102] Mueller B., The quotient category of a Morita context, J. Algebra 28, (1974) 389-407.

[103] Murdoch D., Van Oystaeyen F., Noncommutative Localization and Sheaves, J. Algebra 35 (1975) 500-515.

[104] Nagata H., Local Rings,Interscience Tracts in Pure and Appl. Math., 13, New York, 1962.

[105] Năstăsescu C., Anneaux et Modules Gradués, Rev. Roum. Math.7(1967) 911-931.

[106] Năstăsescu C., Strongly Graded Rings of Finite Groups.Comm. in Algebra 11 (1983) 1033-1072.

[107] Năstăsescu C., Van Oystaeyen F., On Strongly Graded Rings and Crossed Products, Comm in Algebra,10 (1982) 2085-2106.

[108] Năstăsescu C., Van Oystaeyen F., Graded and Filtered Rings and Modules, LNM 758, Springer Verlag, Berlin, 1980.

[109] Năstăsescu C, Van Oystaeyen F., Graded Ring Theory, North Holland, 1982.

[110] Nauwelaerts E., Van Oystaeyen F., Birational HNP Rings, Comm. in Algebra 8 (1980) 309-338.

[111] Nauwelaerts E. , Van Oystaeyen F., Finite Generalized Crossed Products over Tame and Maximal Orders, to appear.

[112] Orzech M., Brauer Groups and Class Groups for a Krull Domain, in LNM 917, 68-90,Springer Verlag, Berlin 1982.

[113] Orzech M., Small C., The Brauer Group of Commutative Rings, M. Dekker, New York, 1975.

[114] Orzech M., Verschoren A., Some Remarks on Brauer Groups of Krull Domains, in LNM 917, 91 - 96, Springer Verlag, Berlin, 1982.

[115] Paré R., Schelter W., Finite Extensions are Integral, J. Algebra 53 (1978) 477-479.

[116] Procesi C., Rings with Polynomial Identities, M. Dekker, New York, 1973.

[117] Procesi C., On a Theorem of M. Artin, J. Algebra 22(1972) 309-315.

[118] Ramras M., Maximal Orders over Regular Local Rings of Dimension Two, Trans. AMS 142 (1969) 457-479.

[119] Raynaud M., Anneaux Locaux Henseliens, LNM 169, Springer Verlag, Berlin 1970.

[120] Razmyslov, Yu, P., On a Problem of Kaplansly, Math. USSR, Izv 7 (1973) 479-496.

[121] Razmyslov, Yu.P., Trace Identities of Full Matrix Algebras over a Field of Characteristic Zero, Math. USSR, Izv. 8 (1974).

[122] Reiner I., Maximal Orders, Academic Press, London, 1975.

[123] Reiner I., Ullom S., A Mayer Vietoris Sequence for Class Groups, J. Algebra 31 (1974)305-342.

[124] Riley J., Reflexive Ideals in Maximal Orders, J. Algebra 2 (1965), 451-465.

[125] Rim D.S., Modules over Finite Groups, Ann. of Math. 69 (1959).

[126] Robson J.C., Non-commutative Dedekind Prime Rings, J. Algebra 9 (1968) 249-265.

[127] Robson J.C., Idealizers and Hereditary Noetherian Prime Rings , J. Algebra 22 (1972)45-81.

[128] Robson J.C., Small L.W., Hereditary Prime P.I. Rings are Classical Hereditary Orders, J. London Math. Soc. 8 (1974) 499-503.

[129] Robson J.C., Small C.W., Idempotent Ideals in PI rings, J. London Math. Soc. 14 (1976) 120-122.

[130] Roggenkamp K., Huber-Dysen V.Lattices over Orders I, LNM 115, Springer Verlag, Berlin, 1970.

[131] Roggenkamp K., Huber-Dyson V., Lattices over Orders II, LNM 142, Springer Verlag, Berlin, 1970.

[132] Roggenkamp K., The lattice Type of Orders, in : LNM 825, 104-129, Springer Verlag, 1980.

[133] Rosenberg A., Zelinski P., On Amitsur's Complex, Trans. AMS 97 (1960) 327-356.

[134] Rowen L.H., On Rings with Central Polynomials, J. Algebra 31 (1974) 393-426.

[135] Saltman D., Division Algebras over Discrete Valued Fields, Comm. in Algebra, 8 (1980) 1749-1774.

[136] Samuel P. Anneaux Gradués Factoriels et Modules Réflexifs, Bull. Soc. Math. France, 92 (1964), 237-249.

[137] Schelter W., Integral Extensions of Rings Satisfying a Polynomial Identity, J. Algebra 40 (1976) 245-257.

[138] Schelter W. Affine P.I. Rings are Catenary, J. Algebra 51 (1978) 12-18.

[139] Schilling O. The Theory of Valuations, AMS Math. Survey IV, 1960.

[140] Silver L., Tame Orders, Tame Ramification and Galois Cohomology, Ill. J. Math. 12 (1968) 7-34.

[141] Small L.W., Localization in P.I.-Rings J. Algebra 18 (1971) 269-270.

[142] Stenström B., Rings of Quotients, Springer Verlag, Berlin, 1968.

[143] Swan R., Algebraic K-Theory, LNM 76, Springer Verlag Berlin.

[144] Turnidge D.R., Torsion Theories and Rings of Quotients of Morita Equivalent Rings, Pacific J. Math. 37 (1971) 225-234.

[145] Van den Bergh M., Graded Dedekind Rings, J. Pure Applied Algebra, to appear.

[146] Van Deuren J.P., Van Oystaeyen F., Arithmetically Graded Rings I, in : LNM 825, 130-152, Springer Verlag. Berlin, 1980.

[147] Van Oystaeyen F., Prime Spectra in Non-commutative Algebra, LNM 444, Springer Verlag, Berlin, 1975.

[148] Van Oystaeyen F., Compatibility of Kernel Functors and Localization Functors, Bull. Soc. Math. Belg. 18 (1976) 131-137.

[149] Van Oystaeyen F., Localization of Fully left Bounded Rings, Comm. in Algebra 4, (1976) 271-284.

[150] Van Oystaeyen F., Zariski Central Rings, Comm. in Algebra 6 (1978) 799-821.

[151] Van Oystaeyen F., Graded Azumaya Algebras and Graded Brauer Groups, in : LNM 825, 158-171, Springer Verlag, Berlin, 1980.

[152] Van Oystaeyen F., Birational Extensions of Rings , in : Ring Theory 1978 287328, M. Dekker, New York.

[153] Van Oystaeyen F., On Graded Rings with Polynomial Identity, Bull. Soc. Math. Belg. 32 (1980) 22-28.

[154] Van Oystaeyen F., Crossed Products over Arithmetically Graded Rings, J. of Algebra 80, 537-551.

[155] Van Oystaeyen F., Generalized Rees Rings, and Arithmetically Graded Rings, J. Algebra 82, (1983) 185-193.

[156] Van Oystaeyen F., A Note on Central Class Groups of Orders and Graded Orders, in : Methods in Ring Theory, Proceedings, Reidell Publ. Comp. Dordrecht, 1984.

[157] Van Oystaeyen F., Some Constructions of Rings, Proceedings E. Noether days, North Holland, to appear.

[158] Van Oystaeyen F, On Clifford Systems and Generalized Crossed Products, J. Algebra, to appear.

[159] Van Oystaeyen F., Verschoren A., Reflectors and Localization, M. Dekker, New York, 1979.

[160] Van Oystaeyen F. Verschoren A., Relative Localization, Bimodules and Semiprime PI Rings, Comm. in Algebra 7, (1979) 955-988.

[161] Van Oystaeyen F. , Verschoren A, Non commutative Algebraic Geometry, LNM 887, Springer Verlag, Berlin, 1982.

[162] Van Oystaeyen F., Verschoren A., Relative Invariants of Rings: The Commutative Theory, Marcel Dekker, New York, 1983.

[163] Van Oystaeyen F., Verschoren A, The Brauer Group of a Projective Variety, Israel J. Math., 42 (1982) 37-57.

[164] Van Oystayen F. Verschoren A., Splitting and Crossed Products in Relative Brauer Groups, Comm. in Algebra 11 (1983) 2375-2392.

[165] Verschoren A., Localization and the Gabriel-Popescu Embedding Theorem, Comm. in Algebra 6(1978) 1563-1587.

[166] Verschoren A., On the Picard Group of a Grothendieck Category, Comm. in Algebra 8(1980) 1169-1194.

[167] Verschoren A., About "On the Picard Group of a Grothendieck Category.", Comm. Algebra 10 (1982)1119-1124.

[168] Verschoren A., Exact Sequences for Relative Brauer Groups and Picard Groups, J. Pure Appl. Algebra 28 (1983)93-108.

[169] Verschoren A. , A Mayer-Vietoris Sequence for Relative Picard Groups, J. Algebra, to appear.

[170] Verschoren A., Global Relative Invariants, to appear.

[171] Verschoren A., On Relative Picard Groups, J. Algebra, to appear.

[172] Verschoren A., Danilov's Theòrem Revisited, to appear.

[173] Verschoren A., The Affine Scheme Associated to a Nonnoetherian pi algebra, to appear.

[174] Villamayor O., Zelinski D., Brauer Groups and Amitsur Cohomology for General Commutative Ring, Extensions, J. Pure Appl. Algebra 10(1977) 198-202.

[175] Wauters P., Jespers E., Asano-orders, Marubayashi-Krull Orders and Strongly Graded Rings, preprint.

[176] Wauters P., Jespers E., What is an $\Omega$-Krull Ring ? in : Proceedings of the E. Noether days, North Holland, to appear.

[177] Webber D.B., Ideals and Modules of Simple Noetherian Hereditary Rings, J. Algebra 16 (1970) 239-242.

[178] Yuan S., Reflexive Modules and Algebra Class Groups over Noetherian Integrally Closed Domains, J. Algebra 32 (1974) 405-417.

[179] Zariski O., Samuel P., Commutative Algebra I, II, Van Nostrand, 1958-60,

[180] Zelinski D., Long Exact Sequences and the Brauer Group, in LNM 549, 63-70, Springer Verlag, Berlin, 1976.

# INDEX